DEVELOPMENT POLICY AND PUBLIC ACTION

Edited by
Marc Wuyts, Maureen Mackintosh and Tom Hewitt
for an Open University Course Team

OXFORD UNIVERSITY PRESS

in association with

1992

This book has been printed on paper produced from pulps bleached without use of chlorine gases and produced in Sweden from wood from continuously farmed forests. The paper mill concerned, Papyrus Nymölla AB, is producing bleached pulp in which dioxin contaminants do not occur.

Published in the United Kingdom by Oxford University Press, Oxford in association with The Open University, Milton Keynes

Oxford University Press, Walton Street, Oxford OX2 6DP

Oxford New York Toronto
Delhi Bombay Calcutta Madras Karachi
Petaling Jaya Singapore Hong Kong Tokyo
Nairobi Dar es Salaam Cape Town
Melbourne Auckland

and associated companies in
Berlin Ibadan

Oxford is a trade mark of Oxford University Press

The Open University, Walton Hall, Milton Keynes MK7 6AA

First published in the United Kingdom 1992

Copyright © 1992 The Open University

All rights reserved. No part of this publication may be reproduced or transmitted, in any form or by any means, electronic or mechanical, including photocopy, recording, or any information storage and retrieval system, without permission in writing from the publisher or under licence from the Copyright Licensing Agency Limited. Details of such licences (for reprographic reproduction) may be obtained from the Copyright Licensing Agency Ltd of 33–34 Alfred Place, London WC1E 7DP

This book is sold subject to the condition that it shall not, by way of trade or otherwise, be lent, re-sold, hired out or otherwise circulated without the publisher's prior consent in any form of binding or cover other than that in which it is published and without a similar condition including this condition being imposed on the subsequent purchaser

British Library Cataloguing in Publication Data

Data available

ISBN 0–19–877336–6
ISBN 0–19–877337–4 (Pbk)

Library of Congress Cataloging in Publication Data

Data available

ISBN 0–19–877336–6
ISBN 0–19–877337–4 (Pbk)

Edited, designed and typeset by The Open University

Printed and bound in Great Britain by
Butler & Tanner Ltd, Frome and London

CONTENTS

THE OPEN UNIVERSITY
U208 *THIRD WORLD DEVELOPMENT*
COURSE TEAM AND AUTHORS

Tom Hewitt, Lecturer in Development Studies, The Open University (Course Team Chair 1991–92)

Ben Crow, Lecturer in Development Studies, The Open University (Course Team Chair 1989–90)

Tim Allen, Lecturer in Development Studies, The Open University

Paul Auerbach, Reader in Economics, Kingston Polytechnic (Part Assessor)

Carolyn Baxter, Course Manager, The Open University

Henry Bernstein, Senior Lecturer in Agricultural and Rural Development, Institute of Development Policy and Management, University of Manchester

Krishna Bharadwaj, Professor, Centre for Economic Studies and Planning, Jawaharlal Nehru University, New Delhi, India

Suzanne Brown, Course Manager, The Open University

Janet Bujra, Lecturer in Sociology, Department of Social and Economic Studies, University of Bradford

David Cleery, Research Officer, Centre of Latin American Studies, University of Cambridge

Kate Crehan, Associate Professor, New School for Social Research, New York, USA

Sue Dobson, Graphic Artist, The Open University

Harry Dodd, Print Production Controller, The Open University

Kath Doggett, Project Control, The Open University

Joshua Doriye, Professor, Institute of Finance and Management, Dar es Salaam, Tanzania

Chris Edwards, Senior Lecturer in Economics, School of Development Studies, University of East Anglia

Diane Elson, Lecturer in Development Economics, University of Manchester (Part Assessor)

Sheila Farrant, Tutor Counsellor and Assistant Staff Tutor, The Open University, Cambridge (Course Reader)

Gerry Farrell, freelance musician and writer

Paul Frenz, Research Assistant, The Open University

Jayati Ghosh, Associate Professor, Centre for Economic Studies and Development, Jawaharlal Nehru University, New Delhi, India

Heather Gibson, Lecturer in Economics, University of Kent at Canterbury

Garry Hammond, Senior Editor, The Open University

Barbara Harriss, Lecturer in Agricultural Economics and Governing Body Fellow, Wolfson College, University of Oxford

John Harriss, Director, Centre for Development Studies, London School of Economics (Part Assessor)

Pamela Higgins, Graphic Designer, The Open University

Lakshmi Holmström, freelance writer

Caryl Hunter-Brown, Liaison Librarian, The Open University

Gillian Iossif, Lecturer in Statistics, The Open University

Rhys Jenkins, Reader in Economics, School of Development Studies, University of East Anglia

Hazel Johnson, Lecturer in Development Studies, The Open University

Sabrina Kassam, Research Assistant, The Open University

Georgia Kaufmann, Research Fellow, Institute of Development Studies, University of Sussex

Andrew Kilminster, Lecturer in Economics, School of Business Studies, Oxford Polytechnic

Patti Langton, Producer, BBC

Christina Lay, Editor, The Open University

Anthony McGrew, Lecturer in Government, Social Science Faculty, The Open University

Maureen Mackintosh, Reader in Economics, Kingston Polytechnic

Mahmood Mamdani, Professor, Centre for Basic Research, Kampala, Uganda

Charlotte Martin, Teacher and Open University Tutor (Course Reader)

Mahmood Messkoub, Lecturer in Economics, University of Leeds

Alistair Morgan, Lecturer in Institute of Educational Technology, The Open University

Eleanor Morris, Producer, BBC

Ray Munns, Cartographer, The Open University

Kathy Newman, Secretary, The Open University

Hilary Owen, Lecturer, Department of Hispanic Studies, University of Belfast

Debbie Payne, Secretary, The Open University

Ruth Pearson, Lecturer in Economics, School of Development Studies, University of East Anglia

Richard Pinder, Training consultant, Sheffield (Course Reader)

David Potter, Professor of Government, The Open University

Janice Robertson, Editor, The Open University

Carol Russell, Editor, The Open University

Vivian von Schelling, Lecturer in Development Studies, Polytechnic of East London

Gita Sen, Fellow (Professor), Centre for Development Studies, Kerala, India

Meg Sheffield, Senior Producer, BBC

Paul Smith, Lecturer in Environmental Studies, The Open University

Ines Smyth, Senior Lecturer, Institute of Social Studies, The Hague, Netherlands; Research Associate, Department of Applied Social Studies and Social Research, University of Oxford

Hilary Standing, Lecturer in Social Anthropology, School of African and Asian Studies, University of Sussex (Part Assessor)

John Taylor, Head of Centre for Chinese Studies, South Bank University (Course Reader)

Alan Thomas, Senior Lecturer in Systems, The Open University

Steven Treagust, Research Assistant, The Open University

David Treece, Lecturer in Brazilian Studies, Kings College, University of London

Euclid Tsakalotos, Lecturer in Economics, University of Kent at Canterbury

Gordon White, Professorial Fellow, Institute of Development Studies, University of Sussex

David Wield, Senior Lecturer in Technology Strategy and Development, The Open University

Gordon Wilson, Staff Tutor in Technology, The Open University, Leeds

Philip Woodhouse, Lecturer in Agricultural and Rural Development, Institute of Development Policy and Management, University of Manchester

Peter Worsley, Emeritus Professor, University of Manchester (External Assessor)

Marc Wuyts, Professor of Applied Quantitative Economics, Institute of Social Studies, The Hague, Netherlands

The Course Team would like to acknowledge the financial support of Oxfam and the European Community in the preparation of U208 *Third World Development*.

The editors wish to acknowledge the cooperation of the Population and Development Programme at the Institute of Social Studies, The Hague in writing this book. They also acknowledge the support of the United Nations Population Fund (UNFPA) Global Programme.

Countries and major cities of the world

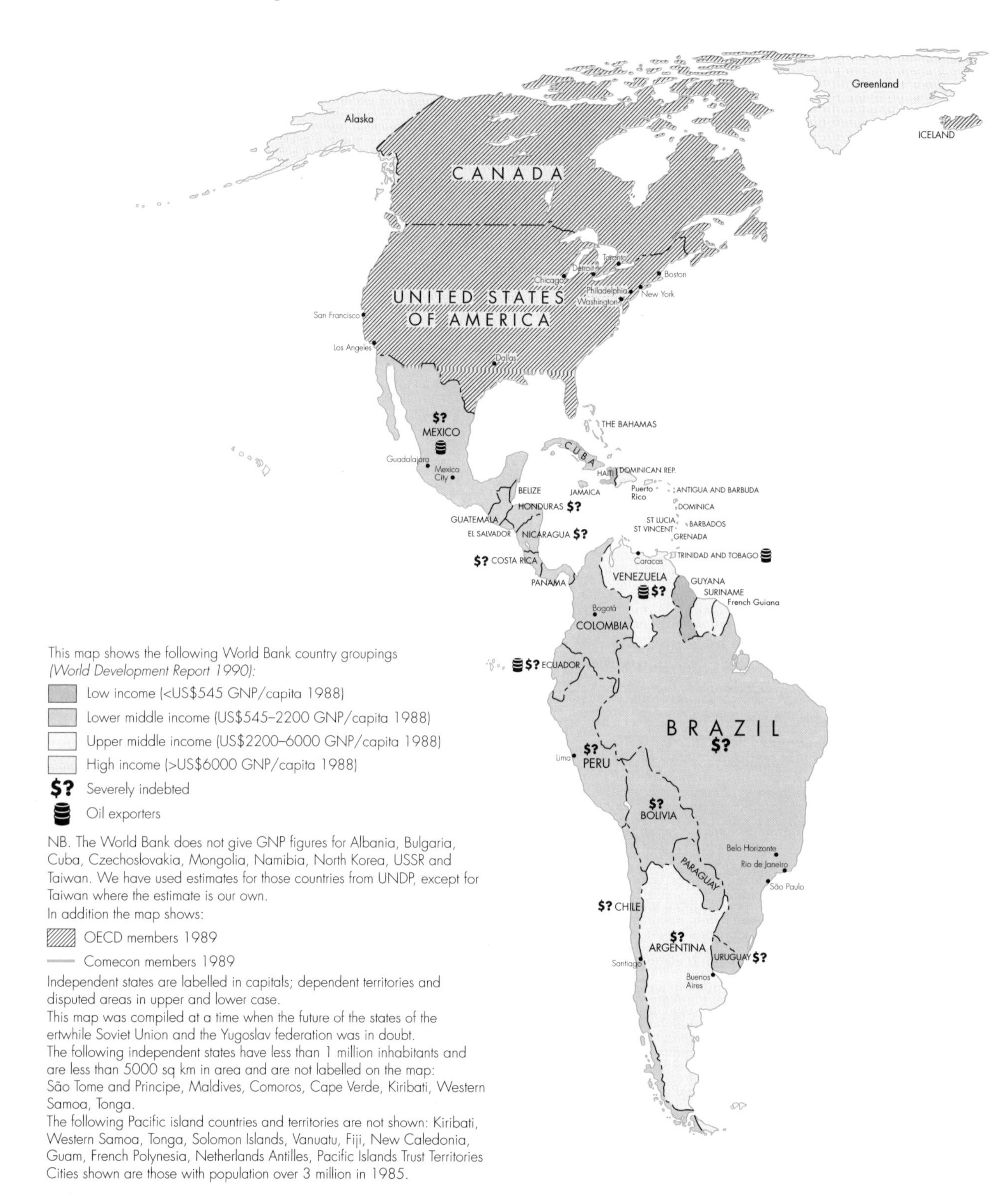

This map shows the following World Bank country groupings *(World Development Report 1990)*:

- Low income (<US$545 GNP/capita 1988)
- Lower middle income (US$545–2200 GNP/capita 1988)
- Upper middle income (US$2200–6000 GNP/capita 1988)
- High income (>US$6000 GNP/capita 1988)
- **$?** Severely indebted
- Oil exporters

NB. The World Bank does not give GNP figures for Albania, Bulgaria, Cuba, Czechoslovakia, Mongolia, Namibia, North Korea, USSR and Taiwan. We have used estimates for those countries from UNDP, except for Taiwan where the estimate is our own.
In addition the map shows:

- OECD members 1989
- Comecon members 1989

Independent states are labelled in capitals; dependent territories and disputed areas in upper and lower case.
This map was compiled at a time when the future of the states of the ertwhile Soviet Union and the Yugoslav federation was in doubt.
The following independent states have less than 1 million inhabitants and are less than 5000 sq km in area and are not labelled on the map: São Tome and Principe, Maldives, Comoros, Cape Verde, Kiribati, Western Samoa, Tonga.
The following Pacific island countries and territories are not shown: Kiribati, Western Samoa, Tonga, Solomon Islands, Vanuatu, Fiji, New Caledonia, Guam, French Polynesia, Netherlands Antilles, Pacific Islands Trust Territories
Cities shown are those with population over 3 million in 1985.

USSR
NORWAY
SWEDEN
FINLAND
St Petersburg
Moscow
UNITED KINGDOM
DENMARK
NETHERLANDS
IRELAND
London
POLAND $?
Katowice
GERMANY
BELGIUM
LUX
CZECH.
HUNGARY $?
Paris
AUSTRIA
SWITZ.
FRANCE
Milan
ROMANIA
YUGOSLAVIA
ITALY
Rome
BULGARIA
Barcelona
Istanbul
PORTUGAL
Madrid
SPAIN
Naples
ALBANIA
GREECE
Athens
Ankara
TURKEY
MALTA
TUNISIA
CYPRUS
LEBANON
SYRIA
ISRAEL
Palestine
JORDAN
Baghdad
IRAQ
Tehran
IRAN
KUWAIT
BAHRAIN
QATAR
UNITED ARAB EMIRATES
SAUDI ARABIA
OMAN
YEMEN
MONGOLIA
Beijing
Shenyang
Tianjin
NORTH KOREA
JAPAN
Seoul
SOUTH KOREA
Pusan
Tokyo/Yokohama
Osaka
CHINA
Kashmir
AFGHANISTAN
Lahore
PAKISTAN
Tibet
BHUTAN
NEPAL
New Delhi
Wuhan
Shanghai
Karachi
Ahmedabad
Calcutta
Dhaka
BANGLADESH
INDIA
MYANMAR
Guangzhou
Hong Kong
TAIWAN
LAOS
Hanoi
Bombay
Hyderabad
THAILAND
Bangkok
VIETNAM
CAMBODIA
Bangalore
Madras
Manila
PHILIPPINES $?
Ho Chi Minh City (Saigon)
SRI LANKA
MALDIVES
BRUNEI
MALAYSIA
SINGAPORE
INDONESIA
Jakarta
East Timor
PAPUA NEW GUINEA
$?
MOROCCO
ALGERIA
LIBYA
EGYPT
Cairo
Western Sahara
MAURITANIA
MALI
NIGER
CHAD
SUDAN
Eritrea
$?
SENEGAL
THE GAMBIA
GUINEA-BISSAU
GUINEA
BURKINA FASO
BENIN
NIGERIA
DJIBOUTI
SIERRA LEONE
COTE D'IVOIRE $?
GHANA
LIBERIA
TOGO
Lagos
CAMEROON
CENTRAL AFRICAN REP.
ETHIOPIA
SOMALIA
EQUATORIAL GUINEA
$?
GABON
CONGO
UGANDA
KENYA
RWANDA
ZAIRE
BURUNDI
TANZANIA
SEYCHELLES
Cabinda
ANGOLA
ZAMBIA
MALAWI
MOZAMBIQUE
ZIMBABWE
MADAGASCAR
MAURITIUS
NAMIBIA
BOTSWANA
SWAZILAND
SOUTH AFRICA
LESOTHO
AUSTRALIA
Sydney
Melbourne
NEW ZEALAND

INTRODUCTION

MAUREEN MACKINTOSH

This is a book about public policy for development, with a particular (though not exclusive) focus on economic policy. There have been many such textbooks; but this is an unconventional one in at least two ways. First, we treat development policy making as an activity of many different types of public institution: not just governments, but also aid agencies, non-governmental organizations, community groups, collectives and political movements. Second, we are interested in the whole process by which policy is made and remade. This process includes the public agencies which compete over the social definition of needs, and continues through the patterns of public participation in policy making, to the sources of changing policy ideas and their implications for future action.

In other words, this book sees development policy as embedded in the wider sphere of public action in society. The purpose of this introduction is therefore:

- to explain how we are defining key concepts such as the 'policy' and 'public action' of the book's title
- to explain why we have chosen these less conventional definitions
- to introduce the reader to the line of argument in the book.

What is development policy?

We understand development policy to mean deliberate action by public institutions seeking to promote development. These institutions, as just suggested, can be of many different types, and they certainly act on competing definitions of what might be meant by 'development'.

The experience of the 1980s has made this broad definition of development policy essential. That decade was a period of very sharp conflict about what sort of development policy was needed, for example to promote industrialization (*Hewitt* et al., *1992*). And it also called into question who should make policy and for whom. Could governments make policy on behalf of others, or did policy only work if it came 'from below' (Crow, in *Bernstein* et al., *1992*)? The earlier consensus, which saw a central role in development for an activist, interventionist state, had broken down. A stronger policy-making role, though a contested one, was promoted for agencies as different as popular organizations, charities, multilateral aid agencies and associations of industrialists.

This situation, of policy made through competing ideas and agencies, seems to invite a number of questions. For example:

Q Why is policy so contested?

Q How do conflicts of ideas relate to the interests of the agencies which put them forward?

Q How have certain past policies influenced subsequent public policy and debate?

In other words, we are invited to consider the process of policy making. How does policy emerge from the public institutions and social conflicts in a society (local or global)? And how does it feed back via social change into new policy? All these questions are much too general to answer in this form. But the aim of this book is to explore 'policy' from this angle.

Policy as social process

Once we start to think of policy as a social process, then clearly the nature of the policy-making institutions matters a great deal. Indeed, some policy ideas cannot be pursued, or even really be formulated, until institutions emerge to put them into practice. One example is social security systems in industrialized countries. These were predicated not only on a recognition by the better-off of the potential benefits to themselves, but also on the development of a competent civil service and effective information-gathering. Once established however, institutions can develop powers and interests, and hence opposition, of their own. Out of institutional conflict emerges policy in practice, intended or unintended. In the process, people campaign and bargain, co-operate and cheat — and can be observed by policy analysts!

This book seeks to do some such observing and analysing of the policy process. So one preoccupation is with changing institutions: the incentives they provide, and those to which they respond; their probity and their effectiveness. This development policy textbook therefore examines a whole string of issues not conventionally thought of as 'policy'. Here are some examples.

- How do groups of civil servants behave in times of economic crisis, in pursuit of their own survival? And how does this influence the way that public services are delivered to users (Chapter 4)?
- How do active women's organizations negotiate between a government agenda that specifies which are the proper economic and social activities for women, and their own evolving ideas about women's interests, needs and capacities (Chapter 6)?
- How and why do different agencies, governmental and non-governmental, come into conflict over the proper way to respond to severe social deprivation, and with what consequences for the vulnerable (Chapter 2)?
- How do political and social movements influence both government provision of public services, and their effective organization and take-up (Chapter 10)?

Each of these questions is answered in a specific context: a country, a region, a historical period. This is another necessary aspect of the 'policy as process' framework of the book. Examining policy processes in historical contexts allows us to identify cumulative patterns of change in-

Box 1 A definition of 'policy'?

'Policies' are purposive actions undertaken by the state (governments and their employees), or by other institutions (such as voluntary organizations), with an avowedly public purpose. That is, they are actions conceived as (or defended as) serving some wider public objective such as social and economic development, and not (or not solely) individual private gain. Hence, not all actions by public institutions are 'policy'. Institutions and organizations make public policy in so far as they act or claim to act for the public good.

duced or sustained by public policy. For example, Chapter 4 outlines a process of cumulative decline of public probity in the provision of public services. The decline in this case has been induced by a mutually reinforcing set of economic pressures and policies, and may be worsened by some of the proposed policy responses.

Conversely, Chapter 10 has a more hopeful message. It describes a case where political movements for social reform have shaped and sustained public service provision in a very low-income community. Its message is that virtuous circles are possible, not just vicious circles (as in Chapter 4). For example, the government may be held accountable for effective service provision by active citizens (notably including women), who also ensure that the services are taken up and well used. This idea of cumulative circles of change is further explored in the book's conclusion.

Our conception of policy as a process, emerging from the interaction of ideas and agencies, is an increasingly common one. The burgeoning field of 'policy studies' within social science incorporates this perspective (see, for example, Hogwood & Gunn, 1984). It is in sharp contrast to what may be called the 'rationalist' approach to policy, which still dominates textbooks on policy for economic development.

Policy as prescription

We can describe the rationalist approach as follows. In reply to our question, 'What is development policy?', the most common textbook answer would be along the following lines. Development policy consists of the actions of *governments* to promote development. The study of development policy therefore encompasses the prescription, analysis and evaluation of governmental action.

In other words, development policy analysis in this more conventional framework is defined by the further question, 'What should the government do?' Answers to that question begin by assuming that the government and civil service do indeed aim to promote development for the benefit of the country's citizens, and then ask, how can the government best do this?

Discussion then proceeds by asking which policies have worked or not worked, and by trying to extract models and lessons from evidence of success. This approach has been reinforced in recent years, especially in economic development writing, by the openly prescriptive approach of international agencies, notably the World Bank. There has been an associated proliferation of 'manuals' which offer a how-to-do-it guide for 'policymakers' (for example, Dornbusch & Helmers, 1988).

This rationalist approach, then, can be summarized as 'policy as prescription'. It is in sharp contrast to the definition of policy used in this book. Though still important in textbooks, the prescriptive approach to policy has been under attack from all sides in recent years. As Chapter 3 explains, neo-liberal critics have rejected the assumption, mentioned above, of the benevolent intentions of governments and state employees. Instead, the state is envisaged to serve the private interests of government officials or influential élites.

While these 'private interest' models are an improvement on policy as prescription, they are narrower than the 'process' framework used here. Our definition of policy allows us to suppose that public institutions may or may not act in the interests of a wider constituency including the poor, and to explore the question of why they do or do not do so in particular historical contexts.

Public action

We have already, in the above discussion, used the concept of the 'public' many times: 'public services', 'public institutions', 'public purpose'. But what do we mean by 'public'? There is no simple answer to this, because the idea of the public, and the distinction between 'public' and 'private' spheres of life, have always been complex and contested. And the association of 'public' with the state or government has been particularly at issue during the 1980s.

In some contexts, we still tend to associate 'public' with the state. For example, the 'public sector' is still used to mean activities undertaken

Box 2 Two models of development policy

1 *Policy as prescription* proposes appropriate government policy for development, based on a set of assumptions about the benevolence of government. It is largely ahistorical.

Typical question: What public services should the government provide?

2 *Policy as process* seeks to explain the actions of public institutions, governmental and non-governmental, and their effects, as outcomes of social processes. It takes a historical and evolutionary approach.

Typical question: How can non-governmental action improve public service provision?

by the state or by state-owned agencies. Discussions of 'public policy' frequently refer only to the decisions of governments and the actions of government employees (the combination of which we are calling the 'state'). There was a time when the phrase 'public services' was generally taken to mean services provided by the state, but that association has been weakened recently by deregulation, privatization, and the emphasis on a 'mixed economy of welfare' including private service provision.

Public action and the public good

However, the adjective 'public' of course also has far wider associations than this. 'The public' is often used to mean the citizens of a country, at least in so far as they participate in political and social processes. The 'public sphere' conjures up the wider stage of politics and social movements, campaigning and organizing. The 'public good' is something to which charities and voluntary organizations claim to contribute, and the phrase carries connotations of need and social purpose.

Out of this overlapping set of associations, comes a concept of 'public action' which is considerably wider than the actions of the state. A recent influential book put the point this way:

> "Public action is not ... just a question of public delivery and state initiative. It is also ... a matter of participation by the public in the process of social change."
>
> (Drèze & Sen, 1989, p.259)

Despite the play on the word 'public', Drèze and Sen clearly envisage here a broad sphere of public action by a variety of organizations for developmental ends. But what organizations and actions are we to consider as falling within this concept of 'public action'? Drèze and Sen appear to have in mind:

- voluntary associations of citizens seeking (their own definition of) the public good
- locally based groups seeking community development
- political parties and campaigning activities.

But what of, say, organizations (including some political parties) which are dedicated to the promotion of sectional interests? Are they acceptable only if they promote the interests of the disadvantaged?

Drèze and Sen, while examining the negative as well as the positive effects of state action, tend to concentrate, when discussing non-state public action, on examples of co-operative action, or publicity and campaigning, for benevolent ends such as the relief of famine. They seem to slide towards an implicit definition of public action as action for the public benefit.

But such a circular definition is not satisfactory. Many 'public' organizations, as we have already argued, do act in the service of 'private' ends, that is, in the sole interests of their members. Do we include those in our concept of public action, or do we exclude them, conceptualizing them rather as private clubs? If we do the latter, are we not drawing the boundaries between private and public action solely on the basis of our own (private) prejudices?

Box 3 Public Action

By *public action* we mean purposive collective action, whether for collective private ends or for public ends (however defined).

By the *public sphere* we mean the arena of such public action from parliaments to public demonstrations, passing through the media, trades union activity and voluntary associations for mutual assistance.

Public action and private ends

In this book we have sought to sidestep this problem by using a wide definition of public action. We have included in the public sphere the actions, not only of states and organizations which claim to seek to promote the public good or assist the disadvantaged, but also clubs of the powerful who seek to influence the public sphere for their own ends. Such clubs could include political parties, and also private combines (such as collusive traders) who seek to manipulate market structures to their own advantage. Collective, purposeful manipulation of the public environment constitutes public action, whether through legislation, lobbying, self organization or rigging the market! This makes the concept of public action considerably wider than 'policy' as defined above.

These definitions distinguish 'public action' from, for example, trading on a market, farming to support a household, marrying, caring for children or saving money for one's old age. These are examples of private actions for one's own or other private individuals' benefit. Over time, our ideas about the proper division between the public and the private spheres of life shift and change, as do our concepts of the proper organization of both spheres.

Our definitions, therefore, are closely related. We see development policy as emerging from, and influencing in turn, the public institutions of society and their activities.

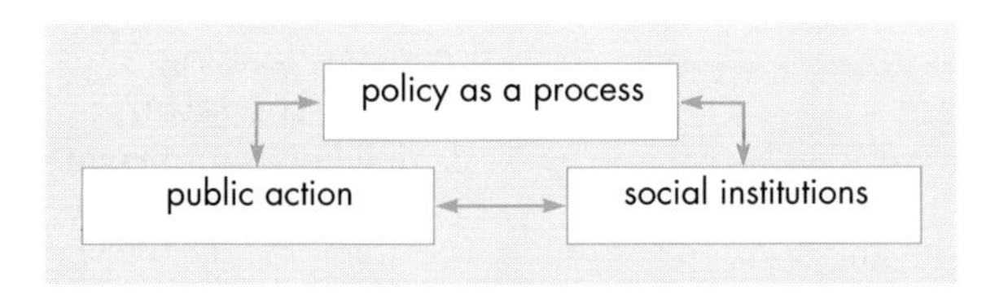

The incomplete market

Our choice of topics for our exploration of development policy and public action has been structured by two themes. Both will be, we believe, central themes for development policy in the 1990s.

The role of the market

The first theme is the proper role and limits of the market, and the associated question of policy towards markets. This theme is forced upon us by the legacy of the 1980s, which saw an upheaval in ideas about development drastic enough to be called a 'counter-revolution' (Toye, 1987). The theorists of this counter-revolution were neo-liberal economists who questioned the very idea that development should be led by governments (Chapter 5 of *Hewitt* et al.*, 1992*). Instead, they proposed that private action — individuals pursuing their private interest in the marketplace — would more effectively promote economic development.

This neo-liberal agenda came in the 1980s to dominate development policy. Its most important principle was that markets should provide the yardstick of public policy. Abstract economic models were used to argue that markets would work efficiently, where allowed to do so. According to these models, state intervention tended to worsen market functioning and should therefore be minimized. Hence market deregulation (removing legal and regulatory constraints on markets) and 'rolling back the state' by reducing its size and scope dominated the policy agenda (Chapter 3). State policies were ranked by the extent to which they extended the scope of markets (Chapter 8).

'Orthodox' economic theory

Another way of making this same point is to say that a rather narrow version of economic theory came to exercise a considerable ideological predominance in development policy in the 1980s. This economics, with its strong assumptions about economic individualism and the benefits of competitive markets, we shall call 'orthodox' economics at various points in this book. Its predominance spread outwards during the decade from industrial and agricultural policy to social policy and administration, and the language of competition, markets and business began to pervade the worlds of government, welfare and charity.

This orthodox theory formed the basis for the single dominant set of policies faced by lower-income countries in the 1980s: the package called 'structural adjustment' (Chapter 7). In a highly prescriptive approach to development policy, this package mixes a reduced role for the state (including a reduction in public services), with a liberalization of both domestic markets and of trading across national boundaries. Much criticism has focused on the undifferentiated application of this package, with a lack of attention to local diversity of needs and institutions (Chapter 8).

The predominance of abstract economic theory over consideration of historical and institutional diversity in recent development policy is one reason for the emphasis in this book on economics and economic policy. Different chapters in the book seek to explain the major policy arguments and to examine how they have worked out in specific contexts, including the responses of people and institutions to the pressures set up by these reforms (Chapters 2, 7 and 9). As market-based reforms spread to Eastern Europe and the previously socialist low-income countries, we include an examination of economic reform in China (Chapter 9).

The role of non-market institutions

One paradoxical conclusion stands out. The efforts in the 1980s to reduce the power of government and to 'free' markets have served in part to remind us of the importance of non-market institutions. This perception has been as true for the neo-liberals as for their critics. Neo-liberal economic reforms, while reducing state economic activity for example by selling state enterprises, have tended to strengthen the state in other areas such as policing: the 'free economy and the strong state' are often seen as the twin hallmarks of Thatcherite Britain. Furthermore, the neo-liberals have a strong critique of the state, based again on orthodox economic models, which has led them to seek not only restriction but also reform of its economic and welfare activities on a market model (Chapter 3).

So neo-liberal reformers, by contesting the organization of the state and the nature of non-state public action across the world, have put these issues firmly in the centre of the policy agenda. In the World Bank's policies for Africa, for example, a decade which began by rolling back the boundaries of the state, ended with a focus on 'capacity building' within state structures. This book examines some of the background to this apparent turn-round, using our perspective on policy and public action to examine some of the contradictions thrown up by the neo-liberal reforms. (World Bank, 1981 and 1991b)

One theme, then, is that markets, while important, cannot be the exclusive focus and yardstick of public action. As the economist A.K. Sen (in Helm, 1989) argues, '... the market mechanism is an essentially *incomplete* specification of a social arrangement'. In other words, if we are to understand how the market functions, we need to know about the legal and social structures within which markets are embedded. Markets, we shall argue, are never truly 'free'; and the way they are regulated matters greatly especially to more vulnerable participants (Chapter 1). The proper role for public action by non-market institutions will be a development debate for the 1990s.

Poverty and politics

If markets and economic efficiency were one dominant theme of the 1980s, poverty and deprivation formed another. Images of famine in

Africa and destitution in Latin America marked the news programmes of the decade. While some countries and people in both the West and the Third World grew richer, simultaneously the scale and levels of poverty worsened (*Allen & Thomas, 1992*). The heightening sense of economic conflict was focused by the debt crisis and the years of economic recession, which dominated the 1980s and drained the resources of many both stronger and weaker economies in the Third World (*Hewitt* et al., *1992*).

As a result, alongside the language of business and efficiency, there has been another level of debate concerning morality and need. Advocates of market-led development have justified their approach not only as efficient, but also in the language of morality: choice, freedom, self-respect, individual rights. Neo-liberal theorists have established the 'moral standing of the market' (the phrase again is from A.K. Sen) as an issue of debate, and critics, including Sen, have taken up the challenge.

To this moral language of freedom and responsibility, the critics of the neo-liberals have counterposed ideas of needs, capacities and participation. To the 'negative freedom' from constraint by the state, which is the basis of neo-liberal arguments, they counterposed a concept of 'positive freedom'. Such positive freedom was originally defined by Berlin (1958) in terms of a capacity for making one's own decisions, a measure of conscious self-determination. This has been interpreted to imply the need for access to minimum resources and capacities — sufficient food, minimum education — for such conscious participation in society (Dasgupta, in Helm, 1989).

But even among the advocates of public response to need, the nature of that response has remained contested. The 1980s saw strong criticism of what might be called the 'welfarist' approach to needs provision: the idea that the state could and should 'count, cost and deliver' in response to identified 'basic needs' (Streeten, 1981). Both neo-liberals and others argued in the 1980s that direct state welfare provision tended to be wasteful, inefficient, and inequitable in practice (Chapters 3 and 10). In practice neither rising incomes nor relatively high total state expenditure on public services guarantee lessening deprivation (Chapter 1).

What, then, does constitute successful public action against deprivation? We begin the book with this question (Chapter 1). We argue that the answer does not lie in any individual policy by an individual institution, state or non-state. Instead, successful public action against deprivation implies a set of mutually reinforcing activities, involving both state and non-state public institutions. Chapter 1 (following Drèze & Sen, 1989), calls this mix a 'strategy'.

Such a strategy is not something a state, or the World Bank, or any other body can decide to 'implement'. It involves the internal structure of the state and its accountability; the structure of the market, based in the pattern of public regulation and market institutions; the pattern of public participation in politics; the extent of social division; and no doubt many other factors. Particular policies, such as a land reform, can influence the overall strategy only if other institutions are supportive, or can be simultaneously reformed. One aspect of these interactions, the relations between economic reform and political reform, is now being much debated, and will be another major development policy theme for the 1990s. This book illustrates these issues by tracing some patterns of feedback between different types of public action in particular contexts (Chapters 9 and 10).

The structure of the book

Section 1 — Public need and public action

Section 1 takes us straight into the issues of the social definition of need and the question of 'strategy'. The first chapter defines 'poverty' and 'need', not as characteristics of people, measurable on some straightforward set of indicators, but as contested issues within the sphere of public action. People's vulnerability to destitution, it argues, cannot be relieved by markets alone. But the response of public organizations is a matter of conflict and a problem for historical analysis.

The second chapter illustrates this argument with a grim story of conflicting public responses to the millions of street children in Brazil. The author shows that, while the scale of the children's deprivation is not in doubt, both the nature of the 'problem' to be solved, and the appropriate public actions for dealing with it, are matters of extreme conflict. He also gives a historical dimension to public conflict over children on the street.

Section 2 — The state: a crisis of governance

The second section of the book turns its attention to the state, and addresses the debate over the proper role of the state in public action. Again, both chapters take a historical approach to conflicts of ideas and policies. Chapter 3 argues that a convergence of neo-liberal arguments with other critiques of the state in the 1970s and 1980s has fatally undermined the 'rationalist' model of state action outlined above, and brought to centre stage the question of explaining the behaviour of the state itself. It then traces the conflicting ideas about the reform of the state and their impact so far.

This theme of state reform is then taken up in Chapter 4. The author attacks the idea, common in some circles, that all African states are inherently corrupt, and in its place traces an unhappy story of the breakdown of the probity of the state under economic pressure from the mid-1970s. He argues that converging pressures and policies have driven the public service into a situation of breakdown from which it will be very difficult to extract it.

Section 3 — Development from below: non-governmental action

The third section of the book takes up the question of non-governmental public institutions (NGOs) and their role in development policy. Chapter 5 takes a critical look at the claims of NGOs to promote development 'from below' with and on behalf of the deprived. It concludes that while there are NGOs which do fulfil this 'empowering' role, other non-governmental aid agencies are operating in a top-down manner which extends foreign influence while fragmenting rather than consolidating local initiative.

Chapter 6 then takes a closer look at one less formal kind of non-governmental public action: women's organizations in Latin America. Drawing on women's own views and voices, the chapter explores the extent to which women have been able to develop and put into effect their own distinctive policy agenda, balancing their immediate needs with their longer term and more strategic aims.

Section 4 — Structural adjustment

Section 4 tackles the policies of 'structural adjustment'. Chapter 7 explains the economic logic of the policies enforced. It then goes on to look at some of the problems and contradictions which emerge in practice, as the policies are 'filtered' through the most important institutions structuring our private lives: households, markets and the relations between men and women. It ends by arguing that 'adjustment' will continue to be impoverishing and destructive of long-term development unless the policy process itself can be democratized.

Chapter 8 then focuses on one specific and highly important aspect of structural adjustment, the deregulation of food markets. Its central argument is that the economic models underlying 'adjustment' are too abstract and simplified to provide a good guide to policy. Markets have been deregulated in ignorance of their institutional structure and functioning, and often did not work as expected. The poor and vulnerable who depend on markets have rarely benefited as predicted. Where successes have been achieved, this is generally because of attention to local conditions and the development of new and better forms of regulation. Free Trade is not the answer.

Section 5— Policy as process

The fifth and final section draws our arguments together. It consists of two case studies of public action: the market-oriented reforms in China, and the development of public services in Kerala in southern India.

Chapter 9 examines the impact of market-based reforms in rural areas of China, which abolished the commune system, upon the governmental structures and the public services. It argues that so far, China has succeeded in switching to market-led economic growth without destroying social provision. But the strains on the public services are showing and may become worse over time.

Chapter 10 turns from a huge country to a small Indian state which has a remarkable record in health and education given its low incomes. The chapter argues that this success can only be understood as the outcome of Kerala's history of social reform movements and the unusually strong position of women in society. The level of public participation has held the state accountable for the services it provides, and ensured that the services were effectively used. It concludes, 'The definition of the public sphere must needs incorporate both government and people.'

PUBLIC NEED AND PUBLIC ACTION

1

DEPRIVATION AND PUBLIC NEED

MARC WUYTS

1.1 Introduction

> "The problem of the poor is to stay alive; the problem of poverty is a problem for the rich."
>
> (De Swaan, 1988, p.14)

This opening chapter examines the problem of public action in the face of need and deprivation. It argues that economic growth and the spread of markets do not necessarily bring with them a lessening in human deprivation. Instead, we observe great variation in human health and welfare between countries with very similar levels of income per head.

Why is this? Part of the variation seems to arise from the fact that as markets spread, large sections of the population can become more vulnerable to poverty. When economic growth slows down or goes into reverse, as it did in Africa and Latin America in the 1980s, that vulnerability produces sharp increases in deprivation. So new forms of security are needed if the most vulnerable are not to suffer the full brunt of economic risk.

In some countries, even very poor ones, the state has offered some security through the provision of public services: health, education, subsidized food. In some countries this has worked well to relieve destitution, while in others it has not.

But these successes and failures in turn require explanation. Why is it that only certain needs, certain forms of deprivation, become the focus of public attention? Why is it that in some societies the state provides goods and services to address certain needs, while in others it does not? Why are some needs, and only some, the focus of charitable action or of mutual self-help?

Public definitions of need change over time. Here we will define 'public needs' as those forms of deprivation and vulnerability which, at any historical moment in particular countries, become the subject of public action.

Defining and responding to such public needs involves conflict. The conflict may be between state and non-state bodies. Or there may be conflict within the state itself. The way the 'problem of the poor' is defined, and the nature of the political processes which define it, will influence both public action and its outcomes. We cannot, from some supposedly independent standpoint, define the gaps in the market and look to the state to fill them. Instead, we have to seek to understand the complex determinants of public action.

1.2 The health and wealth of nations

Q To what extent is human welfare dependent on the level and the growth of income?

One influential view argues that economic growth in itself is the most direct way to alleviate poverty. Wealth must be created, it is said, before it can be spent on providing for need. Moreover, as the benefits from economic growth 'trickle down' throughout the economy and society, the poor will have a greater margin to fall back on, and a family's own insurance will be enhanced. Finally, a premature concern with enhancing social security through state spending may well end up reducing the growth potential of an economy.

This argument, besides being prominent in the development literature and in political discourse, appeals to our common sense. We would expect the welfare of a country's population to bear some relation to its economic strength. But is it the only thing that matters?

Q Are poorer countries totally unable to deal with deprivation because of the lack of wealth?

Q Do richer societies cope with deprivation merely because average income is high?

Much can be learned about these questions by looking at the nature of the association between key indicators of the *health* and the *wealth* of nations.

Wealth means health?

Does wealth bring health? To answer this question we can start by looking at some indicators of wealth and health in a variety of countries (Box 1.1).

Why is the comparison of these indicators useful? Although they are all average measures,

Box 1.1 Indicators of health and wealth

The *gross national product* (GNP) is a measure of the overall size of a country's economy. It includes wealth produced elsewhere but bought into the country. For example, millions of migrants earn money abroad and send part of it back home to support their families. Such incomes will be included in the GNP of the receiving country. Conversely, profits repatriated by foreign private investors will be deducted from the national income of the country in which such profits originate.

GNP per capita is the total GNP divided by the number of residents of a particular country. It is a measure of average wealth, but clearly it says nothing about the actual distribution of wealth in a country.

The *infant mortality rate*, measured as the number of deaths in the first year of life per 1000 live births, is often used as a key indicator of socio-economic development. This may appear odd because it concerns a death rate, and moreover, one that only applies to infants. However, we can look at it as an (inverse) indicator of health or, better still, of survival. Furthermore, the fact that it only applies to infants should not worry us. Firstly, infant mortality is quite strongly, and positively, correlated with adult mortality. And more importantly, young children are most vulnerable to adverse socio-economic or environmental conditions. In this sense, the infant mortality rate is a composite measure of 'a genealogy of hazard, in the form of low family income, lack of sanitation, ignorance, discrimination, crowding, high fertility, or exposure to toxic substances. Many of the direct and indirect causes of death in the very young interact, so that it is difficult to pinpoint a single fatal factor' (Hancock, in Ekins,1986).

Life expectancy at birth is another indicator of mortality. Like infant mortality, it reflects the average standard of living of a society, or, more precisely, the average conditions which secure the survival of people. However, because it is based on both childhood and adult mortality rates at a given point in time, it avoids bias from the effects of the age structure of the population. That is, if we compare a population where most people are still fairly young with a greying population, we may find that the death rate in the latter is higher merely because there are so many more old people. The measure of life expectancy at birth avoids this error.

which hide variations between social classes and between regions of a country, the health indicators nevertheless prove to be far more sensitive to deprivation and hardship than the indicators which measure average income. So, indicators such as life expectancy or infant mortality give a better reflection of the extent of satisfaction of basic needs than an economic measure such as the gross national product (GNP) per capita.

Life expectancy and GNP

Here, we shall use life expectancy (expressed in years of life) and GNP per capita (expressed in US dollars) as indicators, to evaluate the relationship between national wealth and levels of health. Figure 1.1 provides us with a visual display, a scatter diagram, of this relationship for 110 countries for which data were available for 1989 in the *World Development Report.*

As we can see, life expectancy increases rapidly at first as GNP per capita rises, but tends to level out for the richer group of countries. The shape of the scatter, therefore, tends to confirm that a strong positive association exists, on average, between national wealth and levels of health. However, a more careful look at the graph also reveals that some points in the scatter appear to jump out of the general pattern. But it is difficult to see the pattern clearly because a large number of the countries in our sample, and in the World at large, have relatively low average incomes, in comparison to the fewer richer countries. In fact, 73 countries (or 66% of all points in the scatter) have an income per capita below US$2 500; these data points are all stuck together against the left side of the box.

To unravel the patterns within the data we shall begin by making a close-up display of the far left side of the scatter. To do so, we make a separate plot of all countries with incomes below US$2 500 in 1989. This is done in Figure 1.2. The picture becomes clearer now, and also more interesting. All the more so, because some of the data points are labelled.

According to the World Bank's classification, low-income economies are countries with incomes below US$500 per capita in 1989. For these countries, the average life expectancy

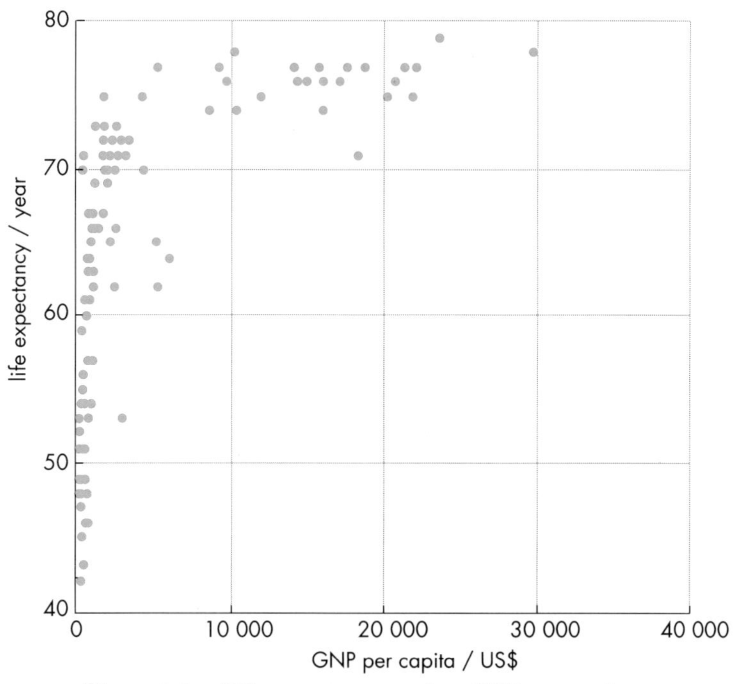

Figure 1.1 Life expectancy against GNP per capita, 1989 [Data Source: World Bank, 1991].

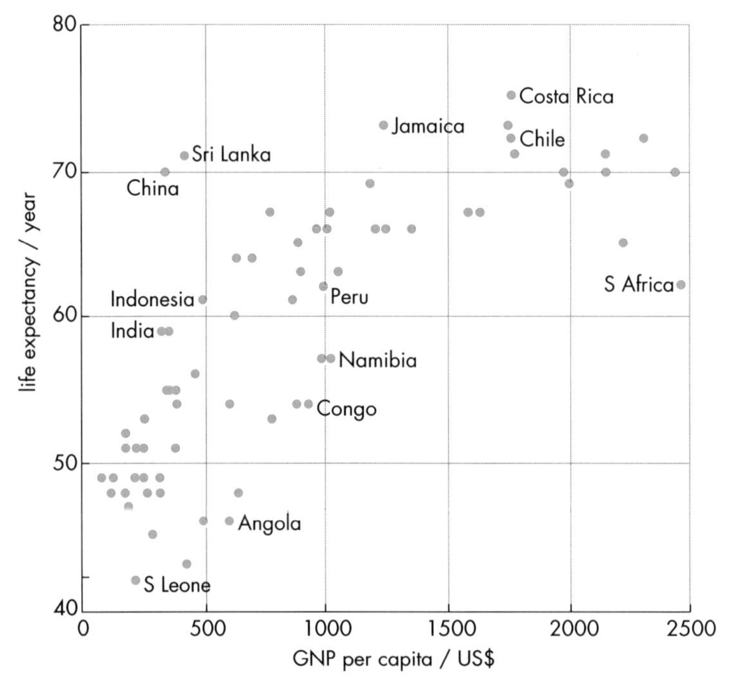

Figure 1.2 Life expectancy against GNP per capita (73 countries with GNP per capita less than US$2500).

(without China and India) was 55 years, but a significant number (15 out of a total of 33 in our sample) have a life expectancy below 50. Against this background, the performances of China and of Sri Lanka are truly remarkable. Both countries have very much higher life expectancy than we would predict from their average incomes.

Looking again at Figure 1.2, we also note the sharp contrast between China and India. Both countries have a comparable level of income per capita, but life expectancy in China exceeds that of India by 11 years. Average indicators, however, often hide considerable variations within a country. This is certainly true for India. The different states vary greatly in terms of public health. One state in particular, Kerala, deserves special attention (see Chapter 10). It is one of the poorer states of India, but its life expectancy is similar to that of China. Interestingly, for women in Kerala both life expectancy and adult literacy are well above those of China. China has a female adult literacy rate of only 56% as against 71% in Kerala. Furthermore, China's female life expectancy is 64.1 years as against Kerala's 67.6 years (Drèze & Sen, 1989).

Turning now to the lower-middle income countries on Figure 1.2, Jamaica, Chile, and Costa Rica also stand out as providing relatively good 'health' in relation to their 'wealth'. This is in sharp contrast to South Africa, where life expectancy is low relative to the level of GNP per capita. This may not surprise us given the apartheid policies of the South African state. In this case again, the average indicators hide a great deal of inequality in the levels of health of its population, as the following quotation illustrates.

> "Among whites, 13 infants die in their first year for every 1 000 live births. Smith notes that there is no such figure available for blacks because many births and deaths are not recorded, 'but a figure well in excess of 100 per 1 000 live births is likely'. Soussan and O'Keefe have culled the following figures from studies published in South African medical journals: infant mortality for settled urban blacks, 82 per 1 000; for 'illegal' or 'unsettled' urban blacks, 107; for 'migrant worker' families, 227; and for families permanently restricted to 'homelands', 282 per 1 000 live births.

If this latter figure reflects the case accurately throughout the 'homelands', then child mortality is much higher there than in the poor black African nations: Burkina Faso (210), Sierra Leone (206). The 1984 *World Population Data Sheet* gives South Africa as a whole an infant mortality rate of 95, higher than such nations as Kenya (86) and Botswana (82) and about the same as Uganda (96). Yet South Africa — based on per capita GNP, a meaningless figure in a nation which keeps 85% of the population poor to keep 15% wealthy — is much richer than those black African nations. South Africa has a per capita GNP of $2 670, compared to Uganda's $230.

Robert Coles, a child psychiatrist who has been conducting research in South Africa for 10 years, offers the following grim conclusion: 'Black infant mortality is 190 per 1 000 live births ... Life expectancy is 15 years less than for whites; and 55% of black deaths occur between the ages of one and four, compared to 7% among whites.'

Adult blacks are not healthy either. In 1979–80 there were 17 reported cases of tuberculosis (mainly a disease of poverty) per 100 000 of the white population (which doctors and hospitals were told about) — and 1 000 among blacks. Blacks have poor if any access to modern health care; there are 300 black doctors, a ratio of one per 90 000 (WHO considers one to 10 000 too low), and few black hospitals; entire black communities must make do without any medical assistance at all."

(Timberlake, 1985, p.183)

This quotation suggests one explanation for these divergences between health and wealth: the pattern of state spending on health care. We return to this issue below.

Figure 1.3 gives us an overview of countries with income levels per capita of US$2 500 and above. Among this group are a number of developing countries which experienced rapid growth in income per capita in recent decades. For some (such as South Korea, Hong Kong, Singapore

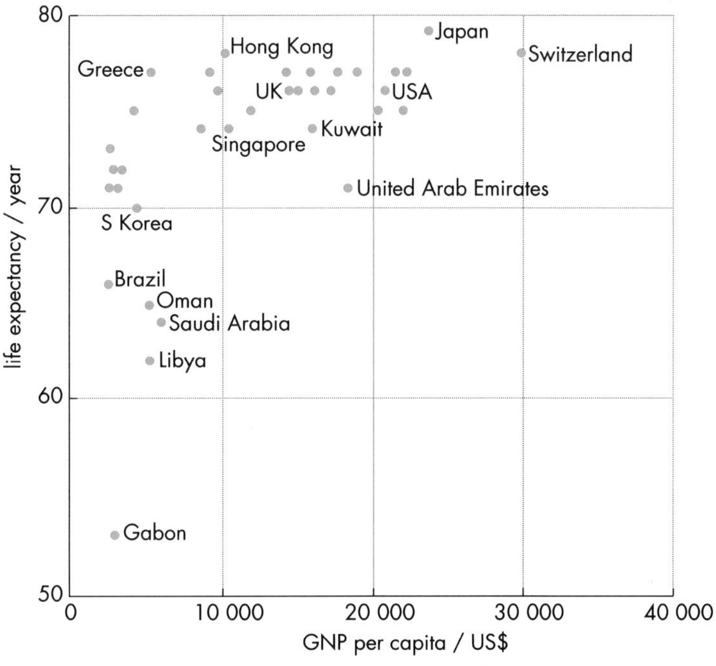

Figure 1.3 Life expectancy against GNP/capita (37 countries with GNP per capita more than US$2 500).

and Brazil) this was due to rapid growth in the volume of output. For others (such as Gabon, Oman, Libya, Saudi Arabia, Kuwait, and United Arab Emirates) the principal factor in income growth was the rise in the price of oil in recent decades. Gabon, in particular, stands out on the graph because it has a relatively small population (1.1 million), combined with an income from oil exports which constitutes about 70% of its total export earnings.

Compare, for example, Kuwait and Oman. Both countries are characterized by high (and unequal) incomes. But, in Oman life expectancy is low for such a relatively wealthy country, while Kuwait had a more impressive record, at least for its own nationals. Similarly, the experience of South Korea is quite different from that of Brazil. In the period 1965–87 both countries witnessed high growth rates in GNP per capita: respectively 6.4% and 4.1% on average per annum. But South Korean life expectancy is substantially higher.

Also on Figure 1.3 are the rich industrialized countries, all of which have levels of life expectancy of 75 years and over. Note, however, that at this upper-end of the scale, *small* changes of 1 or 2 years are quite significant. The reason is simple: as life expectancy increases and approaches old age, it becomes progressively more difficult to add on more years. Given its level of income per capita, the USA, for example, performs significantly worse than Japan or Sweden. Spain, with a GNP per capita of less than half of that of the USA, has a life expectancy of 77 years as against 76 in the USA. One reason might be that in the United States large sections of the population are effectively excluded from access to medical care.

Some conclusions

1. On average, a *positive association* exists between health and wealth. This does not mean, however, that wealth *causes* health. There are good reasons to doubt such a simple conclusion.
2. Some countries (especially Sri Lanka, China, and Kerala State in India) managed to improve health considerably at very low levels of wealth. These experiences are important because, even under the most optimistic assumptions of economic growth, low levels of wealth will unfortunately remain with us in many parts of the globe.
3. Some countries at a much higher level of wealth still have a relatively poor health record. This may be related to inequality in income and in access to health services.

The rest of this chapter explores some of the reasons for these divergences between wealth and health. We start by considering more closely the relation between the spread of markets and the changing nature of poverty and need.

1.3 Markets and vulnerability

So far we have treated deprivation, in the form of ill health and poor life expectancy, as a measurable feature of people's lives. From this perspective, poverty can appear to be an objective characteristic of certain people: those who lack sufficient income to cover a minimal acceptable standard of living. If people have too little income, so the story goes, then they are poor, and the state (or charity) may intervene to supplement their incomes if they can. So it looks as if unequal incomes or poor state services might explain poor average health.

We now turn to considering more closely the causes of poverty. Economic development tends to be characterized by a changing pattern of vulnerability to poverty and destitution, generating the need for new forms of security among the poor.

Poverty and vulnerability

Concepts (and words) matter when we try to organize our thoughts. The word 'poverty' projects a particular image: lack of money, living below the breadline, being in need, being helpless or powerless. This image of poverty focuses our attention on some questions and deflects our attention away from others.

Most of these 'poverty' questions have to do with *measurement*. How many people are poor? But how do we decide who are the poor? Where do we draw the dividing line? Should we focus on 'absolute' poverty or 'relative' poverty? In the former case, how do we decide on the cut-off line; the breadline or, more formally, the 'poverty datum line'? Is poverty a question of inequality in income distribution only? Going one step further, we may want to pinpoint, and measure, certain characteristics of the poor: their education and skill levels, their occupations, their rural and urban location, their fertility and mortality patterns and household composition, etc. All these questions, and many more, propel us to come to grips with measuring the incidence and the structure of poverty at a given point in time. And we can repeat this exercise to get a sense of how the incidence of poverty is evolving and to pinpoint key aspects of any structural changes.

Many of the studies of poverty in different parts of the world have followed this approach. While they have taught us a great deal, they also deflect our attention from less measurable aspects of deprivation. To see the point, let us compare the term poverty with one of its close relatives, 'impoverishment'.

The term 'impoverishment' projects a different set of images: the peasant household, ruined by debt, which is forced to sell its land to the money lender; the clerk or cleaner who gets the sack as public expenditures are cut under structural adjustment programmes; a rural household running away from war and famine into refuge; the smallholder sugar producer confronted with a sharp decline in the price of sugar; etc. Our attention shifts towards the question as to why people become deprived and destitute. We discover that the deprivation of a family is as much related to its *ability to cope with risk* as to its income. The following quotation makes this point:

> "A family with a lower income but with more assets to meet contingencies may be better off than a family with higher income but fewer assets. Families whose assets are mainly productive are especially vulnerable to impoverishment, since disposal of them to meet a contingency will reduce the family's productive or earning capacity. Government programmes, however, tend to overlook the implications of this point. The Integrated Rural Development Programme in India is an example. It is targeted to households below the poverty line and designed to raise them above it in income, usually through a subsidized loan to acquire an asset. But the asset itself may constitute an element of vulnerability. Milch buffaloes are often provided but they are large and indivisible, that is, they cannot be sold in less than single major units, and if they die all is lost. Poverty, in income, may be reduced while vulnerability to impoverishment is increased. In contrast, recognition of the importance of assets which are small or divisible, which spread risks, and which can be disposed of readily without a conspicuously distress sale, points towards smallstock (goats, sheep, pigs, poultry, rabbits, guinea fowl and so on) and trees, which can usually be cut and sold at any time of the year. With these, income may be raised and vulnerability reduced at the same time."
>
> (Chambers, in Ekins, 1986, p.315)

This example shows that an increase in income can go hand-in-hand with a greater vulnerability to impoverishment. Chambers is stressing that vulnerability, instead of being reduced by the extra income, will increase if there are no assets which can be realized to help cope with times of trouble. What matters is not just the quantity of assets, but also the ability of those with assets to diversify the risks which face them.

Destitution has many dimensions, of which poverty, in the sense of lack of income, is only one. Worse perhaps than a lack of income now, is losing the ability to cope with risks over the longer term. Insecurity, vulnerability, risk: these are all important aspects of deprivation.

Markets and vulnerability

One aspect of economic development which can sharply increase the vulnerability of the poor is

Figure 1.4 Bangladeshi children's pictures of floods. After the 1987 floods, there was a 'competition' throughout the schools in Bangladesh, in which the children were asked to paint pictures expressing how they felt about the floods. The chosen pictures were exhibited in Spring 1988.

the spread of markets. Think, for example, about the following argument:

> "In 1981 the economist A.K. Sen published an important and controversial book. In *Poverty and Famines* he argued that the normal working of markets is an important factor in the creation of hunger and famine. As markets spread through and transform rural areas, so individuals come increasingly to depend on the workings of markets for survival, by selling goods or their own labour to buy food. The net result is an increase in the *vulnerability* of many people, especially those who own few resources bar their own labour. Small farmers, pastoralists, labourers, crafts workers become vulnerable not only to drought and pests but also to changes in prices and quantities on volatile markets. Previous payments in kind are transformed into cash: 'more modern perhaps, more vulnerable certainly'. Old methods of insurance against disaster weaken or disappear.
>
> As a result, famines can be caused, or more often sharply reinforced, by the normal working of the market. The loss of crops and animals in a drought is compounded when prices of animals fall for lack of demand, and remaining food leaves a region for better prices elsewhere. Or rising prices in a boom can tip vulnerable people in poorer areas into famine. As Sen says bleakly of the 1974 famine in the Wollo region of Ethiopia, 'The pastoralist, hit by drought, was decimated by the market mechanism.'
>
> Sen is making the point that these disasters are not the result of markets working badly. Markets respond to demand backed by cash, not to needs alone. Sen therefore directs us to consider the non-market determinants of the ability to command goods on the market: ownership of resources and the terms on which people come to market and which influence their ability to trade. Not people's lack of income, but in Sen's phrase, 'how come they didn't have that income?'."
>
> (Mackintosh, in Bernstein *et al.* 1990, p.43)

The central point here, again, is that while the development of markets may raise incomes, it also undermines existing patterns of security: mutual aid organizations for example. Old forms of security become ineffective or obsolete. The poor become more vulnerable to shifts in demand and changes in prices originating at a distance, and less able to cope when they occur (Platteau, in Ahmad *et al.*, 1991). Market integration can both increase productivity through specialization and at the same time increase vulnerability to impoverishment and disaster. Famines are only the extreme of this process. The growing individual vulnerability is widespread.

These points are important because, by the start of the 1990s, most people lived and worked in market economies. Most of the previously socialist countries have been swept by market reforms. The dominant economic philosophy of our time emphasizes the benefits of efficiency, productivity and freedom of choice. It tends to reduce the problem of poverty to one of 'purchasing power' (World Bank, 1986). This approach ignores the vulnerabilities just outlined and looks to market-led growth to resolve the situation.

Section 1.2 already put a question mark over market-led growth as a solution. It showed that the benefits of rising average wealth do not necessarily trickle down to the poorest. Now we have a further criticism of market-led growth: even as incomes rise, vulnerability may rise too. If markets cause vulnerability, then new mechanisms are needed to enhance security in the face of market fluctuations. In seeing what this might involve, it will help to consider the concepts Sen uses in developing the argument just summarized above.

The need for security

People are secure if they can gain access to the goods they need to survive. Sen distinguishes between *endowment* and *entitlement* (Box 1.2)

Failure of entitlement

Entitlement failure includes both vulnerability as an aspect of deprivation and poverty as lack of income. We can define it as the inability of a

Box 1.2 Endowment and entitlement

Endowment refers to ownership of assets and an individual's labour power. Assets can be of varied type: cash balances and other financial assets; productive assets such as land, equipment, buildings, or livestock; a house, consumer durables, and stores of food or other products; etc. The ability to work is an important asset in its own right.

The following quotation illustrates who might lack endowment.

"In land-rich societies the very poor are characteristically those who lack access to the labour needed to exploit land — both their own labour (perhaps because they are incapacitated, elderly, or young) and the labour of others (because they are bereft of family or other support). In land-scarce societies the very poor continue to include such people but also include among the able-bodied who lack access to land (or other resources) and are unable to sell their labour power at a price sufficient to meet their minimal needs. The history of the structural poor in Western Europe during the medieval and the early modern periods turns on this distinction. Until perhaps the twelfth century Europe was a land-rich continent which nevertheless contained many structural poor, who were predominantly the weak, especially those bereft. 'The poor of North Italy, in the tenth century', it has been written, 'are the unfortunate, the disinherited, widows, orphans, captives, the defeated, the infirm ...'." (Iliffe, 1987, pp.4–5)

Entitlement is the command that people can exert over goods in two ways; by using their own resources in direct production and/or by using them to buy and sell on the market. It captures the combined effect of owning resources and being able to use them, in production or in trade, to command goods. Someone may own assets and have labour power (i.e. have endowment), but will have no entitlement if the assets cannot be sold or if paid employment is unavailable.

To illustrate: people who grow their own food are entitled to what they have grown, adjusted for any obligations they may have (e.g. to money-lenders). They can sell, if they want, a part of the product for cash to buy other goods and services, and all the alternative commodities they can acquire in this way are part of their entitlement. Similarly, wage labourers' entitlement is given by what they can buy with their wages, if they they can find paid work. (Drèze & Sen, 1989)

household to preserve a minimal acceptable livelihood with the resources it commands and the production and market conditions it confronts.

How do markets play a role in entitlement failures? Low-income wage labourers in the Third World have little else but their ability to sell their labour. An 'entitlement failure' in this case can come about in three ways:

- a steep rise in the price of necessities, especially food
- a drop in wages
- the loss of a job.

When people spend 60% to 80% of their wages on food, they will be hard hit if the price of their staple food increases sharply. Indeed, the purchasing power of their wages, the quantity of goods they can buy with their cash in hand, will drop accordingly. A fall in money wages has the same effect. In both cases, hunger and deprivation may result.

The loss of a job, however, is much more dramatic and sudden in its impact. Wage labourers unprotected by unemployment benefits or any other form of social assistance are potentially in a deeply vulnerable situation. Note that, in the event of the loss of a job, it is the labourer's only asset, labour power, which has become an element of vulnerability.

Entitlement failures in the case of small farmers, fisherfolk, pastoralists, or crafts workers are also market-related. Typically these self-employed people all own productive assets, engage in direct production, and sell at least some part of their produce. They depend on markets to buy basic necessities, as well as the materials needed for their productive activities. As in the case of wage labourers, entitlement failures for these categories of producers may result from adverse movements in prices. The price of the commodity they sell may drop, or the prices of things they need may increase. Both cause in a fall in real income, but the former tends to have a worse effect.

Apart from adverse movement in prices, entitlement failures can result from a drought or a flood. But, as pointed out above, the normal workings of the market may reinforce the impact of disasters. For example, a drought may hit pastoralists twice: they lose their cattle and meat becomes cheaper relative to other foodstuffs. This type of vulnerability, however, is not a factor which distinguishes small producers from wage labourers. For example, a flood may cause landless labourers losing their jobs at a time when food prices increase sharply.

Preserving assets

What distinguishes small producers from wage labourers, however, is that they own assets, particularly productive assets. A disaster or an economic crisis may provoke distress sales of assets to avoid hunger, or simply to pay a debt. But, as we have seen, disposal of an asset in a crisis may reduce a household's productive or earning capacity. In worse cases, it may threaten a household's whole way of life. This threat to the very basis of one's livelihood is often foremost in the mind of small producers when faced with a crisis. Consider the following passage:

> "During some of Africa's most terrible famines, famine victims have chosen not to consume food in order to try to preserve

Figure 1.5 'The remarkable thing is not how many animals die during famine but how few are sold.' — Herding cattle in Sudan.

their livestock and plant their fields. A few examples will suffice. Mesfin Wolde Mariam, who has greater first-hand familiarity with famine in Ethiopia than probably any other social scientist, devotes a long discussion to this issue. He argues that we have to understand farmers' and herders' fear of continued destitution after the famine in order to understand why they do not sell or slaughter their animals or dispose of their land during famine. The remarkable thing is not how many animals die during famines but how few are sold. The anthropologist Glynn Flood worked among the Afar people in Ethiopia, who were perhaps the worst hit by the famine of 1973. But in his analysis the question he asks is not: 'Will Afar people preserve their lives?' but instead: 'Will they preserve their way of life?'

My own work on the 1984/5 famine in Darfur, Sudan found that, right at the nadir of the famine, people were spending only a proportion (sometimes as little as one-tenth) of their income on food. Although they were hungry and many people around them were dying, nevertheless they could buy food; instead they were spending their money on maintaining their animals, buying seeds for their farms, hiring labour, etc. Their potential income from selling assets was even higher, but they chose not to sell assets whenever possible, and spent much of their money on preserving assets. In particular they were paying for fodder and water to sustain their animals. These were not relatively rich people unaffected by hunger: though it may be difficult to believe, these were ordinary villagers who by any description were 'famine victims'. They had lost much of their entitlement to staple food, but this was not their central concern during the famine."

(De Waal, 1990, p.475)

What we see here, de Waal is arguing, is a response by poor people to acute entitlement failure which focuses, not on their current income (or rather lack of it), but on their desire to

Figure 1.6 Children teach each other school lessons in the midst of famine. Korem relief camp, Wollo region, Ethiopia.

prevent the collapse of their whole way of life. For them, this means trying to retain the assets which have secured that way of life for many years. In other words, this is an example of people attempting to sustain some future security at the cost of present hunger.

Non-market support needed

Though we have used famine as our example, the lessons of this section are wider ones. The development of markets tends to create economic upheaval, breaking people away from older sources of security, and leaving many people dependent on an individual search for wage labour. However rich the country, a household without jobs or assets will be in trouble. Only non-market interventions, whether by state benefits or by help from neighbours, will save it.

We have established that deprivation is not merely a lack of income. Underlying deprivation is a problem of vulnerability to impoverishment which markets cannot solve alone and may well exacerbate, for two reasons:

1 Markets respond to demand backed by cash, not to need. People's ability to provide for themselves through the market is, therefore, limited by their assets and by the market conditions confronting them as buyer or seller.

2 The normal workings of the market entail the vulnerability of many people, especially those who own few resources but their labour. Hence markets by themselves cannot resolve impoverishment. For this, we have to look, as the summary of Sen's argument above suggested, to the public institutions which surround and shape markets: the non-market determinants of the terms on which people come to market.

1.4 Health, wealth and public provision

How can non-market institutions reduce poverty and vulnerability?

There are many forms of non-market provision which can reduce deprivation. They can be summarized by the concept of *public provisioning* (Box 1.3).

Public provisioning can enhance security in two distinct, but interrelated, ways: such expenditure protects vulnerable groups against misfortune, and it may enhance people's ability to secure a livelihood. Thus, state provision of health care helps to guard a family against crippling expenditures associated with illness or disease. Social assistance to those whose access

Box 1.3 Public provisioning

Public provisioning refers to the various ways in which society organizes the provision of goods and services through non-market institutions. It includes:

- State provision of goods, services and social security benefits. Examples in Britain are health care and education, unemployment and child benefit. In Third World settings, possible examples include basic health care and education, child feeding programmes, subsidized food and rural water supplies. The provision of these goods requires public expenditure financed mainly by compulsory taxation.
- Charitable and voluntary provision of certain goods and services financed through charitable donations, mutual contributions, and possibly, supplemented by state grants. Examples in the Third World include mission hospitals or schools, self-help schemes and rural water schemes provided by overseas charitable aid agencies.

to labour is severely impaired can protect them against utter destitution. Access to education may improve one's chances for a better livelihood, just as state supported training schemes and active employment policies can be an effective way to ensure broad access to gainful employment.

Variations in public provision

Most discussions of public provisioning in poor countries focus on state provision. Let us start, therefore, by asking to what extent the variations in state provision explain the variations in the 'health' of nations at different levels of 'wealth' which we discovered in Section 1.2.

Let us begin by considering the case of Sri Lanka. Drèze & Sen, in an important study on which this section draws extensively, attribute the success of Sri Lanka shown in Figure 1.2 to state provision of public health, education and cheap food:

> "Judged in terms of life expectancy, child mortality, literacy rates, and similar criteria, Sri Lanka does indeed stand out among the poor countries in the world. Sri Lanka's experience is particularly worth studying not only for the exceptional nature of its achievement, but also for its timing. Large-scale expansion of basic public services began early in Sri Lanka. The active promotion of primary education goes back to the early decades of this century. The sharp increase in public health measures took place later, but still as early as the middle 1940s. The radically innovative scheme of providing free or heavily subsidized rice to all was introduced in 1942. The fruits of

Figure 1.7 Chinese communes. A painting for the Peasant art movement which began in Huhsien County and spread over China. Peasants painted their own lives and aspirations: building a water tank.

this expansion were also reaped early, and by the end of the 1950s, Sri Lanka was altogether exceptional in having an astonishingly higher life expectancy at birth than any other country among the low-income developing countries."

(Drèze & Sen, 1989, pp.227–229).

This quotation attributes the remarkable success of Sri Lanka primarily to a well chosen set of state expenditures schemes. The authors note that this early progress in developing public health has slackened in recent decades. Furthermore, since the late 1970s, the decline has been associated with a marked shift in economic policy: the emphasis has been put on economic growth, and there has been a noticeable retreat from some of the earlier public intervention schemes.

The case of China is equally remarkable. Here, success was due not only to the state initiating a particular set of public expenditures schemes, but also to a much wider social transformation. Of particular importance in the case of China was the role of the communes: a great deal of public provisioning was managed and partly financed within the context of these local institutions (Mackintosh & Wuyts, in Fitzgerald & Wuyts, 1988). For example, the communes financed 30% of the budget of education and health care (apart from medicines) as against 60% by the government budget and 10% privately, although a substantial part of the cost of medicines was borne privately (World Bank, 1983b). Subsequently, in the 1970s, not unlike Sri Lanka, a programme of economic reforms was embarked upon which implied some disengagements from its earlier course of development (see Chapter 9).

In our discussion of India in Section 1.2 we noted that there was great internal variation, and that Kerala stood out as having a high life expectancy rate. What is especially remarkable about Kerala's experience is that its public expenditures per capita on health care are less than average for all India (Open University, 1985). In Chapter 10 in this book, Gita Sen develops an argument about this remarkable success story. She shows that state expenditures have been important in Kerala but only in a social context which renders them effective. Furthermore, in Kerala this effectiveness resulted from protracted historical processes, in which various strata and classes of society participated in the development of public action against deprivation. Of particular importance was the role of women's autonomy and participation.

This argument about Kerala, in turn, raises interesting reflections about Sri Lanka. Why did state expenditures work so well there? We all know that merely spending money on a particular problem does not necessarily resolve it. State spending, it appears, only improves people's health and welfare in certain contexts: where people can use the state resources provided; where public pressure ensures that state expenditure schemes recognize and respond to real needs of the deprived rather than being diverted into the hands of the better off.

It seems likely then, that the advances in public health and education in Sri Lanka were characterized by widespread popular involvement and participation. Such large-scale promotion of primary education and public health schemes require popular participation for success. That is, they demand broader public action to sustain state action. And it seems likely that the escalation of civil strife and the resulting civil war, by destroying the cohesiveness of society, have undermined the state and hence its commitment to public health.

Turning now to the lower-middle income countries, Jamaica, Chile, and Costa Rica were identified as countries with relatively good health records, and all of these have historically relied on similar state expenditures schemes. The case of Chile may come somewhat as a surprise, but Chile had a long tradition of public action for the improvement of living standards; a tradition with which even the 'monetarist experiment' of the Pinochet regime did not manage to break.

The failure to break down public provision can be compared with the situation in South Africa, a country with comparable average income levels. What is remarkable about South Africa is the extent to which the apartheid regime excluded blacks from the realm of public provision.

Box 1.4 Support-led security and growth-mediated

"One approach is to promote economic growth and take the best possible advantage of the potentialities released by greater general affluence, including not only the expansion of private incomes but also an improved basis for public support. This may be called the strategy of 'growth-mediated security'. Another alternative is to resort *directly* to wide-ranging public support in domains such as employment provision, income redistribution, health care, education, and social assistance in order to remove destitution without waiting for a transformation in the level of general affluence. Here success may have to be based on a discriminating use of national resources, the efficiency of public services, a redistributive bias in their delivery. This may be called the strategy of 'support-led security'."

(Drèze & Sen, 1989, p.183)

Note that a strategy of growth-mediated security is not identical with a strategy of economic growth *per se*. The latter offers no real guarantee that the benefits of economic growth will trickle down into the smallest pores of the economy and society. Instead, it can merely signify a strategy of 'unaimed opulence' or, more graphically, 'casino capitalism'.

The politics of racial segregation, employment policies based on a colour bar, the organization of health care and education, and access to land were systematically structured to enhance the welfare of the white population at the expense of black communities.

Support-led versus growth-mediated security

With the exception of South Africa, the cases we discussed above all concerned countries which relied extensively on public provisioning to achieve often impressive levels of health at relatively low levels of wealth. These countries, to a greater or lesser extent, all pursued policies characterized by what J. Drèze and A.K. Sen refer to as a strategy of support-led security. In contrast, some countries, which we shall discuss below, pursued a strategy of growth-mediated security. Box 1.4 describes the distinction between these two strategies.

To illustrate these distinctions, compare again, for example, Kuwait and Oman. Both countries are characterized by a very inegalitarian distribution of income. But, in Oman the poor are left mostly to their own devices. Kuwait had, before the war, a more impressive record of public provisioning, although, admittedly, one which focused primarily on its own nationals, leaving the large immigrant community on the fringe.

Similarly, the experience of South Korea is quite different from that of Brazil. In South Korea, the rapid expansion of employment opportunities played a significant role in increasing security. This was based on the encouragement of labour-intensive export industries and the often ruthless preservation of highly competitive labour markets, coupled with an active policy of education, skills diffusion and training, and supplementary public works programmes. Furthermore, while South Korea has, until recently, eschewed the idea of large-scale welfare programmes, it had a long-standing concern for the prevention of acute destitution. More recently, the government's commitment to social policies has rapidly increased, in part, no doubt, due to the vocal political discontent and frequent outburst of public protest.

In contrast, Brazil emerges as a country where the poorest 20% of the population have to get by with as little as 2% of the national income and where public services are persistently sacrificed at the altar of economic growth. As Drèze & Sen (1989, p.188) commented dryly, 'unaimed opulence, in general, is a roundabout, undependable, and wasteful way of improving the living standards of the poor'.

Finally, our discussion of the experience of rich industrialized countries in Section 1.2 suggests that it is social security, and not just high average real incomes, which prevents the most abject forms of deprivation in the West. A clear illustration of this point is provided by looking at the expansion in longevity in England and Wales during this century, as listed in Table 1.1.

Table 1.1 Longevity expansion in England and Wales

Decade	*Increase in life expectancy per decade (years)*	
	Male	*Female*
1901–11	4.1	4.0
1911–21	6.6	6.5
1921–31	2.3	2.4
1931–40	1.2	1.5
1940–51	6.5	7.0
1951–60	2.4	3.2

[Source: Drèze & Sen, 1989, p.182.]

Drèze and Sen comment as follows on the patterns revealed by the data:

> "[The table] presents the increase in life expectancy at birth in England and Wales in each of the first six decades of this century (starting with life expectancy no higher than that of most developing countries today). Note that while the increase in life expectancy has been between one to four years in each decade, there were two decades in which the increase was remarkably greater (around seven years approximately).
>
> These were the decades of the two world wars, with dramatic increases in many forms of public support including public employment, food rationing and health care provisions. The decade of the 1940s, which recorded the highest increase in British life expectancy during the century, witnessed an enormous expansion of public employment, extensive and equitable food rationing, and the birth of the National Health Service (introduced just after the war)."
>
> (Drèze & Sen, 1989, pp.181–182).

So, in conclusion, public provision is an important factor in explaining the variations in deprivation found among countries at similar income levels. But this may not be a question of state expenditure alone.

Instead, what appears to matter most is the way in which state spending is embedded in wider patterns of public action, and in different market structures. Wealth creation matters, but *casino-type* wealth creation is a poor strategy to attack deprivation, as the Brazilian example suggested. 'Growth-mediated security' of the South Korean-type, though not without its tensions, succeeds in associating growth with smaller differences between rich and poor and with reduced vulnerability of the poor.

In the poorest countries, there is some evidence to suggest that non-state public action is important in making state expenditure effective.

These conclusions focus our attention on the political processes which define and sustain public action against deprivation. How do deprivation and vulnerability become matters of public concern and the focus of public action? The next section explores that issue in more depth.

1.5 Public need and public action

The continuing existence of poverty and vulnerability, which markets in themselves may change in form but not abolish, provides a clear case for public action. But how do political processes translate deprivation into a public cause? Abram de Swaan formulates the following questions

Q Why, if people fall individually into distress, should remedies not also be individual?

Q Why do some forms of distress become matters of public concern calling for public action by the state or by other forms of collectivity?

De Swaan answers these questions as follows:

> "The interdependence between the rich and the poor, or between the strong and the powerless, is central to the collectivizing process. In feudal times, the poor represented both danger and opportunity to the established in society: the threat of violent attack on the person and property of the rich, and, at the same time, the opportunity to use the 'sturdy poor' as workers and soldiers in the power struggles among competing elites. In later phases of state formation and capitalist development, the poor were seen as a threat to public order, to labor harmony and also to public health, while at the same time they constituted a reserve of potential laborers, recruits, consumers and political supporters.
>
> But on their own, those established in society could not ward off the threat that emanated from the poor, nor could they individually exploit the opportunities that the presence of the poor also afforded them. The external effects of poverty and its potential benefits collectively affected the established in society. And this created a dilemma ... any joint effort on the part of the rich to control the 'externalities' or to exploit the opportunities the poor offered, might also benefit those among the established ranks who had not contributed to it.
>
> Thus, as the indirect consequences of poverty increasingly affected those established in society, they also intensified interdependence among the rich themselves. To the established in society, the problem of poverty represents a problem of collective action. The dynamics of the collectivizing process in poor relief, health care and education stems largely from the conflicts among the elites over the creation of collective goods and the distribution of costs among them."
>
> (De Swaan, 1988, p.3)

This takes us back to the quotation at the start of this chapter. Public action, in this perspective, is the product of tensions within society: tensions between the threats and the opportunities represented by the poor to the rich; tensions between a demand for emancipation on the part of the poor and the drive towards containment and preservation of the status quo on the part of those established in society; tensions among the rich as to how the costs of public action are to be distributed.

This argument in turn suggests that we can understand public response to deprivation as a process of conflict both over the definition of the problem, and over the way in which the problem will be addressed. Political conflict serves to define and redefine a category of public need (Box 1.5).

Box 1.5 Public need

Public need consists of forms of deprivation which become identified as problems in the public sphere: problems requiring public action.

Public need arises from tensions and conflicts set up in society by economic and social deprivation. It is defined by political processes, and changes with the changing forms of deprivation in society.

The character of social and economic deprivation, and the public perception of it, changes as markets spread and grow. As a result, pressures build up in society for public response to these changing needs.

Public need, therefore, is not just an identifiable 'thing', a feature of people's lives. Rather, public need is socially constructed: it springs from economic and social deprivation, but is filtered through political processes which reflect tensions and conflicts in society.

Box 1.6 Public goods

The use of the concept 'public goods' in this chapter, and in this book as a whole, is different from its definition in orthodox economic theory.

In our view, public goods are socially *defined* and *constructed*: the outcome of complex political processes which evolve around the definition of public need in response to poverty and deprivation in society. Public goods, therefore, result from public action prompted by these perceived public needs. The character of public goods will differ depending on the specific complex cooperative and conflicting relations within society in the past and at present.

Orthodox theory, by contrast, defines public goods as specific types of goods or services (such as law and order, defence, the use of a road or a bridge, etc.) which markets find it hard to deal with. The reasons for this market failure (which will be discussed in detail in Chapter 3) are often stated in economic jargon, but in essence they are fairly simple to grasp.

Take, for example, the case of the use of a ordinary bridge or road. Two characteristics of the road imply that there is no effective market for its use. The first is that one person's use of the good does not impede its use by another person (provided the bridge or road is not utterly congested). This is what gives the good its 'public' character. The second is that it is virtually impossible, or far too costly, to prevent a person using the bridge or the road once it is built. That is, the service is generally available irrespective of whether or not a price is paid. Both these reasons imply that public goods cannot be provided through buying and selling, but need to be financed with public money.

Hence, in orthodox economic theory, public goods are defined solely with respect to the inherent characteristics of the goods and services concerned. The question as to which goods qualify as public goods is, therefore, only technical in nature.

Hence, how we view public need depends partly on the society in which we live, and partly on who we are within it. The social processes which identify needs also serve to develop views about which needs can be addressed and how.

This is the second of de Swaan's themes above. Through public debate some goods come to be defined as goods which people should have. At the same time, indeed through the same social processes, collectivities and public institutions come to be created in some countries to ensure their distribution.

However, there is one problem with de Swaan's argument. It appears to suggest that in all industrializing and urbanizing economies, the growing interdependence between the rich and the poor, or between the strong and the powerless, inevitably leads to public action on the part of the rich to create the public goods to satisfy public need. In this chapter, and in this book generally, we shall question this link, while adopting the basic tenor of de Swaan's argument.

Our argument then, is that, in all societies, two processes tend to occur in the public sphere. First, some needs are defined as 'public needs', matters of public concern, while other forms of distress are excluded from the public sphere. Second, different forms of public action, whether by the state or other forms of collectivity, respond to these perceived 'public needs' in complex ways. Both these processes lead to the creation of public goods (see Box 1.6).

Understanding these interrelated processes requires us to come to terms with three basic questions:

Q Who defines public needs?

Q Who pays for public goods: how are the costs of public goods distributed among the various strata in society?

Q Who manages and controls public goods?

Answers to these questions, as the rest of this

53537

Figure 1.8 (Top) Construction of London sewer, 1859. (Bottom) Digging a sewer in Dhaka, Bangladesh, 1987.

book will show, depend not in any simple sense on 'state policy', but on complex cooperative and conflicting relations within society.

Urban sanitation as a public good

An example, taken from De Swaan (1988), may help to clarify this argument. Consider the development of urban sanitation as a public good in 19th century Western European cities. In brief, the story runs as follows. Cities like London and Paris grew at an unprecedented rate, due to massive rural-to-urban migration propelled by rapid industrialization. As a result, established town dwellers and newly arriving migrants, men and women, rich and poor alike found themselves thrown in a turmoil which grew incessantly and anarchically. City life was full of hazards such as 'crime, mob violence, rebellion, the unpredictable cycles of the urban industrial labour market, and mass epidemics'. Cholera, in particular, became the symbol of urban interdependence. Not surprisingly, for those who could afford it, the dominant strategy was to move to wealthier, cleaner, roomier, and safer areas; 'it was a strategy of individual isolationism'. In fact, at the time, most people, even without much medical knowledge, saw that poverty and overcrowding were in some way or other associated with epidemics and illness. Those with money, therefore, left the inner cities and sought refuge in better neighbourhoods.

This process of spatial segregation based on class and status opened up the way for the private provision of water and sewer networks to the wealthier citizens who had moved to these neighbourhoods. Private companies invested in the development of these networks in response to the demand originating from these wealthier neighbourhoods. This required voluntary action by members of a particular neighbourhood to participate in the financing of local services. Indeed, the success of the operation necessitated that all living in the neighbourhood would take part. It involved, therefore, the voluntary pooling of financial resources by the local community to finance the provision of these services.

The construction of these networks, however, not only required the installation of pipes and channels within these neighbourhoods, but also, at the heart of these systems, the construction of a large facility such as a water basin or a sewage mill which serviced the various neighbourhoods connected with it. Obviously, once large investments in developing these central facilities were undertaken, private companies sought to reduce costs by getting as many neighbourhoods as possible within the system.

> "Gradually, the city became saturated with pipes and channels until only a few areas, usually the older and poorer zones and those that were harder to reach, remained unconnected: 'Class divisions thus had hydrological dimensions', writes Berlanstein."
>
> (De Swaan, 1988, p.142)

Up to this point, the process was driven by private demand informed by people's concern for public hygiene, at first in their own neighbourhoods, but increasingly in the city at large. Eventually, the cost of linking up the few remaining areas to the system became relatively low (since the heavy investments in the core of the system were already made), while the benefit of doing so was considerable since these areas continued to endanger the health of the citizenry in its entirety. It is at this point that a coalition of engineers, health experts and administrators, supported by public opinion, managed to push through the necessary legislation to subsidize the water supply and sewerage for all remaining areas. De Swaan concludes: 'only in this last phase did the system become a true public good, extending to all citizens, from which no one could either be excluded or exclude themselves'.

This example is just one story of the genesis of a public good. It does not mean, however, that similar processes of high urban growth propelled by rapid industrialization in a different social context will inevitably lead to similar outcomes.

Exclusion from public goods in Johannesburg

In the case of Johannesburg, the main mining and industrial town in South Africa, the response of the state to rapid urban growth in this century was totally different from that in 19th century London. Here, the effective policy was to remove the black population from the slums in the inner city to newly built townships or to the homelands. Hence, in the 1930s, a first major attempt was made to address the problem of the overcrowding of the black communities in the urban slums of Johannesburg through exclusion from, rather than incorporation in, public need. The 'solution', which did not prove successful in its own terms, was to clear the slums and to build a massive township to house 80 000 Africans 15 kilometres away at Orlando. The main result, however, was an even greater overcrowding of the black population in the fewer remaining slums. The real process of removals only started in earnest from the 1950s onwards. As Iliffe put it: 'this required not merely urban reform but almost total urban rebuilding'. At its base was the Group Areas Act of 1950 which gave the state power to relocate population groups. In its wake, virtually the whole of the African population in Johannesburg was forcefully removed to Soweto, 13 kilometres to the south-west of Johannesburg. This left Alexandra as the one remaining squalid and overcrowded settlement to the North of the city (Iliffe, 1987).

The physical removal of the black communities from the slums of Johannesburg and other South African cities, therefore, illustrates graphically the virtual exclusion of blacks from public need. There was, however, one interesting exception: the question of subsidies for public transport. The reason is easy to understand. The South African economy is virtually entirely dependent on black labour. The African townships, located at the edges of cities and white towns, were often far from most places of work. Moreover, the policy of township development was increasingly redirected to areas in the bantustans insofar as

Figure 1.9 A water standpipe in Soweto, Johannesburg's largest township.

Figure 1.10 Squatter Camp near Cape Town, South Africa.

the latter were considered to be within 'commuting distance' from the place of work. In fact, from 1967, residents of any township within 50 kilometres of a bantustan could be relocated to that bantustan, and be expected to travel up to 160 kilometres a day to get to work and back. In the period 1968 to 1980, 670 000 people were moved on this basis. By the early 1990s there were close to one million commuters from the bantustans, many of whom lived in squatter settlements just within the bantustan borders (Van Ryneveld, 1989).

Obviously, as most of these frontier commuters, and most of the black workers in general, were relatively unskilled and low-paid workers, few were able to bear the cost of commuting. Not surprisingly, public action in protest against these measures ensued. The highly successful bus boycotts in Alexandra and Evanton in the 1950s forced the apartheid regime to provide substantial subsidies for public transport. The state granted monopolies over the provision of public transport to private, white-owned bus companies, which were then assured of 'normal' profits by means of state subsidies.

These two examples provide an interesting contrast, and warn us against making easy generalizations. Each one illustrates a particular set of social processes concerning the definition of public need, the consequent management, finance and organization of public goods. The interaction between wealthy and poor is propelled by economic necessity, but takes place within a social context and through specific political processes. Public needs are not given things; they are socially constructed and reflect the processes which shaped them. Which needs are catered for and how they are responded to varies sharply depending on the social context. The social context in turn defines the realities of public goods.

Street children in Brazil

Chapter 2 in this book, by Tom Hewitt, provides another example. It examines the range of conflicting public responses to the social problem of street children in Brazilian cities in a context where the needs of the poor, of which these street children constitute an important group, are largely excluded from the public sphere. The chapter questions the concept of 'abandonment' of these children. It explores alternative explanations of what is happening including the development of survival strategies among the children themselves. And it describes a context where public responses to this problem, though clothed in moral language, promote more often than not, the continued exclusion of the children from public need and the curtailment of the children's own survival strategies in the face of hunger and oppression.

Famine relief in Africa

For a final example, let us return to the case of famine relief in Africa. De Waal's analysis, quoted in Section 1.3, suggests that famine victims are most concerned with safeguarding their assets so as to preserve their way of life, i.e. their autonomy as producers. In contrast, many donor agencies involved in famine relief see the problem more in a short-term perspective, as one of safeguarding human lives rather than preventing the death of a way of life. De Waal argues, however, that it is precisely the breakdown in the way of life of famine victims which makes the difference between famines which cause hardship, and famines which kill. Here we have a clear problem of the definition of need: who defines it? Who is made dependent on whom in the process? How is finance provided and who imposes what restrictions on the use of funds?

Social context

In conclusion, these examples show that public goods develop within a social context, emerging from a particular state and a particular configuration of political processes. Only some states act benevolently in this regard on some occasions. Public action is the result of conflict and compromise within the public sphere, and these processes determine both the character and the effectiveness of public goods.

Our examples in this chapter have illustrated this argument. Public action, we have established, is an essential response to deprivation and vulnerability. But the examples have displayed a bewildering variety of strategies and great variation in the public goods provided. The rest of this book explores the determinants of different types of public action, in different contexts, and examines and seeks to explain their effects.

Summary of Chapter 1

1 Public need is socially constructed. It springs from economic and social deprivation, but is filtered through political processes which reflect tensions and conflicts in society.

2 Deprivation has many dimensions of which poverty, in the sense of lack of income, is only one. Vulnerability to loss of livelihood also matters. Markets imply insecurity, and their development influences the nature of public need and the pressure for public action.

3 For deprivation to decline, public action must be effective. Even very poor countries can greatly reduce poverty by enhancing security.

4 Economic growth and the spread of markets alone cannot alleviate distress. The effects of wealth on deprivation vary greatly depending on the character of public action.

5 Determining public action and developing public goods are not technical matters but social processes. The effectiveness of public provisioning is determined by the interaction between markets, public institutions and political processes.

2

CHILDREN, ABANDONMENT AND PUBLIC ACTION

TOM HEWITT

2.1 Children in development

> "Brazil endures a famished prosperity. Amongst the countries which export food, it is in fourth place. Amongst the countries which suffer hunger, it is sixth. Now Brazil exports armaments and cars as well as coffee, and it produces more steel than France; but Brazilians are shorter and weigh less than twenty years ago.
>
> Millions of children with no roof over their heads wander the streets of São Paulo and other cities, hunting food. Buildings have become fortresses, janitors have become armed guards. Every citizen is either assaulted or assailant."
>
> (Eduardo Galeano in *O Século do Vento*, 1988)

The street 'urchin' is a common cultural image. Since Dickens' novels, street children have caught the public imagination. The characters of Dickens were drawn from the real world of the nineteenth century industrial revolution. Even before this — in the seventeenth and eighteenth centuries — gangs of vagrants were a common phenomenon in Europe. Many of them would have been children.

In Brazilian cities today, street children are very much in evidence. 'Abandoned' children they are often called, conjuring up images of despairing families pushing children out into the streets. But we should be cautious of such simplistic images. Chapter 1 suggested that societies develop explanations of deprivation, such as that suffered by these children, through contested ideas and competing social pressures. And public response to acknowledged need emerges also from social conflict, the outcomes of which are not predetermined.

One purpose of this chapter is to explain more clearly what we mean by these generalizations, by questioning the 'abandonment' of these children. The chapter examines alternative *interpretations* of what is happening including the development of survival strategies among the children themselves. It then examines the range of conflicting public *responses* to this situation: the authoritarian response of the state; public fear at the presence of children on the streets; the community and church groups working with the children on the street; and the children's own responses, their allegiances and gangs.

This is why we have included a chapter on abandoned children in a book about development policy and public action. Abandoned children are a major group of deprived people: an easily identifiable and priority social concern which has its roots in an uncontrolled and inequitable social and economic system. And the plight of children in developing countries has been very much in the public eye. Since the International Year of the Child in 1979, there has been consistent coverage of the plight of poor children in developing countries. The most recent statement of concern for the rights and welfare of children is contained in the United Nations Convention of the Rights of the Child (1990). This convention dates back more than thirty years to the UN Declaration of the Rights of the Child; rights which range from children's entitlement to have security, care, a name, education and play, to protection from neglect, cruelty, exploitation or any other form of discrimination. Just how little the articles of the declaration have any bearing on the experiences of children in the world today will become evident in the course of this chapter.

In parallel to such well intentioned although inherently limited international initiatives, there has been extensive coverage of Third World children in the media. Sunday supplements, TV documentaries, cinema and news broadcasts provide audiences with a steady stream of helpless, malnourished faces engendering an affluent pity which, often as not, sends people reaching for their cheque books or credit cards. A donation to 'help' the poor. When charitable donations to salve the conscience are not possible, moral outrage takes its place.

A common feature of both international initiatives and media coverage has been that they are led by Western perceptions of childhood. That is, for example, Western adults expect their children to be educated in schools at least until their mid-teens, they do not expect children to need to seek employment (beyond the paper round type) before this age and they expect children to a have a 'home' to go to whether in a family or an institution. Such notions of childhood (despite the growing incidence of young homelessness in Europe) make it difficult to accept other realities as anything but wrong. However, the notion of 'childhood' is not unproblematic, as Ben White warns:

> " 'Childhood' itself is not a universal or absolute category; its definition varies from one society and from one time to another, and also according to both class and gender. Research must therefore deconstruct the category of childhood ... by identifying the social, economic and political factors contributing to its changing definition and to the activities defined as suitable for children."
>
> (White, 1982, p.468)

Children who beg, work, borrow or steal on the streets of Third World cities are particularly emotive signs of deprivation. For those living at closer quarters to such injustice, such as in São Paulo, Recife or Rio de Janeiro, public perception of poor children swings between pity (or moral outrage) on the one hand and fear on the other; the former leading to a sense of helplessness, the latter often leading to violence and, increasingly, to cynical murder.

Figure 2.1 About half Brazil's population is aged under 16.

Brazil has some 53 million children under the age of sixteen or close to half the country's total population. Of these, it is estimated that between seven and twenty million work and/or live on the streets of cities. These are 'guesstimates' so we should not jump to the conclusion that these are abandoned children. Some argue that these figures may be severely exaggerated. Ennew & Milne (1989) forcibly make the point that available statistics are frequently guesses which have entered official documents and become 'facts'. Nevertheless, pervasive urban poverty forces many children (and adults for that matter) to seek out a living on the streets, even though those abandoned are a small proportion of the total.

This chapter is about these children. The aim is to look behind the images of children sleeping rough, sniffing glue and getting involved in petty theft and prostitution, to show that they are the visible tip of a story of urban deprivation.

Abandonment and public action

Abandonment of children is a phenomenon which has repeated itself through the history of the development of capitalism. Public responses to it have also been remarkably similar. Pity and fear have been at the extremes of these responses. That is, either pity driven by charitable altruism or fear for property, life and social disintegration have engendered responses which skim the surface rather than attacking the roots of the problem. Both reactions are dangerous as they take no account of the reasons for social deprivation and, as a result, they do not help inform the kinds of public action more appropriate to alleviating the plight of the children themselves.

Street children are often depicted in an inevitable downward spiral. From rural idyll (as far as raising children goes) to urban misery of low-paid, low-productivity work, then theft and the slide into marginality, criminality and drug addiction. Such images feed into people's fears and prejudices.

But a moral approach to the analysis of street children and child labour is limited. Instead we want to look at the interdependencies between households, the labour market, survival strategies, child labour and abandonment. This puts the image of abandoned children into a socio-economic context; puts them into the wider surroundings and lets us examine available options. Further, it allows us to take a more critical approach to public action intended to alleviate the plight of poor children.

To understand public action we first need to grasp the *private actions* of the children themselves. Private actions in the face of extreme poverty, as faced by these Brazilian children, are survival strategies. The survival strategies of the poor take many forms. We want to focus here on poor children because in the league table of poverty they are often the bottom of the pile. They are powerless but not necessarily helpless.

Public action, furthermore, is not necessarily carried out to reduce poverty but rather to contain and control it. This is the argument made in Chapter 1 Section 1.5. As such, it may not be in the interests of the children who are the object of public action. It is in this broad framework that we will analyse the notion of abandonment.

The first sections of this chapter examine the scale of the problem (Section 2.2); different interpretations of the problem (Section 2.3); and what lies behind the lives of street children and what drives them to the streets in the first place (Section 2.4).

The second half of the chapter deals with responses (Section 2.5). It looks at street children through the lens of public action. The emotive and fearful responses to street children share one thing in common: that something must be done about the situation. But what kinds of public action (whether state action or from other quarters) have been employed to alleviate this visible social problem? We will examine the state's response to street children and the reasons behind that response and we will examine other, non-state and community-based, responses. But first let us see the extent of the problem in Brazil.

2.2 Working children and street children in Brazil

It has been estimated that up to one third of 5–15 year-olds in Latin America are economically

Figure 2.2 Children pirouetting at traffic lights in Mexico City. They then 'pass the hat'.

active (Ennew & Milne, 1989). Of these, only some are street children and even fewer could be described as abandoned children.

The large majority of poor working children live permanently in households, even though they are economically active. In other words, child labour is not synonymous with abandonment. The second category, *children on the streets*, are those who spend their days (and some nights) on the streets. Many of these are the street traders of Latin America who use their earnings to contribute to household income and to help pay for school. The final, smaller, group are *children of the street* who have no contact with families and relatives and have no shelter.

The following extract from Alec Fyfe's book on child labour illustrates that urban child labour in Latin America can take on a wide variety of forms and conditions:

> "Child work in an urban context is highly diverse, ranging between intra-family work, work to apprenticeship outside the family, domestic service, wage labour, odd jobs and errands, and independent activities in the street. In urban areas the spectrum of the child/adult relationship can embrace, at one end, working with parents in such income generating tasks as making hand-rolled cigarettes, weaving straw and packaging home-made craft products. At the other end of the spectrum, the working child may be almost invisible, as in the case of poor migrant girls sold into virtual slavery as sweatshop workers or as prostitutes. Less dramatic, but even more pervasive, is the plight of child domestic workers, often sent by their parents in the rural areas to distant 'relatives' in the cities, where they remain unpaid or poorly paid in return for their room and board. Between these two extremes are the more typical urban child workers living with their families on construction sites, working in factories, shops and restaurants. Then, there are the petty,

so-called 'marginal' activities, of shoe shining, looking after and washing cars, and other casual means of scraping a meagre living on the street."

(Fyfe, 1989, p.96)

Since around 75% of Brazil's population is urban, it is not surprising that the majority of the country's working children live and work in cities. This is not to say that child labour does not exist in the countryside. On the contrary, rural child labour, often in slave-like conditions, has been well documented (Lee-Wright, 1990; Whittaker, 1987). Children work on plantations, in logging and mining. Official statistics show that four and a half million Brazilian children

Figure 2.3 Child labour is prevalent in both rural and urban Latin America. (Top) Cutting cane in Brazil. (Bottom) Carrying someone else's shopping, Rio de Janeiro.

worked in agriculture in 1975, 68% of whom worked more than a forty hour week But it is in urban agglomerates where there is a high incidence of child labour. The evidence seems to indicate that this has increased during the recession years of the 1980s (Myers, in Bequele & Boyden, 1988).

The journalist Jan Rocha describes the conditions of street children:

> "Only a minority [of Brazil's 7 or 8 million street children] are totally on their own — orphaned, abandoned or without any contact with their parents. Most maintain some contact, however tenuous with their family. On the streets, home is a shop doorway, a bench in a square, a hot air duct outside a restaurant, a bonfire on the beach, the steps of a railway station. Bed is a piece of cardboard, an old blanket, newspapers. Some sleep alone, others huddle together for warmth or protection. They never know when they might be woken up by a policeman's boot, a jet of cold water from a street cleaning truck, or even a bullet from a vigilante group or gun-happy officer of the law ... During the day, the street children's main concern is survival — food. To get it they beg, pick pockets, steal from shops, mug tourists, look after parked cars, shine shoes or search litter bins. Frequently glue takes the place of food. They sniff it from paper bags and for a few glorious moments they forget who or where they are."
>
> (Rocha, in Dimenstein, 1991, p.2)

2.3 Interpretations of abandonment

In Brazil, many children work in and out of the household in both rural and urban areas. This work, as we have seen, takes many forms. A related social concern is abandonment. In Brazil, *abandonados* has become something of a catch-all phrase, even for those children who are not abandoned by their families. Some are orphaned and alone in the world. Some are left to their own devices for much of the day. Some leave their families in search of a better life. In all these cases, the term 'abandoned children' is applied.

It is worth looking at the notion of abandonment because the way in which it is perceived is at the root of the different ways which interventions are structured. Here we will look at the various interpretations of the term 'abandonment'. Later, in Section 2.5, we will see what effects these interpretations have had on public action.

Charity and the needy poor

In Section 2.1 there was mention of western perceptions of childhood, and the warning from Ben White that this is not an absolute definition, but one which varies by society and through time. In Europe, for example, there was a gradual change in attitudes towards childhood some 200 years ago. At this time, childhood became increasingly segregated from adulthood.

That children should be protected from the adult world has become a strong and emotive notion in European societies. Charities use this to great effect. For many years, fund-raising by charities has been based on images of children in distress. At the extreme, images of children suffering famine are used to engender pity for those in need. They have been relatively successful in raising funds for famine relief, but have done little to give a fuller picture of the lives of the children they portray.

The exposure through the media of abandoned children is a more subtle example of how notions of a 'proper' childhood can be exploited. Films of street children such as *Pixote* and *Salaam Bombay* to some extent reflect the conditions of life of street children in Brazil and India, but their appeal to European audiences has been to engender pity for those in need (of a childhood) rather than recognition of the more immediate needs (of food, shelter, friendship and hope).

Whether charitable organizations intend it or not, the way that abandoned children are portrayed is that they are victims of circumstances and deserve help. This view is accurate but limited on two counts. First, it is caught in notions of what childhood should be rather than what it is for many of the world's children. Second, abandonment itself is treated uncritically. How children got to be abandoned in the first place and the options available to them are little discussed. Still less are the views of the children themselves taken into account.

Fear of the delinquent

At the opposite end of the spectrum of interpretations of abandonment lies the notion of the vagabond. In place of the deserving poor we find the dangerous delinquent.

This is a view which has a long history. From the middle ages in Europe when groups of homeless and landless roamed in search of sustenance through to the present day in many large cities of the world. Take the following point of view as an example:

> "There are no dangers to the value of property or to the permanency of our institu-

Figure 2.4 Parallels with the past. (Top) Mining in Columbia circa 1980. (Bottom) Mining in Britain circa 1840.

tions so great as those from the existence of a class of vagabond, ignorant, ungoverned children. This 'dangerous class' has not begun to show itself, as it will in eight or ten years, when these boys and girls are matured ... Then let society beware, when the outcast, vicious reckless multitude of New York boys swarming now in every foul alley and low street, come to know their power, and use it."

(The Reverend Charles Brace writing in 1853, cited in Fyfe, 1989, p.97)

This statement was made well over 100 years ago and may be shocking to the liberal sensitivities of a late 20th century social consciousness. The parallels with the past are unmistakable. In the middle ages in Europe and the early days of the the Russian revolution, gangs of roaming children were a common sight (Agnelli, 1986). During the Industrial Revolution in Britain in the 19th century, street children were a commonplace in cities.

Ironically, it is those close at hand who are more likely to take the view that children are a threat. From a distance, it is easy to recognize that street children are not getting the best deal in life and to understand how petty crime is one of the very few options available for survival on the streets. Pity at a distance can rapidly turn into fear at closer quarters. When a perceived threat of loss of property and/or violence against individuals overcomes sympathy, then the way that abandonment is viewed becomes quite different.

Abandoned children become a social problem to be alleviated not because the children themselves deserve a better chance in life but because they are a threat to the functioning of the rest of society. It is a short step then to considering street children as somehow outside society and therefore not eligible for the same rights and respect that other citizens expect. As we shall see in Section 2.5, once abandoned children are seen to be on the margins of society, the reactions to them can also become less than human.

Abandonment as exclusion from public need

Neither of the two above interpretations get very close to explaining abandonment. Both, however, hint that the common denominator in understanding abandonment is that such children have been *excluded* from the social and economic system. How and why this exclusion comes about needs some explanation. The rest of this section, and the next, argues that abandonment can best be explained as the most visible aspect of urban deprivation.

The mechanisms by which public needs are defined are argued in Chapter 1. Why should abandoned children be excluded from the process of social provisioning? The short answer is that children are less powerful than other sections of society. Because the causes of abandonment are intricately tied up with the process of development itself, particularly the persistence of poverty on a huge scale, then to alleviate the plight of abandoned children means tackling the root causes of poverty. In the case of Brazil, as with many other countries, alleviating poverty has not been high on the agenda of development objectives. This argument is elaborated in the following section.

Box 2.1 Interpretations of abandonment

Three views of abandoned children have been sketched out:

- as the deserving poor
- as a threat to society
- as excluded from the socio-economic development.

2.4 Poverty and abandoned children

Q What is it that drives children to the streets of Brazil's cities?

To answer this question we need to examine the wider socio-economic context of poverty in Brazil.

Brazil's economic development in the post-war era has been impressive by any standard. But its social welfare provisioning has been quite unimpressive. In crude quantitative terms, economic development has resulted in a relatively high GNP per capita of US$2 160. But this figure is a bit like the statistician who, with one foot in a bucket of ice and the other in the fire, says 'on average I am feeling very comfortable'.

Per capita national income is merely a measure of total national income divided by the total population. The way that this cake is divided up in Brazil is amongst the most inequitable in the world. According to official statistics, the share of income going to the poorest 20% of the population is 2.4% and the share going to the richest 20% is 62.6% (World Bank, 1990a). Brazil is rich, but few have access to its wealth. As a result, most social indicators in Brazil are worse than many countries with much lower per capita GNP.

Figure 2.5 'Retirantes' (the migrants) by Portinari 1944.

Brazil's under-five mortality rate in 1988 was 85 per 1000, which is almost twice that of Sri Lanka where per capita income was one fifth of Brazil's. Life expectancy in Brazil was 65 years and literacy rates in 1985 for males and females were 79% and 76% respectively.

Part of the problem is income distribution. Another part is that public resources do not reach the poor. Although Brazil's social spending is not as low as in some other developing countries, the benefits go disproportionately to the urban employed (i.e. to rural areas or to the urban unemployed or to large unregistered informal sector). Spending on health and nutrition is already low (2.2% of GDP) but the bulk of this spending (78%) goes on high cost, curative hospital care. Similarly, education spending is disproportionately skewed towards higher education (UNDP, 1990).

In Chapter 1 (Box 1.4) the same point is made. Economic growth does not necessarily mean human development. In the case of Brazil, the emphasis on economic growth at the expense of social provisioning has been aptly phrased 'unaimed opulence'. Wealth has been generated on a huge scale and yet it has not been used to build up the country's social welfare infrastructure. The rapid urbanization of Brazil has exacerbated and concentrated this problem.

Figure 2.6 One view of 'unaimed opulence'. A 'favela' under a São Paulo flyover.

Migration and urbanization

The pattern of land ownership is a crucial factor in the pattern of migration and urbanization in Brazil. Less than 0.1% of landowners own more than 15% of Brazil's land while over 50% of smallholders own less than 3% of the total land. Given the size of Brazil (147 million people in 1989 in an area bigger than Australia) it is remarkable that 75% of total population now lives in cities. The population of São Paulo alone is projected to reach 23.6 million by the end of this decade, making it the second largest city in the world after Mexico City.

The environment of many Brazilian cities mirrors that of many cities in developing countries. They are the centre of economic development and tend to concentrate the wealth of a nation. Thus São Paulo had 10% of Brazil's population in 1980, contributed 25% of the net national product and more than 40% of total manufacturing value added (UNDP, 1990).

Alongside this wealth, there is a concomitant increase in poverty. People were pushed from the deteriorating conditions of the countryside on one side and lured by the 'wealth' of cities on the other. Since the 1960s, rural to urban migration has increased rapidly. A vicious circle set in, as the UNDP Human Development report points out:

> "Urban economies cannot absorb all the rural poor. The persistent problem is that attempts to tackle urban poverty directly — by creating jobs and providing public services unavailable in rural areas — simply attracts more of the rural poor and wipes out any gains."
>
> (UNDP, 1990, p.86)

Even when migration slows down, the present size of urban agglomerates and the cumulative

Figure 2.7 Sleeping rough in Rio de Janeiro.

effects of unchecked expansion are such that the urban environment has become intolerable for many. Poor drainage and the risk of flooding are combined with incomplete water and sewage systems, unreliable electricity supplies, badly maintained roads and inadequate public transport. The lack of all these basic urban amenities hit those in the poorer areas the hardest. Housing itself is in crisis. Squatter settlements (*favelas*, as they are known in Brazil) are the norm for many city dwellers. These are precariously built 'houses' usually quite lacking in the above amenities. It has been calculated that the formal housing sector produces less than 20% of the new housing stock in developing countries. In São Paulo alone there are some 1600 *favelas* built on any available land — hillsides, under bridges and on the banks of rivers and reservoirs (Rocha, in Dimenstein, 1991).

Brazil's 'unaimed opulence' is apparent even to the casual visitor to any large Brazilian city. Ostentatious wealth (luxury cars, large houses surrounded by high walls and guarded gates, shopping centres with all imaginable consumer goods) sits side by side, and seemingly oblivious to, abject poverty (shanty towns sprawling behind luxury apartments, women and young girls handwashing clothes in stagnant, polluted waterways, ragged children running parking lots and carrying someone's shopping).

These are the scenes which prompt outrage. Such juxtaposition of wealth and poverty cannot fail to do otherwise. But this hides the wider picture. Away from the richest parts of town, there is a less obvious (or less visible) social deprivation. This is a structural poverty reflected in all aspects of life: poor housing, poor health facilities, poor education facilities, overcrowded transport to travel many kilometres to poorly paid work. In short, a picture of an urban environment which suffers from rapid growth and a municipal government which cannot cope and anyway does not appear to care except at those infrequent moments when promises of sanitation or electric supply are traded by politicians for votes.

The informal sector

How do the majority of the urban poor earn a living? It is estimated that one in four urban dwellers in Brazil works in the so-called informal economy without official working documents (and, consequently, without access to social welfare provision) (Jaguaribe, 1990).

The 'marginal' or 'informal' economy are terms which have been used to describe the income-generating activities of large parts of the urban poor. It is not an unproblematic term and there has been considerable controversy over its use. One problem is that it is a catch-all term for activities ranging from the very poorest (e.g. shoe shiners and the rubbish tip scavengers) through to productive enterprises with several employees. As such, it is a residual category for any economic activity which is not formal. In practical or policy terms it is not, therefore, a very useful or manageable category.

The *informal sector* as used in this chapter will be defined as 'any way of making a living which lacks a moderate degree of security of income and employment, whether productive or not, working for oneself or others, legally or otherwise' (Foster-Carter, 1985). This is the definition adopted by various authors and which has the advantage of encompassing a variety of forms of income-generating activity without falling into the debates which attempt to categorize these various activities.

Informal sector: activities by which people make a living, legally or illegally, without having any security of income or employment.

In this chapter, we are concerned with the bottom end of the informal sector: the shoe shiners and the supermarket bag carriers. These activities tread the line between illegality and criminality (real or perceived). They are the marginal activities which many, particularly the young, are forced into. For it is here that many of Brazil's street children are to be found and it is here that the results (positive and negative) of public action are most visible.

What characterizes such people's livelihoods (or survival strategies) is that their income-generating options are invariably arduous, low-productivity work which forces individuals to work long hours for a small return. This, in turn, means that children are forced to work in an effort to maintain bare subsistence. Some of this work is done in the household but a substantial part is done on the street.

Child labour is by no means solely an urban phenomenon. But there is somehow a fixed notion that there is a big difference between children 'helping out' on the rural family plot of land and urban children working for a 'wage'. Whereas the rural children's work is deemed acceptable, the urban children's work is not (White, 1982). This difference is related to a debate on population growth (*Allen & Thomas, 1992*, Chapter 4).

Large households in rural areas undoubtedly provide security for parents in old age. They can expand household income through child labour; directly through wage labour or indirectly through children tending siblings to allow adults to do waged work. In cities, however, large households can become more of a liability than a help. Housing is scarcer and access to food becomes more reliant on cash incomes. The more mouths there are to feed the higher the income required. Child labour is a recourse taken up by many poor urban households. Some of this will take place in the household but much of it is carried out elsewhere, including on the street.

If we accept that child labour is a necessary part of the survival strategy of poor urban households, the critical issue is the conditions of that labour. Children's social powerlessness invariably means that their labour can be exploited more intensively than adult labour. Children earn less, have fewer skills and can be made to work longer hours. But for adults and children alike, options are limited due to generalized conditions of poverty and under-employment.

Amongst these options, children may choose to leave their families, like the eight-year-old in the streets of Rio who says:

> "I live here because my stepfather doesn't like me. He drinks *cachaça* (sugar-cane rum) and rows with my mother. He used to hit me, pull my arm, pinch me ..."
>
> (Rocha, in Dimenstein, 1991, p.3).

So for some 'abandonment' may be a better, or the only, option. This is not necessarily abandonment of children by parents although this certainly happens. The converse also occurs. Children make a rational decision to leave. Children, very probably already working, will make calculations and decisions about whether or not to stay with the household. Adults unable to provide for children have little control over the situation. It may not be a one-off decision but something which develops over time. In the process, children will build up multiple sets of allegiances (e.g. commonly to gangs) or will be controlled by relations of exploitation (e.g. as apprentices, factory workers or domestic servants).

Box 2.2 Meanings of abandonment

Depending upon the individual circumstances of children and on the level of analysis, abandonment can have a variety of meanings. In this section, three meanings have been pinpointed:

- Children may be abandoned by their families (either by being left as orphans or simply by being left to their own devices).
- Children may decide to abandon their families in the conviction that going it alone or making allegiances with other children in similar circumstances will better their position.
- Whichever of the above is the case for individual children, a common denominator for street children is that they appear to have been abandoned by society. In this sense, street children have been excluded from definitions of public need.

In short, children may decide that they will be better off if they go it alone or make an allegiance to a gang on the streets than if they stay with their family or relations. The street offers freedom and escape from days locked up in *favelas* while parents are out at work.

Still others have no choice, such as the girl who found violence at every turn:

> "I started living on the streets when I was seven years old, when I lost my mother and couldn't survive. Me and my brother were hungry and then a friend of mine took me on the street ... I became a prostitute. Then I started to steal because as a prostitute I couldn't support my family. When I was thieving the men came to arrest me, they beat me up and did a lot of things. They put me in jail where I had to eat bread and water and spent three days in isolation, before I was taken to the FEBEM (Foundation for Child Welfare) and beaten up again. I used to spend one or two months in prison and then I would escape to the street again. I used to take drugs and start stealing again, and then get arrested and beaten up. Always stealing and getting beaten up. When I got fed up with being beaten I went back to being a prostitute but then the bastards slapped me around if I didn't want to have sex with them. If I was to tell that to the policemen they would just put me in jail and beat me again."
>
> (in Dimenstein, 1991, p.22)

2.5 Public action and abandoned children

So far, it is clear that child labour in Brazil is common and that the incidence of children taking to the streets is high despite sometimes

Box 2.3 Reactions to street children

Public action toward street children can be divided into three distinct areas:

- government action and institutionalization
- public reactions and death squads
- community and children's responses.

The reactions from these different quarters of Brazilian society are driven, in part, by the different perceptions of abandonment set out in Box 2.1.

exaggerated estimates of abandonment. We have seen how different people have interpreted the reasons why children take to the streets. These different views have generated different responses, although no one is in any doubt that some kind of action is needed. However, the action differs with the perception of the problem.

What have been the responses to the growing numbers of street children in Brazil? Clearly, violence is a commonplace. Valdemer de Oliveira Neto, a human rights campaigner, states it this way: 'We are a very violent society. Slavery was formally abolished just 100 years ago. Brutality has been the pattern of behaviour since then'. The threatened children of Brazil's cities would no doubt agree. Violence seems to reflect a desperate reaction to a social situation which is perceived as uncontrollable.

Unhappily, irrespective of whether the motive for action is a desire to 'clean up the streets' or to give the socially deprived a better chance in life, the scale of the problem is such that any response will be an uphill struggle. The argument of this section is as follows:

So far we have seen that the failure of social welfare provisioning on the part of the state has led to serious socio-economic problems. Growing poverty and inequality in urban centres has forced growing numbers of people onto the streets to eke out a precarious existence. Poor children, as one of the most powerless of society's groups, have suffered at the bottom of the pile.

In the past, the state has been trying to contain the outcomes of the problem rather than tackling its roots. Other agencies (community, church, NGOs and children themselves) have tried to alleviate the problem of urban deprivation and the plight of street children. Their initiatives have, by necessity, been localized and small-scale but they have been much more effective than the policies of containment instigated by the state. In this last section, we will examine different types of public reaction to street children.

Government action and institutionalization

We saw in the last section that, for children, leaving the household partially or completely is one means of obtaining some degree of autonomy. Eking out a living on the streets is precarious. Frequently, illegal and semi-legal activities are the only options for survival. Nevertheless, the mere presence of individuals and gangs on the streets is viewed as vagrancy by the police and law enforcement agencies.

The frequent round-ups of 'vagrant' children in Brazil and, in official language, their 'enrolment in schools' has not been successful in remedying the situation. The consensus is that institutionalization has made things worse by merely perpetuating a cycle of marginality and crime.

The National Foundation for Child Welfare (FUNABEM) was a welfare institution set up by the military government in 1964. Along with many of the military government's institutions it was considered a question of 'national security' and therefore immune to criticism (Galheigo,

Figure 2.8 'Vagrant' children in Salvador, Bahia.

Table 2.1 Children taken in by FEBEM in Greater São Paulo, 1979–85

Year	*Number 'assisted'*	*% cases of abandonment*	*% cases of law violation*
1979	26 851	74.6	25.4
1981	29 503	69.6	30.4
1983	30 487	70.2	29.8
1985	38 394	61.3	38.7

[Data source: Macedo, in Cornia *et al.* 1988]

1991). Its brief was to mobilize the the community in seeking solutions to the problems of children. Its role was to take in children abandoned on the streets, given up by parent(s) who could no longer afford to raise them. The institution also took children involved in crime into custody. It was supposed to find means of housing, educating and providing health care for all these children.

Table 2.1 gives official figures on the number of children institutionalized by the state of São Paulo. Growing numbers of children went through the system in the 1980s. An increasing proportion of these were taken into care for violation of the law. In practice, however, there is little distinction made between law breaking and abandonment. As we shall see below, the view that all street children are criminal is prevalent.

As it turns out, the institutions of FUNABEM are little different from the workhouses of industrializing Britain in the 19th century. For the children, FUNABEM means incarceration and fear. The institutions ostensibly set up to protect children in need, in reality rob them of the little autonomy they may have gained when they abandoned their households. More likely, what these institutions do is protect citizens from the real or perceived threat of gangs driven by poverty. Children get the blame for a rise in criminality and are labelled as delinquents. This, argue Ennew & Milne, is simply finding a scapegoat.

> "Most [poor children] are trying to have a childhood, many of them are struggling to survive a mockery of childhood, usually in ways less dramatic than resorting to prostitution or crime. But political propaganda and the media try to convince the public that up to 80% of all offences in Brazil are committed by juveniles. It is a scapegoating of the next generation which results in public demands for closed institutions for punishing child offenders. Children are criminalized by reputation, imprisoned as if they were criminals and, like adult prisoners everywhere, robbed of their human rights. Thus the full force of mainstream society is ranged against a group of children who are particularly powerless."
>
> (Ennew & Milne, 1989, p.152)

Official responses have been to remove the perceived danger of children and lock them up in the modern day equivalent of the poor houses. This has tended to exacerbate, not alleviate, the problem. As one author has described it:

> "FUNABEM is concerned with 'adjusting' the minor to society. The 'needy' or 'abandoned' minor finds in this institution, outwardly dedicated to his protection, a most efficient school of violence, crime and death."
>
> (Machado Neto, 1982, p.536).

This process is illustrated in the following passage:

"There is solid evidence that those institutions which are supposed to take care of the children, only succeed in turning them into hardened criminals. The child ends up joining one of the gangs of the institution. Moreover, the use of violence in these places as a means of discipline only encourages violent behaviour amongst the inmates. In no time at all, the children are back on the street, far worse than when they were taken off it."

(Dimenstein, 1991, p.40)

So the failure of the state to provide adequate institutional care for children in need exacerbated rather than alleviated social problems. In an era when criticism of the government was almost impossible, a scapegoat had to be found. As economic conditions worsened through the 1970s and 1980s, more adults and children took to the streets to make a living. Poor children themselves became at fault. Rather then recognizing that state institutions for children were breeding violence and alienation, poor children themselves were blamed. Street children and delinquency were strongly associated in the public mind. Children have become the targets of violence and brutality.

Public reactions and death squads

The number of children killed by 'death squads' in Brazil reached an average of approximately one a day in 1989. Estimates are that there are over 400 murders of all types a month in Rio alone of which a significant proportion are of children. This compares to 160 deaths a month in New York and 15 in London (Dimenstein, 1991).

In recent years there has been substantial media exposure of violence against children on the streets of Brazil's cities. The death squads, and their documented connection with the police force are not new. During the twenty years of military dictatorship from 1964, death squads operating against 'political' targets were common. What is new is the increasing numbers of young so-called marginals who are becoming the victims of death squads.

In 1990, the human rights organization, Amnesty International, published a report on the violent activities of the police and 'death squads'. It documents a level of brutality which defies belief: many cases of arbitrary arrest, torture and murder. Some 80% of violent deaths are of males aged between 15 and 18. A growing number of murder victims are children under the age of 15.

Amnesty cites a report compiled by the human rights organization, IBASE, and the National Street Children's Movement which concluded:

"These facts ... show a marked presence of organized actions for the elimination of people, in this case children and adolescents, with a view to 'cleaning up the streets', 'removing witnesses' or 'guaranteeing the security' of a given area ... these

Figure 2.9 Boy pointing to the number of children killed by death squads in different northern Brazilian states between 1984 and 1989.

groups are acting in the whole country, without their activities being properly investigated or punished."

(cited in Amnesty International 1990, p.11).

The journalist and writer Gilberto Dimenstein has written a moving account of the extent of murders and violence committed by death squads against so-called vagrants in the streets of Brazilian cities. In it, he documents numerous cases of violence which have frequently ended in murder. Both civilians and the police are implicated, but they are seldom brought to justice (Dimenstein, 1991).

A common theme of these and other publications is that criminality is on the increase while the judicial system is in crisis and cannot cope with what appears to be a massive social problem. The police and, increasingly, ordinary citizens have taken the law into their own hands to 'clean up the streets'. There is a chilling case from Rio de Janeiro where a boy was strangled and a note left on his body which read: 'I killed you because you don't go to school and have no future'. What is behind these private death squads? The long-time campaigner, Bishop Mauro Morelli, explains it this way:

"In my understanding, the killing of children in our streets, behind the killings there is a concept of cleaning, these children make the place disgusting and dangerous. Children running in the streets are seen as a danger to society and to the tourist business. The killings are done by what we call extermination groups, formed by policemen and bandits and paid by businessmen and industrialists who are looking for safety and protection of their own goods and property."

(Dom Mauro Morelli, Bishop of Duque de Caxias)

In other words, fear seems to be the motive. Fear of theft and of damage to person and property. This view is not very different from that of the reverend Charles Brace cited earlier. In recent years the perception that street children are dangerous criminals has spread. As another human rights campaigner explains:

"The death squads say they are cleaning up the city, helping society by killing criminals. But there is another aspect, the population supports the death squads activities. Criminality is so big that the population has no other way of getting security."

(Valdemer de Oliveira Neto, human rights campaigner)

Such fear is breeding increasingly violent responses. Just as the government programme of institutionalization was an attempt to throw perceived dangers behind locked doors, so too the public response to street children has been to strip them of their humanity and then stamp them out. As Jan Rocha explains:

"What must be as hard to bear as the fear and the hunger is the hostility and indifference that street children face from the general public. Brazilians are warm and hospitable, and they love children, but most of them do not see street children as children. What they see in these ragged, dirty kids, drugged or lively, cheeky or sad, is a threat. A threat to their property, a threat to their lives. That is why a few years ago in the centre of São Paulo a lawyer savagely kicked and stamped a 13-year old boy called Jesus to death when he grabbed a woman's gold necklace. A crowd stood round him watching and egging him on. Only two young office girls protested at the brutal act."

(Rocha, in Dimenstein, 1991, p.4)

In sum, successive governments have failed to confront the pressing social problems of poverty in Brazil. Instead, they have tried to make them 'disappear' behind locked doors. Failing in this too, the public appear to have taken the law into their own hands. Condoned and often aided by the police, death squads are practising systematic killings of youngsters who are deemed delinquent by their mere presence on the streets.

Community and children's responses

The state at both national and municipal level has been singularly unsuccessful at tackling the growing social problems of abandoned children in Brazilian cities. Locking them up and hoping that the problem will go away has not only failed but has bred a level of violence and resentment which is quite out of control. The police and death squads have run amok. Citizens treat the children as sub-human and the children themselves trust no-one but their companions on the street. They fear, with justification, that the whole of society has turned on them.

Fortunately, there is more to the story. Disgusted by the state of affairs, many small but significant initiatives have got off the ground in recent years. They are frequently run by a small number of individuals on shoe-string budget. The church has also been involved. A few examples will give a feel of the possibilities. The stress of government policies has been institutionalization and containment. Non-governmental initiatives, frequently with the backing of international agencies such as UNICEF, have emphasized a community based approach. This has involved both emergency help such as food and shelter, but also education projects and work schemes. These latter two are of interest because they function around the existing activities of street children rather than trying to remove them to new surroundings.

Recife, which has one of Brazil's largest concentrations of street children, is in the poor North-east region of Brazil. Here the incidence of violence against children is perhaps the worst in the country. A lawyer, Ana Vasconcelos, appalled by the lives of girls as young as 10 forced by poverty to become prostitutes, opened the Passage House. The house, which provides temporary shelter and food, can only cope with some 60 of Recife's estimated 30 000 street girls.

The Republic of Young Traders in Belém, a city at the mouth of the River Amazon, is a community based initiative which is subsidized by UNICEF. Many children are involved in shoe shining, car washing and petty trade. Some 500 congregate each day at noon for a subsidized lunch. They then set to work. They recycle old goods (fridges, chairs, bicycles, etc) to raise funds for the project. The goods are refurbished by members of the project.

Figure 2.10 The Passage House, Recife.

The Salesian Centre for Minors, sponsored by the Catholic Church, was aimed at shanty towns in general but in practice 80% of its participants are street children. The Centre acted as an employment agency providing messenger boys, carriers, market packers and so on to different businesses. As a result nearly 1000 youngsters have full working documents, rights and benefits, plus an official minimum wage. These are all unusual not only for working children but also for many adults working in the informal sectors. As one observer has noted on this project:

> "By its insistence on dignified wages and working conditions from the companies, and on high quality work and dependability from the [children], this programme demonstrated that it is not necessary to accept either economic exploitation or condescending charity as the price for creating

> employment for poor youngsters. The model was sufficiently replicable that the programme expanded its work to several Brazilian cities."
>
> (Myers, in Bequele & Boyden, 1988, p.130)

The largest non-governmental initiatives is the National Movement of Street Children set up in 1985. Its aim was to expose violence against children in making official complaints to the authorities.

One of the important issues to be raised by the national Movement of Street Children is that of self-advocacy, in other words, the idea that children are not just the passive accepters of 'help' from adults. The movement, itself led by youngsters, promoted a national conference of street children from all over Brazil in the capital Brasília in 1986. 432 delegates aged between 8 and 16 years of age spent three days sharing their experiences of work, education, violences, family, political organization and health. Themes were chosen by the children. Amongst the most 'popular' were: 'Why do we girls have to have sex with policemen to get out of prison?' and 'Why do police have the right to beat minors?' (cited in Fyfe, 1989).

The discussion, as reported in extensive press coverage, was both painful and liberating. Since the conference, exposure of the conditions of street children has become widely disseminated.

Despite the activities of many non-governmental organizations, it is the National Movement of Street Children which brought international attention to the problems of street children and thereby jolted the authorities into action. In the last few years there has been unprecedented media coverage of street children both in Brazil and internationally. The Brazilian government has set up a string of enquiries and legislation, at least, is gradually changing.

However, as Jan Rocha notes in the introduction to Gilberto Dimenstein's condemning book, *Brazil: War on Children*:

> "These are all good signs, but the indignation that should be caused by the killing, rejection or abandonment of any child is still lacking. Action is being taken because the world has thrown up its hands in horror, not because that horror is shared by the authorities, the congressmen, the judiciary, or the general public. Furthermore, the reasons behind the existence of so many street children are still only being timidly discussed."
>
> (Rocha, in Dimenstein, 1991, p.15)

Hopefully this chapter has began to uncover some of these reasons. There are no simple answers, of course. But if one thing has become clear, the scale of social deprivation for Brazil's street children is such that it will not go away without a concerted effort. More important, this chapter, along with many other publications, has shown that the plight of street children will only be ameliorated by attacking the roots of poverty in Brazil.

To date, the development process in Brazil has been one of unaimed opulence. So tackling the roots of poverty is an ambitious task, confronted by formidable political opposition. But anything less will only mean continued deprivation for many thousands of youngsters. For some it is already too late.

Summary

1. This chapter sets out to examine interpretations of abandonment and reactions to abandoned children in the light of these interpretations.
2. Child labour is pervasive in Brazil's cities. Up to 8 million children are calculated to spend much of their life on the streets.

3 Abandoned children are perceived in different ways: as deserving of charity; as a threat to property; as a group excluded from any definition of public needs.

4 Brazil's pattern of development based on rapid economic growth and minimal social provisioning is at the root of children taking to the streets.

5 In this sense, abandonment has several meanings:
- Children may be abandoned by their families.
- Children may decide to abandon their families.
- The state and society at large appears to have abandoned street children.

6 The state's repressive response to street children has been to contain and institutionalize them. This has been done in a most brutal fashion leading to a vicious circle of violence.

7 Violence, characteristic of the Brazilian state, has spread to society as a whole with death squads taking matters into their own hands.

8 With state action completely discredited, non-governmental initiatives have become the most hopeful response for the future. There are signs that the state is now changing its attitude as a result of public outcry.

THE STATE: A CRISIS OF GOVERNANCE

3

QUESTIONING THE STATE

MAUREEN MACKINTOSH

3.1 Some initial questions

Q What is wrong with the state?

Q How should the state be reformed or controlled?

These are two of the dominant political questions of our time. This chapter sets out to survey conflicting answers, comparing recent debates in industrialized countries with changing ideas about the role of the state in economic development.

This chapter begins by tracing common threads in criticisms of the state across the world, showing converging attacks from different political perspectives. Then two more specific questions are raised.

Q What are the ideas about the state which these critics are attacking?

Q What are the assumptions about society and economy which underlie the attacks?

This chapter illustrates the power of changing ideas by tracing the scale of institutional reform which has emerged from the dominant, neo-liberal critique of the state, and from alliances between market-oriented reformers and other critics. The end of the chapter considers some alternative views, and outlines some serious anxieties about the current direction of public sector reform.

The ideas examined here have brought an upheaval in the world of development theory and policy. The very subject of 'development' was built on the idea of the state as the main lever for changing the economy and society. The ideology of 'developmentalism' and the concept of the interventionist state were inseparable in the optimistic post-war beginnings of development theory (White &Wade in White, 1988). The crises and disillusion of the 1970s, and the international politics of the 1980s, brought a dramatic shift to the other extreme. The state came to be seen as the central problem. Other forces, particularly the market and non-governmental organizations, it was suggested, would have to lead economic development.

But the state cannot be abolished. State action will remain an important part of public action for good or ill. Hence, deciding what can be done about the state, will remain an important element in debating what can be done about development.

3.2 What is wrong with the state?

In the mass of critical writing about the state, two sets of ideas can be distinguished:

- the unresponsive but invasive state
- the inefficient but restrictive state.

The unresponsive but invasive state

Here are two critiques of the state. The first was written in 1979 by a group of people who worked for the British state, mainly in professional and semi-professional jobs.

> "The ways in which we interact with the state are contradictory — they leave many people confused. We seem to need things from the state, such as child care, houses, medical treatment. But what we are given is often shoddy or penny pinching, and besides, it comes to us in a way that seems to limit our freedom, reduce the control we have over our lives ... it is not just that state provision is inadequate, under-resourced and on the cheap. ... The way it is resourced and administered to us doesn't seem to reflect our real needs.
>
> State provision leaves a bad taste in our mouths. State institutions are often authoritarian, they put us down, tie us up with regulations.
>
> A distinction is often made between our public and our private life. But even the parts of our life designated private do not any longer, if they ever did, seem to be fully under our control or unaffected by the state and its policies. The state seems at times to penetrate even our closest relationships with each other ... officialdom has a well defined view about the family and what it should be."
>
> (LEWRG, 1979, pp.5–6)

Compare that argument with the following extract. It was written in 1987, on the basis of research and widespread consultation with grassroots women's organizations in the Third World. It discusses 'basic needs' policies, by which states seek to ensure delivery of basic goods and services to the poor.

> "A third aspect of these [basic needs] programmes is their continued use of a top-down approach to project identification, planning and implementation. ... Surely those programmes meant specifically to improve the quality of life of the poor ought to listen to their voices.
>
> Experience has taught us that the absence of local participation in favour of a more bureaucratic approach is not only undemocratic and inequitable but highly

Figure 3.1 Bureaucracy.

inefficient. The Indian experience with anti-poverty programmes (on which the government now spends almost one sixth of public plan outlays) run by bureaucrats in a top-down manner, is that they are poorly coordinated and tend to be insensitive both to poor people's needs and unsuited to local resources, since they are often not accountable to the local population. They are also seen as rife with corruption, and considerable leakage of resources meant for the poor towards the wealthy; furthermore they foster dependency rather than self-reliance, and they engender considerable contempt for the government mechanism."

(Sen & Grown, 1987, pp.39–41)

Figure 3.2 'You have to apply for relief on the prescribed form in triplicate and two witnesses, one a J.P., must endorse it and all applications should be sent by registered post…', Laxman cartoon.

There are many similarities between these two critiques. Both focus on the type of state activity which we generally call 'social services' or 'welfare': direct provision of health, education, housing, water. Both passages emphasize the unresponsiveness of the bureaucracy to people's needs and views, and the inequity which imposes poor services on the most vulnerable and diverts resources away from the poorest. Both point to undemocratic or authoritarian methods of supposedly democratic states with elected governments. Both note shoddy provision, and a resulting response of resentment or contempt by those subject to the programmes.

There are also differences of emphasis. The discussion of India focuses in part on corruption; the critique of the British state emphasizes the invasiveness of the state in investigating and influencing private life. But one could easily find similar passages which take up the theme of corruption in Britain (for example, racism in council house allocation), and of invasiveness in the Third World (for example, top-down population control programmes).

What are the politics of these writers? The emphasis on poverty, need and accountability may have suggested to you, rightly, that the authors would define themselves as of the political left. But the critique of bureaucracy (encompassed neatly in cartoon form by Figure 3.1) was by then widely shared across the political spectrum. And the neo-liberal right would agree with these authors that the state:

- restricts freedom and imposes its view of how people should live
- reduces people's control over their private lives
- fosters dependency rather than encouraging self-reliance.

The inefficient and restrictive state

Now here are two critiques from politically further right, both written by people seeking to reform the state. In the first passage, Mahathir, the prime minister of Malaysia, discusses public (i.e. state-owned) enterprises. In Malaysia, these public enterprises had been seen as important instruments for promoting economic development in general, and the economic advancement of the Bumiputra (Malay) people in particular.

Figure 3.3 The invasive state: this woman is being given a three-monthly injection of the contraceptive Depo-Provera, in Bangladesh.

Hence, the views put forward in this passage imply an important political turnaround.

> "... public enterprises never seem to be profitable or efficient. Even when they are monopolies, they cannot seem to earn their way, much less pay tax or dividends to the owner — the Government. More often than not, a privately owned enterprise which has been making profits and paying taxes, not only ceases to do both on nationalization, but requires subsidies and copious injections of capital every now and then by the Government."
>
> (Mahathir, quoted in Leeds, 1989)

These are familiar political themes: state-owned enterprises as monopolistic and inefficient users of resources, and a drain on the country's Treasury. The equation of inefficiency with subsidy, and the attribution of inefficiency to the fact of state ownership itself, dominate the political arguments for privatization of public enterprises worldwide.

Second, here is an Indian academic, writing in Britain but with experience of the civil service and industrial policymaking in India. Deepak Lal has been an influential voice in developing a critique of post-war development economics, and in particular of the interventionist role it assigned to the state. Here he is arguing against a *dirigiste* role for the state; that is a role which involves the state in planning and organizing production and development, and specifically, attacking the 'basic needs' policies which Sen & Grown were also discussing above.

> "The new gap is between the different goods and services actually consumed by the Third World's poor and those deemed by technocrats to be necessary to meet basic needs. Filling that gap is considered to be a matter of social engineering, which the bureaucracies of the Third World can readily perform. Further support is thereby lent to their *dirigiste* impulses which, in attempting to supplant the price mechanism, have done so much indirect damage to the prospects of the Third World's poor. By not emphasizing enough the inherent limitations of an imperfect bureaucracy at the same time as they castigate imperfect

markets, those seeking to supplant the price mechanism in the provision of basic needs may yet again divert attention from … 'getting the prices right'. …"

"… many developing countries are closer in their official workings to the rapacious and inefficient nation state of 17th or 18th century Europe, governed as much for the personal aggrandisement of their rulers as for the welfare of the ruled. … Yet *dirigistes* have been urging a myriad new tasks on Third World governments."

(Lal, 1983, pp.102 &108)

These views too have echoes of British and US political debates. All states (especially, says Lal, those in the Third World), are inefficient and self seeking bureaucracies, creating more problems than they solve, and serving their own interests rather than those they claim to assist. Worse, the state prevents people from following their own best interests via the market. Technocrats arrogantly but mistakenly believe that they know what people need, and can provide it better than the poor themselves. There is some convergence here with the arguments of Sen & Grown, though in a different language and leading to different political conclusions.

It was precisely the apparent convergence of ideas about the state, across the political spectrum, which opened up the political space for widespread reform. This chapter will argue that the convergence was more apparent than real. First however we must look more closely at what was under attack: as much a set of ideas about the state as the institution itself.

3.3 The 'public interest' state

Implicitly or explicitly, the critics just quoted were attacking certain commonly held ideas about the state. This set of ideas can be called the 'public interest' view.

The influence of this conception of the state as a benevolent institution, can be illustrated by tracing its dominance in orthodox economic thinking about development up to the late 1970s. Until the last fifteen years, economists' working model of the state and the economic policy process could be described as follows.

- There was the *government* as economic policy maker. This was understood as a group of decision makers (politicians and civil servants) who sought to define the public interest, and then to use the economic powers of the state to further it. These economic powers included the capacity to tax, to regulate markets, to spend and to produce goods and services.
- Related to this, there was the *public sector* of the economy. This was defined as the processes of taxation, state expenditure and control, including the provision of welfare services and the control of state-owned industry. Those who worked in this sector

Box 3.1 The 'public interest' view of the state

At its simplest, this has three elements, which we can summarize as follows.

1. It is possible to identify a 'public interest'. Society has some common interests which can be served.
2. The state is competent to identify that public interest.
3. The state will in practice serve (or at least, can be made to seek to serve) that public interest.

Probably nobody has ever held this idealistic view in its pure form! But this general approach to thinking about the state has been an influential one.

were supposed to carry out the decisions of government, in seeking to administer this sector in the public interest.

- The *process* of defining the public interest combined the political judgements of elected politicians (e.g. on matters construed as value judgements, such as the degree of welfare support for the destitute), with the judgements of economic advisers on what were construed as technical matters, e.g. the best way to promote economic efficiency and economic growth.

These assumptions together constituted the conventional wisdom which has now come to be called the 'public interest' view of the state. Oddly, it excluded from economists' professional concerns most of the *internal* workings of the government and its bureaucracy. Those were seen as issues for political scientists.

This apparently rational model of policy-making assigns a clear role to the economic adviser, who assists in identifying the public interest, and economic policies which will best promote it, but leaves decisions to the politicians and implementation to the civil service.

This is not to suggest that before 1970 no one had noticed that politicians can be venal and public sector institutions can be corrupt and incompetent. Far from it. Many economic advisers in the industrialized countries and in the Third World wrote thoughtfully about the limitations and problems of the advisory role. But the fact remains that most economic theorizing was still done as if the position just outlined was a satisfactory conception of the state.

Before we proceed to the demolition job still in progress on this world view, we should examine some elements of the 'public interest' approach more closely. This will help us to evaluate the critiques.

Public interest and market failure

How is the 'public interest' defined?

The short answer to this question is: with reference to the market. Until the late 1970s, the predominant definition of the proper economic role of the state was that it should intervene where markets worked inefficiently. Most of the activity of the public sector was justified on those grounds, including much of health and education expenditure in developing countries. The state might also legitimately, and separately, use its powers to redistribute income or assets to the poor, if society (in the person of political representatives) so decided.

The economic justification for the interventionist, 'developmental' state was thus built on the argument that markets in the Third World tended to be particularly inefficient. We should be clear what 'efficiency' meant here (see Box 3.2).

Such a perfectly competitive economy is not necessarily a *desirable* economy. Those without

Box 3.2 Efficiency and competition

Orthodox economics defines an efficient economy as one where:

- firms and farms are producing at lowest possible cost
- no consumers can be made better off, without making others worse off, by redistributing resources (people, machines, land) between different producers and products.

In probably its single most important theorem, orthodox economics demonstrated that a highly competitive market economy would be efficient in that sense. The extremely stringent conditions under which this is true — which prevent any single firm or consumer (or a coalition) exercising power over pricing — economists call 'perfect competition'.

Box 3.3 'Externalities' and state education

A market works badly when someone who makes a decision is not affected by, and hence does not take into account, its full implications. Economists call this an *externality*.

Here is an example. We can think of money spent on education as an investment in people, which raises their productivity. In a market, such investment will only be undertaken if a return can be foreseen: in this case the additional income of an educated as compared to an uneducated person. But, development economists argued, uneducated people in a poor country may not foresee the widespread benefits from education which lie in the future, when the jobs which need educated people will exist. Worse, if education is not undertaken now, the firms cannot profit by it and so will never create the jobs for educated people.

"The trouble lies in the fact that in an economy where there are as yet no tractors, the supply of drivers is unlikely to be forthcoming and vice versa. ... Left to household decisions, neither the market knowledge, foresight or financial requirements are present which are needed to secure adequate supplies [of educated people]."

(Musgrave, 1966, quoted in Meier, 1984)

Hence the state must step in to develop 'sensible education targets' to support its plans for economic development.

incomes can still be destitute in such an 'efficient' situation. Nor does any such perfectly competitive economy exist. There are too many constraints in the real world: for example, on freedom of movement, and on access to technology and markets and information. But this idea, or 'model', of a perfectly competitive, efficient economy became (and remains) a benchmark for economic policy making.

What this meant was that economists' justification for a leading state role in development was essentially built upon a negative: the concept of 'market failure'.

Market failure and development planning

Post-war development economists argued that developing countries' markets worked badly. They did not transmit accurate information to firms and individuals, so bad decisions were made. Box 3.3 gives an example.

Precisely this same type of argument was used to justify industrial production and economic planning by the state (just the sort of *dirigisme* Deepak Lal deplores in the quotation in Section 3.2). Many of the early development planning theorists were influenced by the experience of the post-war reconstruction of Europe. Here is a summary of the argument by a critic of the time.

"In underdeveloped countries there is little incentive to invest capital in the introduction of modern efficient methods of large scale production in individual industries producing goods for domestic consumption because the markets for the respective industries are too small. Since however the adoption of such methods in any one such industry would increase the demand for the products of the other industries, the incentive would be much greater if investments in a wide range of ... industries were undertaken, or at least considered, in conjunction. The adoption of investment projects which, though unprofitable individually, would be profitable collectively would, it is implied, be a good thing."

(Fleming, 1955, in Agarwala & Singh, 1958 p.272).

Hence, these theorists went on to argue, the state must plan industrial development. Only the state could foresee the growth of the market and the lowering of costs which could result from a mutually reinforcing group of industrial investments, (that is, only the state could predict the externalities of these investments). Hence it should undertake these investments, or at least subsidize the initial excess capacity; later as the external effects came through the private sector

Figure 3.4 Changing ideas about development: state-owned steel plant in Bihar, India, 1969.

could take over. These kinds of arguments are unfashionable now, but in the 1950s they were setting an agenda of debate for development planning.

There are a number of striking features of this post-war development theory. Note the scale of the role assigned to the state. There is great optimism about the state's benevolence and competence. The state is well informed and able to act logically on a complex set of economic forecasts. Industries and services it runs or subsidizes, by implication, will be run efficiently.

How the industries and services will be run, however, is not explored. They replace market activity, hence cannot be run wholly on market principles. Where the market failed there was to be planned production, often apparently assumed to be managed through a simple hierarchy from the government and its advisers downwards. But throughout this literature there was a relative lack of interest in how efficiency was to be ensured, or more broadly, on what non-market principles of economic organization the services were to operate.

Finally, the assumption of hierarchical control reflects the implicit assumptions about accountability. The 'public interest' here is brought into the public sector only through the top, the decisions of politicians. There is no sense of a conflict over differing definitions of the public interest, which might be fought out within the public sector of the economy. Whether the state really acted as an agent of the people was a problem left to political science. Economic policy was recommended *as if* to a benevolent state. Hence, cash in state hands was truly cash to be used for the public good: public sector investment analysis was based on that principle. One might slide unselfconsciously between the government and the country : 'If Ghana were to do ...', meaning the government as representing the nation.

3.4 Class, power and the state

Critics of the concept of the 'public interest' state have attacked its assumptions from two angles. One approach is to deny the first assumption in Box 3.1: the very existence of a single, identifiable public interest. Critics who have taken this line include radical theorists of economic development from Latin America, the Caribbean and parts of Africa. Their views are summarized in Chapter 5 of the second book in this series (*Hewitt* et al., *1992*).

These writers, strongly influenced by Marxist theory, reject the idea of 'the public' as a homogeneous category. Instead, they see society and economy as divided into social classes, the most important division being between those who own the means of production for the market, and those who do not. In the Third World the class structure, they suggest, is further complicated by foreign ownership of large scale production. Hence, the 'national bourgeoisie' of local capitalists is often weak.

The state, then, has to be understood in this class context: as an institution which exercises power (including repression and violence as well as economic and administrative power) in the long-term interests of dominant classes, not in some abstract 'public interest'. Marx himself famously referred to the state, in the *Communist Manifesto*, as, 'but a committee for managing the common affairs of the whole bourgeoisie'.

In the West, such a state might promote indigenous industrial development. It might also create a (partly benevolent) welfare state, both to reduce industrial conflict by responding to some demands of organized labour and to act as an agent of social control (see the LEWRG quotation at the beginning of this chapter).

But in the Third World, these critics argued, the state would be a harsher institution. It would tend to serve the interests of foreign companies who sought cheap labour and resisted local industrial competition. So the state might divert resources from indigenous capitalist development, repress labour organization and block 'welfare' programmes which made labour more expensive, while seeking a niche to benefit its own employees and supporters.

Issa Shivji and Mahmood Mamdani are two African writers who have analysed the state in these terms. Both see state employees as forming a part of the local petty bourgeoisie: a class of small property owners which includes traders and better-off peasants. Mamdani (1976), analysing Uganda after independence, describes the difficult development of the fragmented petty bourgeoisie into a 'post-colonial ruling class' which for a time acted as the 'guardian' of the interests of the 'metropolitan' (external) bourgeoisie.

In consolidating such an 'intermediary' ruling class, control of the economic functions of the state was crucial.

> "Given that it is located both within the state (state bureaucracy) and outside of it (kulaks, traders), the petty bourgeoisie has two alternative methods of accumulation: to use the state to create public property, which the petty bourgeoisie would then control *indirectly* through its control over the state; or to use the state to expand individual private property, which the petty bourgeoisie would then control *directly* through ownership."
>
> (Mamdani, 1976, p.314, emphasis in original)

Both writers analysed the post-independence expansion of the public sector of African economies, through nationalizations and state control of internal and external trade, as an outgrowth of the class interests of the state bureaucrats. With reference to Tanzania, Shivji wrote that, as state economic activity grows:

> " ... on top of the administrative bureaucracy of the civil service-type running the state machinery there arises the economic bureaucracy ... [which] is involved *directly* in running the production process."
>
> (Shivji, 1973, pp.10–11)

This, Mamdani argued, had two advantages for the 'governing bureaucracy'. It gave it an economic base independent of the traders and peasants, and allowed it rhetorically 'to identify its particular interests with the general interests of society' through an ideology of national and 'socialist' (i.e. state-led) development (Mamdani, 1976).

This economic base was reinforced by the mutual benefits found in joint ventures between state and multinational companies. Overseas companies gained funds, from compensation, and 'from agreements involving patent rights; hiring of trade marks and expert personnel, supplying of management consultants, training of citizen bureaucrats etc.' (Shivji, 1973), while retaining management control. Local economic bureaucrats and their allies gained lucrative directorships, salaries and patronage. The local economic bureaucracy and external capitalists formed an alliance, one which bred corruption and the use of public office for private gain (Mamdani, 1976). Hence, 'bureaucrat capital' became an important basis for the private accumulation of capital in rural and urban areas. (Mamdani in *Bernstein* et al., *1992,* see also Chapter 4 of this book).

This, like the 'public interest' arguments, is a simplified picture, but it contains a coherent (and cynical) vision of the state as an institution serving powerful private interests. It also, note, contains a different view of the market. Here is no pure competitive system, functioning, where it works well, in the public interest. Instead, this market is composed of firms exercising economic power in the interests of their owners. The self-interested exercise of power by the state derives from (rather than correcting) the unequal exercise of market power. In this, these writers differ sharply from our next set of critics, who also see the state serving private interests, but ally this to a more favourable judgement of the market.

3.5 The 'private interest' state

That 'public servants' can act in their own private interests at the expense of the poor, is a perception as old as bureaucracy itself. The neo-liberal theorists of the 'private interest' state, whose ideas have come to dominate public sector reform, agree on this point with the last set of critics. But they associate it with a much more

Box 3.4 The 'private interest' view of the state

This approach, often called the 'public choice' school of economics, brings the market into the state. More precisely, it extends the individualist assumptions of orthodox economic theory from those who buy and sell in markets to all those involved with the state. Politicians, voters, civil servants are all assumed to act solely in their own interests: they pursue individual gain, not the public good.

Note that this is not to assume that politicians and state officials are corrupt. On the contrary, the proponents of this view (in contrast to the previous critics) tend to argue that to pursue one's own interests is an acceptable and indeed morally valuable approach to life. We have always recognized this in the market, they say, so why shirk from recognising it when studying non-market decision making? These theories are therefore sometimes referred to as the 'economics of politics'.

Figure 3.5 '... bridge, hospital, roads, water... You want me to promise all these now? Where is the hurry. The elections are still far away!'

favourable view of competitive markets. In effect, the neo-liberals turn the assumptions of orthodox economics against the 'public interest' theorists themselves, focusing their critique on points 2 and 3 in Box 3.1, the competence and motivation of the state. (See Box 3.4.)

This radical rethinking of the state produces a quite new set of ideas about 'the public' and 'policy'. 'The public' here is composed, not of social classes, but of individuals whose preferences should be of equal weight in determining political decisions. And since politicians are assumed to have re-election as their sole aim, the best poli-tical system is therefore one which best forces politicians to respond to majority views.

'Policy' then, is not the pursuit of the public good by benevolent institutions, but the more or less satisfactory response to public opinion by political institutions staffed by self-seeking individualists. The design of institutions is therefore crucial. And governments, like markets, can 'fail' to work well if their organization allows individuals to exercise undue power.

Government failure

'Private interest' theorists have developed some sophisticated arguments about how, given their assumptions, government can go wrong. They have questioned the competence of the state to put right the failures of the market, and catalogued the costs of the attempt. Box 3.5 lists some of their arguments about 'government failure' which have most influenced the debate on the role of the state in development.

Bureaucracy

'Private interest' economists, unlike their 'public interest' cousins, examine the internal workings of the state. They distinguish the motivations and activities of bureaucrats from those of politicians, and have constructed an influential 'model' of bureaucracies.

A 'bureau' here is a tax-financed, monopoly provider of, say, a social service in a local area. The

Box 3.5 A glossary of the private interest state

The economic theory of bureaucracy — the view that bureaucrats exploit their monopoly of information and services; in order to expand their budgets, powers and perks.

The 'Leviathan' state — the view that the state always tends to grow into a monster : monopolistic, too powerful, larger than the citizens would wish.

Rent seeking — the waste of resources resulting from individual pursuit of the income-earning opportunities created by state regulations.

state officials' behaviour is determined not by the aim of providing a good service, but by the pursuit of their own interests. One theorist (Niskanen, 1971) provides the following list of what those interests might be: salary, perquisites of office (perks), public reputation, power, patronage, output of the bureau, ease of making changes, ease of management. He then argued that most of these interests would be served by enlarging the budget, and hence the size, of the bureau.

We can therefore picture the bureau as a monopoly supplier of services seeking ever more funds from another monopoly: the Treasury which supplies funds. The bureau has a number of advantages in this bargaining process; in particular, it has far more information about its activities than anyone else, making it difficult for even the treasury to judge funding needs. Niskanen concludes from these assumptions that state 'bureaux' will tend to grow too large, by exploiting their bargaining power in the interests of employees. And that they will also tend to be more expensive than necessary, since the bureau's monopoly of supply and information insulates it from cost-reducing pressure (from competition or taxpayers) and allows perks to rise.

Leviathan

The expansion of government budgets by bureaucrats is thought to be reinforced by the self-aggrandisement of politicians, and, especially, by alliances between politicians and powerful élites. Powerful lobbies, such as vocal residents in capital cities, can extract state benefits which the more dispersed and less powerful citizens are forced to pay for. The net effect runs against the public good. Here is an influential application of the argument.

> "An alternative approach [to a public interest view] would view them [governments] as agencies that serve private interests. And rather than interpreting government policies as choices made out of a regard for the public interest, it would instead view them as decisions made in order to accommodate the demands of organized private interests. ... Particularly in the area of food price policy, this approach has much to recommend it. Put bluntly, food policy in Africa appears to represent a form of political settlement, one designed to bring about peaceful relations between African governments and their urban constituents. And it is a settlement in which the costs tend to be borne by the mass of the unorganized: the small scale farmers."
>
> (Bates, 1988, p.346)

Food prices in other words are held down in a private mutual accommodation between government and urban dwellers at the expense of farmers. The government budget expands to provide food subsidy, and production falls.

The power of interest groups to bloat the state against the wishes of the majority is often said to be worst in those countries with a heavily censored press and a single dominant party. In these circumstances Lal (cited in Killick, 1989) envisages an extreme version of Leviathan, a 'predatory state' which uses its monopoly of violence, regulation and production of essential goods to squeeze revenues from its populace. State interventions, such as the protection of local industries by tariffs, are understood from this perspective as self-seeking means of extracting revenues. They are not seen as well intentioned, if sometimes problematic, development policies.

Rent seeking

Industrial policies are also thought to have other costs. Private interest theorists attack the argument, described in Section 3.3, in favour of planning. They argue that the state lacks the information to make efficient decisions where the market fails. Worse, people respond to state regulation and control by exploiting the opportunities offered.

For example, if the state requires people to seek a licence for industrial production, then people will tend to waste society's resources by competing (through lobbying, advertising and bribes) for the extra profits which can be made if the licence is obtained. Or, if the state imposes an import tariff, it raises prices, and too many people will move into that now more profitable field. Such diverted activity is called 'rent seeking', because regulation creates 'rents' (higher incomes for some selected beneficiaries) and people waste resources trying to get in on the act. Though some rent seeking is legal, the roots of corruption lie here: regulators and their clients develop patterns of mutual accommodation which run against the public interest.

Reforming the state

Here is the World Bank (1988a) summarizing the private interest view:

> "Although the pursuit of private interests allocates resources efficiently in competitive markets, this generally does not occur when governments use the monopolistic powers of government to their own advantage. Politicians, bureaucrats and many private interests gain from a growing government and greater government expenditure."

From the logic of these arguments follow influential proposals for reform. Where politicians and élites exploit the power of the state, constitutional reform must limit it. If government bureaucracies are institutions run in the interests of their staff, then the staff must be faced with material incentives to act in line with the preferences of citizens. If regulation causes waste and corruption, deregulation will help. And if monopoly allows inefficiency, then Leviathan can be controlled by a good dose of competition.

So just as 'market failure' suggested a role for government, so 'government failure' led back to the market. The conclusion was inevitable. Once you assume that the state is a private institution like any other, then from orthodox economic assumptions, the prescription of competition emerges at once. Market failures may be a problem, but no viable alternative principle of economic organization to the market exists.

It follows that exhortations to work in the public interest are pointless. Only deregulation can prevent urban consumers from exploiting small food farmers via their influence over the state, a proposal examined in Chapter 8. The state must either be privatized or be made to simulate as far as possible the organization of the private market. A recent discussion of British and American academic work on the public sector put this extreme viewpoint clearly: despite possible reforms 'it is clear that elected governments will continue to be rather inadequate copies of markets' (Bennett, 1990).

These 'private interest' assumptions, not always made so explicit, have come to dominate thinking about the reform of the state in the late 1980s and early 1990s. The neo-liberal reformers have exploited the political space opened up by the widespread doubts about the performance of the state, surveyed in Section 3.1. The neo-liberal desire to restrict the state has found echoes among left-wing critics, who also emphasize self-help and non-governmental action. The 1980s saw some unexpected political alliances against the state.

However, these alliances can obscure profound disagreements about reform of the state. A common political vocabulary of 'self reliance' and opposition to bureaucracy can conceal vast differences in intent: private charity or local control of state resources? A common emphasis on small-scale units and decentralization is frequently used to obfuscate the distinction between privatization (described by Bennett (1990) as 'decentralization to the market') and decentralization of power and control within state structures. It is to disentangling some of these confusions that we now turn.

3.6 Squeezing the state

Q Neo-liberal theorists believe strongly that Third World states are generally too large relative to the private sector of the economy. To what extent have such states been cut or their assets sold off, in recent years?

Two forces have combined to reduce the size of the state over the last decade. One has been economic crisis, which has reduced both govern-

Box 3.6 How to squeeze the state

Here are the methods, with examples from Britain (a pioneer).

Cut state activities. Provide fewer goods and services, either directly, for example by building fewer council houses or recruiting fewer soldiers, or indirectly, for example by funding fewer independent advice centres.

Sell state assets and companies (privatize). Sell state owned utilities, such as electricity and water, to private shareholders; sell state-owned land and buildings to private owners.

Buy in more services (contract out). Purchase services such as road cleaning or care for the elderly from private and non-profit suppliers. This *may* be cheaper; it will certainly cut the numbers of people who work directly for the state.

Charge more for state services. This might reduce the use of the services. It will reduce the taxes which have to be raised to finance a given level of services.

Cut benefits. Spend less on cash benefits for individuals such as state pensions, social security and child benefit.

Between them, these methods aim to cut what the state spends, the taxes it raises, its workforce and its output.

ments' tax revenues and their ability to borrow. The other has been deliberate policy. Neo-liberal governments working from 'private interest' premises have sought to reduce the size of the state; while other governments have been pushed into similar reforms by overseas agencies, notably the World Bank. To squeeze the state, governments must either cut the state, or sell parts of it off (see Box 3.6).

In Britain in the 1980s, privatization and contracting out cut state employment; changing the pension rules reduced the growth in benefits. But state spending relative to the size of the economy only fell in the late 1980s.

Figure 3.6 Britain in the 1980s pioneered the wave of privatization of production and service provision: privatized rubbish collection in the London Borough of Wandsworth.

Cutting the state

How much have Third World states been cut? To explore this question, we first need some definitions.

So what do the (not very reliable) public expenditure data on the categories listed in Box 3.7 tell us about changes in the size of Third World states?

First, there is less scope for cuts than in the West. Most Third World states are relatively small. They spend very much less per head than governments of industrialized countries, and (except for some wealthy oil states), generally spend less in relation to GDP (Figure 3.7).

Furthermore, the structure of spending affects where cuts may fall. There are fewer transfers in the Third World, since there is relatively little social security, especially in very low income countries (Figure 3.8b). Only in a few semi-industrialized countries such as Malaysia and Brazil do payments into social security funds exceed 5% of GDP (World Bank, 1988a). In Britain, transfers are 40% of public spending and 20% of GDP. Subsidies, though, are a possible area for cuts (Figure 3.8a).

Otherwise, three areas can be cut. One is the military (politically difficult). Another is investment and other 'developmental' spending. These

Box 3.7 Measuring the state: some definitions

Public expenditure is spending by the state, financed from taxation, from (some) borrowing and from (some) aid funds.

It includes the following four categories of (annual) *current spending.*

(i) *Goods and services* — direct provision by the state of e.g. health services and teachers. It also includes the military. The category is sometimes called 'government consumption'.

(ii) *Transfers* — cash benefits to individuals, for example social security payments.

(iii) *Subsidies* — payments, usually to firms to keep down prices of goods and services (food for example).

(iv) *Interest payments* — for example interest on overseas borrowing.

Public expenditure also includes:

Capital spending — investment by the state, for example in roads, bridges and hospitals.

The *government budget* accounts for the above spending, and for how it is financed. Financing will include taxes, revenue from import and export tariffs, grants (e.g. aid) and borrowing (locally or abroad).

Public expenditure does *not* include most of the output of the *public enterprise sector.* State-owned companies which sell their goods on the market are not counted in state spending, except (sometimes) where supported by government grants or loans.

Finally we can measure public expenditure in *absolute* terms (e.g. how much per head of the population) or we can measure it *relative to Gross Domestic Product (GDP)*: i.e. to the total public and private sector output of a country. And we can measure it in *nominal* (cash) terms; or in *real* terms, that is, in terms of what the cash will purchase, making allowance for rising prices.

categories ('capital investment' and much of 'economic affairs and services' in Figure 3.8) include: physical infrastructure such as roads, railways and dams; agricultural support services; mining; and spending on public utilities such as water and power. The poorer the country, the more of its budget the government tends to try to spend on building up physical infrastructure.

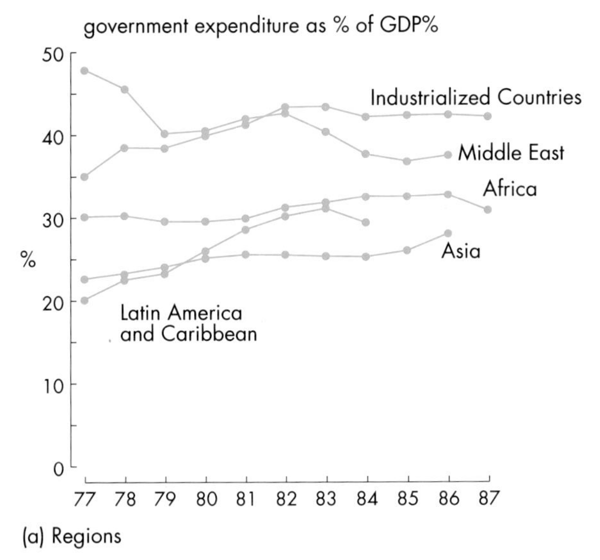

(a) Regions

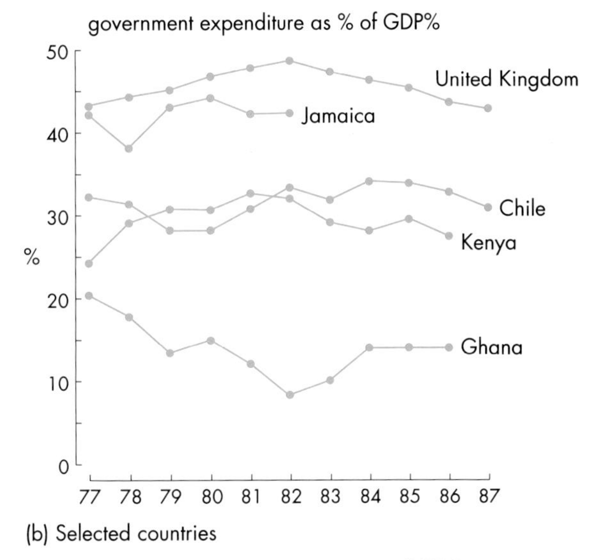

(b) Selected countries

Figure 3.7 The relative size of governments: general government expenditure as a percentage of GDP.

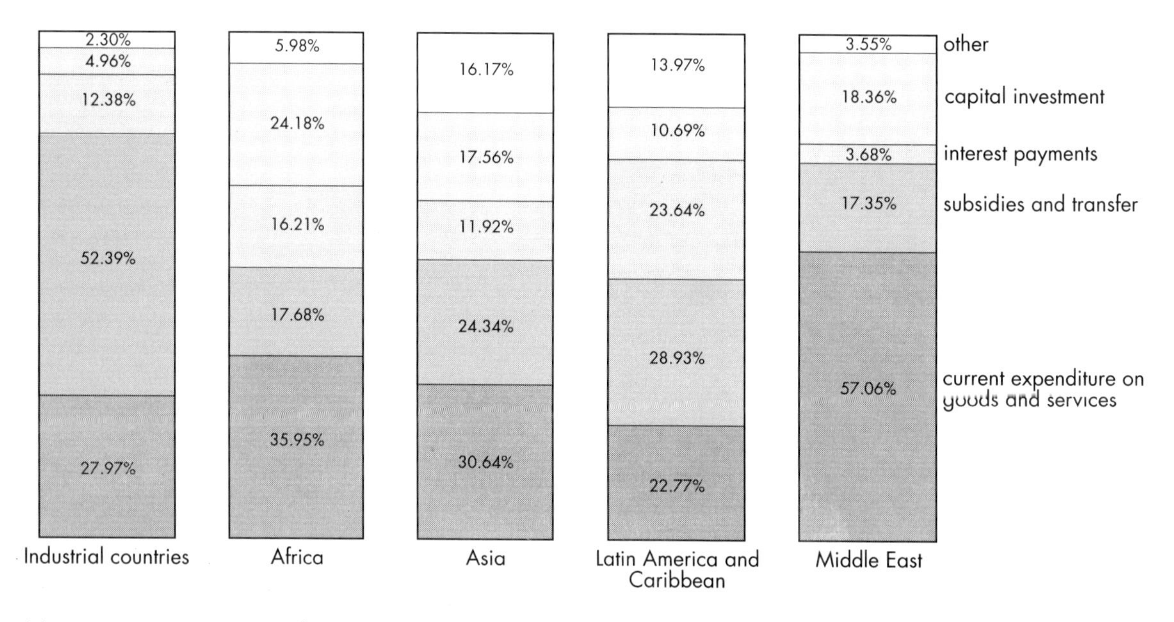

(a) By economic category (% of total expenditure and lending minus repayment)

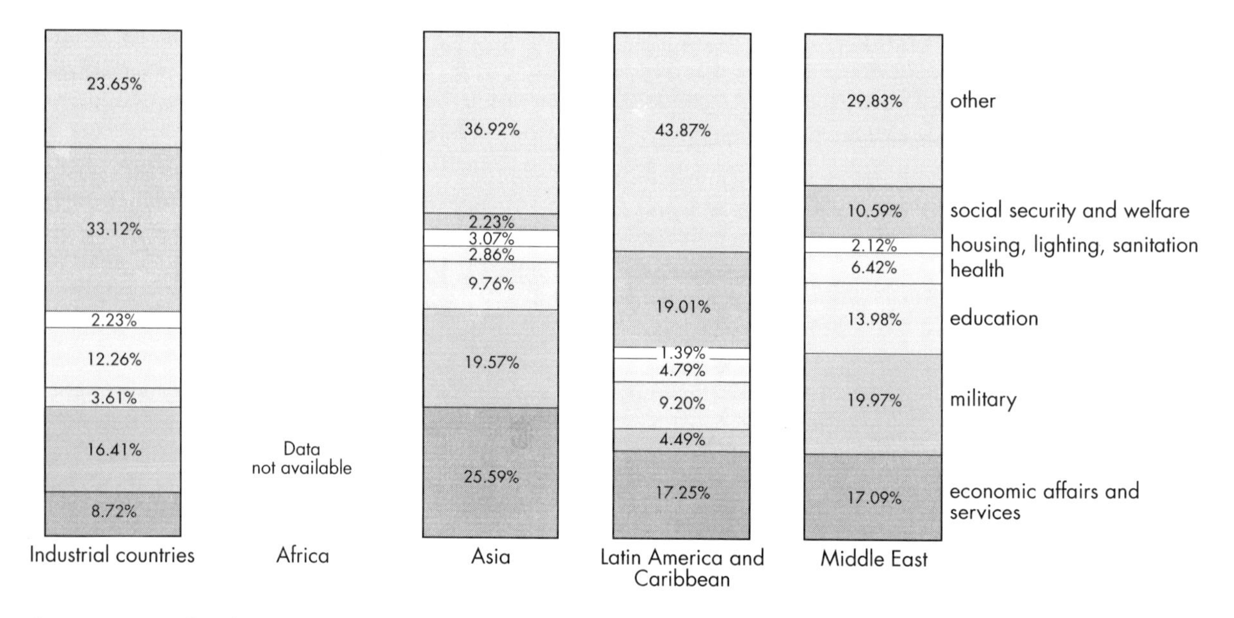

(b) By sector (% of total expenditure)

Figure 3.8 Central government expenditure by category, 1985.

The third area is social services, mainly health and education. Developing countries tend to spend relatively more on education and less on health than the industrialized countries (Figure 3.8b). But absolute spending is generally very low. In 1980, before the worst of the economic crisis, low-income countries' governments spent on average only 8% of their budgets on social services, or less than US$1 per head. Better-off middle-income countries (the classifications are the World Bank's) spent 35%, and industrialized countries 56%.

It is hard to generalize about the squeeze on the state in the very diverse Third World. But it seems that up to the early 1980s, government spending in developing countries had generally grown more rapidly than GDP. This furthermore had seemed reasonable. Developed countries had larger states and, as incomes rose, people want better public services. In particular there was also a widespread belief that a developmentally active state was essential for development (Section 3.3). Even after 1975, when economic growth slowed in most of the world, governments tended to borrow to protect state spending. This was partly because worsening poverty increased public need and partly because governments tried to keep up state employment and services, hoping the crisis would pass.

The crunch came with the debt crisis in the early 1980s. Growth turned into stagnation or decline and this time public spending was squeezed too. Table 3.1 illustrates this turnaround for a number of developing countries. Only in South Korea and Brazil, among the countries shown, was state spending apparently not growing relative to GDP in the 1970s. In the early 1980s, the absolute decline in state spending was especially

Table 3.1 Annual growth rates in selected countries: gross domestic product (GDP) and government spending on goods and services (GFCE)*

	1970–75		*1975–80*		*1980–85*	
	GDP/%	*GFCE/%*	*GDP/%*	*GFCE/%*	*GDP/%*	*GFCE/%*
Africa						
Kenya	4.5	8.3	6.6	8.7	2.3	−0.3
Ghana	1.2	4.6	1.6	7.7	0.5	−4.2
Zambia	2.9	5.0	−0.6	−2.0	−0.1	−3.3
Caribbean						
Jamaica	1.8	10.6	−2.6	2.3	0.2	−1.3
Trinidad & Tobago	3.0	8.4	7.9	9.2	−4.7	−0.7
Latin America						
Brazil	10.7	7.2	6.6	4.6	2.7	1.7
Chile	−2.3	4.0	7.9	3.4	0.1	−0.7
East & South Asia						
S. Korea	9.4	7.6	8.3	6.3	8.2	4.2
Malaysia	10.3	12.0	8.3	8.7	4.7	2.7
Indonesia	8.5	10.9	7.8	13.0	4.2	4.4

* government final consumption expenditure

[Data source: United Nations Statistical Office; National Accounts Statistics] marked in Latin America, the Caribbean and Africa, where debt became a serious burden.

The squeeze was undoubtedly most severe where debts were most serious, and especially where very low income countries faced unsustainable debts. Governments could no longer borrow to cover the gap between (declining) tax revenues and spending. And the squeeze was enforced by the International Monetary Fund and the World Bank under the 'structural adjustment' policies described in more detail in Chapter 7. Both institutions believed that government deficits (expenditure exceeding tax revenues) were a cause of economic crisis, and took the view that Third World states had grown too large and were a drain on the economy.

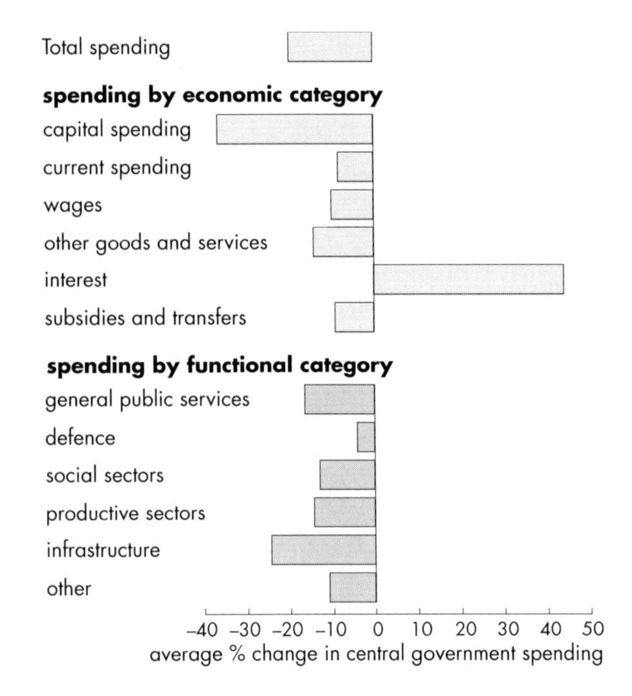

Figure 3.9 Falling real government spending in indebted countries, early 1980s.

As Figure 3.9 shows, in highly indebted countries, real government spending fell in the early 1980s (by 18.3% in the sample shown). And predictably, capital investment in infrastructure, and spending on social services (social sectors) were hard hit. Capital investment is always the easiest element of spending to cut, as in Britain in the early 1980s. Health and education, and subsidies (to food prices for example), came next. They were forced out of government budgets by rising interest payments, and the desire of many (not all) governments to protect military spending from real decline (World Bank, 1988). The weaker the economy, the more the 'developmental' part of public spending tended to be squeezed (Tanzi, 1991).

The effects have been eloquently documented. In the early 1980s, some Latin American countries saw huge cuts in health and education budgets: e.g. public health spending was cut by 75% in Bolivia, and 20% in Chile. But the worst absolute impact was in the much poorer African countries, where schools, hospitals and village dispensaries lost staff and much of the supplies which they needed to function: drugs, books, paper. The long-term effects were huge in some countries. The prolonged crisis in Ghana, for example, cut real health expenditure per head by 1982/3 to 20% of the 1974 level (Cornia *et al.*, 1987). Chapter 7 discusses the implications of this collapse in social provision.

Selling the state

And what of the other arm of the British squeeze, privatization? Seeking to accelerate development, many Third World governments built up a large state enterprise sector after Independence: public utilities, nationalized mining and agricultural enterprises, state manufacturing firms set up to lead industrial development or to bail out ailing private enterprise. They now face pressure to privatize these firms.

The IMF, the World Bank and USAID have all made privatization an explicit aim of their lending programmes. British consultancy firms seek to sell their privatization skills abroad. But, while large numbers of firms have been privatized in developing countries, most of these firms have been quite small, and the overall size of the public sector has changed little. The main exceptions are Chile and Bangladesh, where large parts of the public sector have been denationalized; in Chile by an extreme neo-liberal government and in Bangladesh because of dependence on the US (Cook & Kirkpatrick, 1988).

Figure 3.10 Visible effects of the public spending squeeze. (Top) A barely equipped health clinic, Mpongwe, Zambia. (Bottom) A school in Burkina Faso.

Privatization in the Third World is difficult. Many firms are unprofitable and therefore unsaleable. Local capital markets are small, and governments may not want ownership to move wholly abroad. So privatized firms may dominate local markets and behave as uncompetitively and inefficiently as state monopolies And there is opposition within the country: for example a Malaysian trade union president described privatization as 'a craze ... a capitalist concept to boost the riches of the élites at the expense of the workers and consumers' (Ragunathan, quoted in Leeds, 1989).

Even Chile's privatizations, which began in the mid-1970s, ran into trouble. Local buyers were few, and funds to buy firms hard to come by. As a result, the privatizations of the 1970s created small numbers of large financial–industrial conglomerates, dominating the local market, very reliant on bank loans and with relatively little equity capital. In the economic crisis of the early 1980s, in a country where most tariff protection for local industry had been removed, a number of these conglomerates became insolvent. To prevent financial and economic collapse, most of the important privatized companies had to be taken

Box 3.8 Creating the competitive state

Here are the characteristic neo-liberal reforms designed to bring the market into the state.

- *Enforced tendering* for state-financed activities, which forces public providers to compete with the private sector.
- *Deregulation and liberalization*: opening up areas hitherto state monopolies (like primary education in some developing countries) to private competition.
- *Creating autonomous agencies*: separating off parts of the state as autonomous units able to manage their own affairs (like urban development corporations or polytechnics in Britain); where possible constituting such agencies as *public enterprises*.
- *'Cost recovery'*: a switch from tax financing to charging fees for the services of public agencies.
- *Profit and cost centres*: breaking state agencies down into small units for management purposes, with control of budgets at unit level.
- *Internal competition*: trying to create competition between state suppliers to provide a given state service: for example the 1991 reforms of the British National Health Service.

back into government control by early 1983 (Marshall & Montt, in Cook & Kirkpatrick, 1988). Since then, new privatization programmes have been designed to try to produce stronger companies.

Selling the state will go on, but more slowly than the proselytizers of the 'new development orthodoxy' would wish. Neo-liberals continue to press for privatization, despite the discovery that the efficiency record of public enterprises is very mixed. Pressure for cuts continues, despite a failure to find any significant negative (or positive) association between the size of the state and the rate of economic growth in the Third World (World Bank, 1988a), and despite growing concern about the social and economic costs (Chapter 7). Meanwhile, neo-liberal reformers are increasingly turning their attention to internal reform of the state bureaucracy.

3.7 Splitting the state

If the state cannot be squeezed as fast as the neo-liberal reformers would wish, perhaps it can be made more competitive, and hence, the reformers would argue (Section 3.5), more efficient?

All of the reforms listed in Box 3.8 follow the logic of the 'private interest' view of the state. If the state consists in any case of self interested agencies and individuals, then the state should be reconstructed to provide those individuals and agencies with the right (material) incentives to perform the jobs the government wants done. If monopoly power shields bureaucrats from challenge, then the answer is to create a situation as close to competitive market pressure as possible.

Figure 3.11

Q In Britain, the Conservative governments have brought in all the above reforms in the last decade. How far have they also been tried by Third World governments, willing or reluctant?

The evidence for recent reform is patchy and anecdotal. At the beginning of the 1980s, a higher proportion of health and education services were already privately provided in the Third World as compared to industrialized countries. The diversity of organization of social services in the Third World is too great, and there are too few surveys, for satisfactory generalization.

However, there is now great pressure on developing countries' governments for such reforms. They increasingly form part of the conditions for aid, including structural adjustment lending, with the World Bank leading the trend. Here is a summary (based on World Bank, 1988a) of the main proposals.

Deregulation

The World Bank urges governments to give up public monopolies such as education, and to welcome private and church schools, subject to minimum standards. Government intervention should take the form of ensuring competition (not the licensing of schools). For private health services, the Bank states, 'This involves spreading information on the prices charged by alternative providers, on the appropriate treatment for various ailments, and on the importance of insurance coverage' (World Bank, 1988a). Neither price control nor quality regulation of the health market are mentioned here, despite known and widespread problems of exploitation of the sick, and high costs, in unregulated health markets.

Decentralization and private management

Where state providers levy charges (for example public utilities such as water and electricity), structural adjustment-related reforms tend to emphasize independent management. Such enterprises can also be leased to private management, or employ private management on a contractual basis. In a number of cases, this has been a staging post to privatization, as in the case of a privatized Malaysian container terminal.

'Cost recovery'

This is the Bank's euphemism for the shift from tax finance to charges for public services. Taxes, the Bank argues, are expensive and difficult to collect, especially in low income countries. To expand public services, additional funds are needed and can be best found from fees.

Here are some elements of the Bank's justification of such user fees.

> "In health, a small charge proportional to the service charge will tend to make clients avoid unnecessary services. ..."
>
> "Modest charges for some services used by the bulk of the population, such as drugs and school materials, also appear to be affordable. ..."
>
> "The budget for primary education in some African countries ...could be increased by more than twenty percent if the stipends paid to higher education students were terminated. ..."
>
> "Charges for universities and tertiary level hospitals have a negligible effect on the poor. ..."
>
> "User charges will improve efficiency if public institutions such as clinics or schools are given greater responsibility for collecting them and choosing how to spend the proceeds."
>
> (World Bank, 1988a, pp.136–138).

That is, fees cut demand, promote efficiency, and can even be redistributive, although why governments which do not redistribute taxes should be more progressive with charges is not explained. In some services including health, the Bank sees fees as a necessary stop-gap until a private insurance market, that is, in their phrase, 'an adequate financial environment' develops. The inequity of private health insurance is not balanced against the pro-market view.

Fees and decentralization are clearly related proposals. Services, the World Bank argues, can be improved if fees are retained locally for local provision adapted to local needs. Hence, where possible, management should be decentralized to local areas, with funds being raised as far as possible from local users by a mixture of taxes and charges. Chapter 7 takes a critical look at these and the next set of proposals.

Targeting

Remaining tax-financed spending should be targeted on the most needy. Subsidies to the better-off can be removed by charging them, at cost, for higher-level services such as higher education. Subsidy can then be concentrated on primary education and primary health care, especially in rural areas. Some vouchers, selective support and scholarships based on income and merit, can allow the poor some access to higher-level services. (Compare this with the argument in Box 3.3.) The voucher scheme is particularly characteristic of the 'private interest' approach, shifting funding from providers to users, in order to create a system closer to a market for public services.

This set of proposals takes non-primary services towards private provision, leaving only primary services to be partly state-supported and possibly state-run.

The extent of actual reform so far is poorly documented. Fee charging has expanded in Africa, and Thailand has experimented with vouchers. But the country, again, which has gone furthest and most deliberately along the lines outlined, in reorganizing its health and education services, is Chile. Private education has been encouraged. Public education spending is paid per student to schools, to force public and private schools to compete. Health and education management has been partially decentralized to municipal level. User fees have been introduced for health services. The World Bank is now holding up Chilean reforms as a model for other countries (World Bank, 1991a).

Elsewhere, replacement of general subsidies with targeted subsidies is widespread, especially for food. Some state monopolies of provision have been removed, for example primary education in Pakistan. Many Third World governments already subsidize voluntary service provision, in effect, buying in services, and this may be growing. Civil service staff have been greatly reduced. And 'contracting out' of professional services is developing, as civil services, in Africa particularly, lose professional staff to work as consultants both for overseas agencies and for their old state employers.

This brings us to the connection between overseas aid agencies, and public service reform (see also Chapter 4). Donor agencies can promote decentralization, and targeted programmes, partly from a desire for a fiefdom under their own control, justifying this by a belief in the inefficiency of government. Decentralization becomes devolution: is this efficient? Or does it create fragmentation: an inefficient diversity cloaked in the language of decentralized efficiency?

3.8 Devolving the state

In proposals for devolution, the 'private interest' reformers meet those concerned with class and power. For the neo-liberal right, the idea of devolution arises from an admission that some public action remains necessary, coupled with pessimism about government failure. The market alone cannot do everything. Therefore if bureaucrats are too self-interested to respond to needs, the answer lies in devolving *decision making* ('policy') as well as action to bodies with reasons to be more responsive. The neo-liberals see two possibilities:

- devolving control of state services to groups of private employers
- devolving response to need to the voluntary sector such as private charities.

The British government for example has transferred a large budget for state-funded training to Training and Enterprise Councils (TECs), private employer-dominated bodies. The hope, presumably, is that private employers will act in their collective interests, and expand training efficiently, rather than in their private interests, using state funds to pay for training they would have done anyway.

Governments may also seek to devolve response to needs, such as the relief of destitution, at least partly to private charity. Hence the neo-liberals have rediscovered that diverse group of organizations sometimes called the 'third sector': the voluntary organizations, not-for-profit trusts, charities, cooperatives, non-governmental organizations. These organizations can draw on charitable funds and on unpaid work, to reduce costs and the call on taxation. And they are argued to be more responsive to need than bureaucrats: more open to scrutiny by their beneficiaries, more flexible, because smaller and less rule-bound.

Left–right alliance

On the 'third sector's' role, the neo-liberals and the left have converged on a policy. Those concerned with class and power, who believe the state serves the powerful, see voluntary organizations as a countervailing force. Such organizations do not only offer responsiveness and diversity. They can also be a means of *empowerment* of those with least power, by acting for, on behalf, and under the control of them. They can do this as campaigning organizations exercising political leverage on behalf of the needy, or by transferring state resources into the control of the least powerful. These complex themes of charity, diversity and empowerment are explored further in Chapters 5 and 6.

This ideological alliance of left and right on behalf of devolution has created immense pressure for the expansion of the role of voluntary or non-governmental organizations (NGOs) both in the West and in the Third World. In Britain they have become involved in replacing state provision, providing services under contract, and developing joint schemes with mixed state and charitable funding sources (NCVO, 1990). In the Third World, aid from governments as well as charities is increasingly channelled through NGOs, rather than through Third World states.

But can non-profit organizations deliver all these benefits? Not all the roles proposed for them are compatible, and the moment non-profit organizations become channels for state funds, and major independent service providers, the contradictions emerge.

Foreign or indigenous control?

In Britain, voluntary organizations worry about becoming business-oriented service providers, and losing their advocacy role (NCVO, 1990). In the Third World, the most powerful NGOs are based abroad. So their expanding role can increase foreign control of public policy, in a way which is more difficult to counter or even monitor than conditions on aid to governments.

The funds involved are large. Africa is thought to have received about US$1 billion in 1986 through NGO channels; the EEC in the mid-1980s was channelling about US$600 million a year of its aid funds in this way, and the Canadian state aid agency put 12% of its funds into NGO projects (Bratton, 1989). Most of this goes to large overseas-based NGOs.

Nevertheless, numbers of indigenous Third World NGOs are growing fast, including cooperative movements, church organizations, trade unions, self-help groups, organizations of local skilled professionals, lobbying groups, and organizations headed by the local 'great and good' set up specifically to attract aid funds.

In this way, devolution of service provision has been forced on Third World governments. Services such as health and education in Africa, hit by the cuts, have been partially re-established in a different form. A patchwork of provision develops, each patch tended by an agency or NGO. The state function is reduced to a coordination, or more vaguely an 'enabling' role. At worst, previously integrated services become an inefficient patchwork, and NGO provision and inter-agency competition takes the policy function almost wholly away from government. (For an argument along these lines for Mozambique, see Hanlon, 1991.)

But in the Third World too, official funding can conflict with empowerment, threatening autonomy, campaigning and responsiveness.

NGOs become businesses, or even bureaucracies, losing grass roots and political 'edge'. And the conflict becomes particularly acute when the grass roots are local and the funds foreign.

Overseas NGOs are frequently torn between an ideology of empowerment, and a belief that they know the answer. Their progressive self-image is rooted in a whole trend of thinking about development which stresses local, small-scale, self-reliant solutions. Sometimes called 'neo-populism' or 'the other development', this allies labour-intensive development and intermediate technology with independence from the state.

But successful local grassroots NGOs often play a more complex role than this. A good example is SEWA; the Self Employed Women's Association based in Ahmedabad in India. This, according to a founder (Bhatt, 1989), began as a trade union, albeit of the very poor self-employed and homeworkers. Trade unions have had a bad press in the 1980s, and in the Third World have been accused of representing the relatively privileged and being too closely tied to the state. But combining self-empowerment with winning concessions from both the state and employers has stood SEWA in good stead.

Similarly, the credit and marketing schemes which are effective in reducing poverty appear to be those which manage to promote viable labour-intensive trades while also acting as advocate, intermediary and negotiator with government and local élites (Tendler, 1989). The more NGOs become devolved arms of the state, or independent service providers, the further from this model of pressure, organization, advocacy and support they are likely to move.

Worse, once NGOs lose an advocacy role, they can turn inwards or collapse. Without a formal constituency, many find it difficult to respond to changing needs, and become subject to charity 'fashions'. These problems have turned the US welfare system, which is highly dependent on charitable provision, into a fragmented, fluctuating and unreliable patchwork. And they have led those concerned with class and power to see a very important role for NGOs, not in taking over from the state, but in changing the state itself.

Figure 3.12 Rubber tappers in Cachoeira village in the Amazon set up and run their own school. Pupils range from to 3 to 60 years of age.

3.9 Up-ending the state

Q Can the state itself be reformed, to operate more as 'public interest' theorists assumed?

The neo-liberal reply of course is, no. But left-wing critics, like those quoted at the beginning of this chapter, concerned with pressures of class and power, are less sure. They see self-organization of the powerless as a method, not just of self-help, but also of identifying public need and pressuring the state to respond. Here these critics' concerns, with accountability and responsiveness of the state, meet up with the activities of a group of reformers seeking to open the state to pressure.

Attempts at reform

These reformers' attempts to change the rules and organization of state services are based on distinctly 'public interest' assumptions. They believe that state employees can and will pursue aims of public service if provided with both the incentives and the conditions for doing so. Far from assuming private interest behaviour, the reforms proposed seek to remove impediments to activity in the public interest.

A good example is to be found in the work of Robert Chambers. Chambers explores the possibility of what he calls 'reversals'; making a state bureaucracy more responsive to the people it is ostensibly designed to serve. This, he argues, requires a move away from 'normal bureaucracy' characterized by centralization, standardization and simplification. There are tasks that such normal bureaucracy does well (e.g. a vaccination campaign) but much that it does badly. Its worst performances are in situations where needs are diverse, information is needed from below, and employees can get away with neglecting poor areas (as they cannot, for example, in mass vaccination). In situations such as these, top-down standardized solutions fail. A good example is research and extension related to poor farmers in diverse, rain-fed, low-productivity agriculture (Chambers, 1988).

Reversing bureaucracy

The reforms Chambers proposes follow from the idea of 'reversing' normal bureaucracy, moving towards decentralization, diversification and complexity. In the agricultural research field, decentralization initiatives include (good) farming systems research, involving farmers in experimentation and analysis of results. At best, experimental staff move away from being experts transferring information, to become catalysts, colleagues and consultants of farmers. Complexity is introduced by researching the mixed systems such as intercropping that are already used by farmers, instead of trying to simplify farming systems through monocropping. Diversity is extended by keeping a wider variety of seeds for farmers to try, and by actively involving farmers in the testing process, hence generating bottom-up information flows.

Chambers points out that most agricultural information systems serve the central management control function: they are not information systems for *farmers*. A bottom-up system, Chambers suggests, requires three elements. First, to *perceive* diversity; secondly 'a reversal of control to *permit* diversity'; and third to *promote* diversity in the place of standardization. In other words, to listen; to open the system to differentiation from below; and then to respond to it. Finally, he recognises that senior staff will tend to resist such power reversals and demands from below. If such approaches are to stick, the internal incentive and reward system has to be reorganized to support it: a process which he regards as difficult, but not impossible.

Decentralization and democracy

Decentralization, aimed at making the state work upside down in the interests of those 'below', the public and especially the relatively powerless, requires both internal reform and external pressure. Some reforms in Britain echo these ideas. Many local authorities have been decentralizing their activities, in some cases setting up local small-area 'forums' or committees with budgets, to decentralize the political decision process down to small localities (Hoggett &

Figure 3.13 Can farmers find their own solution? (Above) Controlling soil erosion, Yatenga Province, Burkino Faso. (Right) Watering trees on an experimental site, Weyle Sidke, Somalia.

Hambledon, 1987). Similar aims are behind coopting more people onto local government committees dealing with services.

There has also been such governmental decentralization in a number of areas of the Third World. Karnataka in southern India is in the process of decentralizing its local government system in a series of reforms along these lines. And in the early 1980s Nepal instituted a decentralization of government with a strong developmental and 'populist' bent in its official definition and justification .

The Nepalese reforms illustrate the political complexity of government-instituted reform. They aimed at integrating the development effort locally, curtailing central control, making government more responsive, and also at outflanking demands for more radical social change. Hence they appear more populist in conception than execution. Programmes continued to be run vertically from central government, rather than devolved and integrated locally. Information systems did not support local control. Worse, political decentralization by strengthening the political position of local élites, may have actually reduced government's responsiveness to the poor (Bienen *et al.*, 1990).

In other words, it is possible to decentralize without 'up-ending' in Robert Chambers' sense at all. There is no necessary connection between localizing the operation of the state sector and democratizing it. Decentralization may well be a necessary condition for greater responsiveness of public services, but it is not sufficient to ensure responsiveness, despite the contrary assumption of the World Bank (1988a).

So 'up ending' the state requires a difficult mixture of reform and empowerment. Internal reform of the bureaucracy must include higher rewards and status for front line staff, and information and control systems which allow and encourage interaction with service users, and local control and allocation of resources. Local units then need to be pressured by well organized external constituencies; that is by organized and active service users who can put pressure on the system from below to continue to treat them interactively, learning as well as supplying.

Left-wing critics of the state know how difficult this trick is to sustain in the face of opposition from the powerful. But they nevertheless put forward a different vision of the state in society. They are sharply critical of the optimism of the 'public interest' theorists, and their undifferentiated 'public'. At the same time they argue that the extreme pessimism, both of the 'private interest' theorists and of some of the left, is unjustified. In effect, they are trying to reconstitute a more realistic idea of how a (partly) public interest state, responsive to the needs of the poor, might be developed and sustained.

3.10 The future of the state?

The 'public interest' view of the state has been terminally undermined. Critics have unpicked its assumptions and refuted its predictions. Confidence in the developmental state, and its leading role in industrialization, has gone down with it. Disillusion with the state produced for a time widespread agreement — on the need to reduce its power, decentralize its operation and devolve some of its activities, including decision on the use of state funds, to non-state bodies.

We have seen that when these critiques and proposals are examined closely, the apparent political consensus vanishes. Underlying it are vastly different aims and assumptions. One set of critics seeks to create a society, including its state structures, which functions as far as possible on the principles of private competition. These reformers believe that the world does operate in terms of their individualist assumptions, and that a state modelled on the market will be efficient and responsive, or at least the best we can have.

There is a gulf between this view and the left-wing critics of the state. The latter believe that state and market are institutions riven by class and power, which reinforce each other in opposition to the needs of the powerless. They see a solution in the self-organization of the poor outside the state. But they also, perhaps increasingly, see this as a vehicle for reforming the state itself.

In the end, these two views are irreconcilable: there is scope for cooperation, but their aims and assumptions diverge. The 'private interest' view of politics can seem satisfyingly cynical and realistic, and has attracted adherents for that reason. But can we live by individualized and self-interested motives alone? There are recurrent fears that it is far easier to destroy community feeling, altruism and mutual support than to create it. The 'Green' movement is only the most recent proponent of the view that solidarity may be necessary for survival. Sociologists, psychologists and anthropologists have been warning economists for years that their individualist assumptions were neither empirically valid, nor a sufficient basis for a satisfying life.

If we accept the argument in Chapter 1 that markets spread vulnerability, and hence imply a need for *non*-market action, then the principles of such non-market action need careful thought. The redistributive social security systems of most industrialized countries are not wholly explicable within the 'private interest' framework, however valid the criticisms of their operation in practice. The importance of mutual support systems in the Third World is now being rediscovered by development economists (Platteau in Ahmad *et al.*, 1991). So how, if some form of social solidarity is necessary, can it be effectively and accountably organized, and what is the role of the state in this?

There are left-wing critics of the state who are fatalists, in that they believe its oppressions are inevitable. But the challenge of neo-liberalism, and the extreme failings of some states and governments, have forced those who disagree with market-oriented reforms to turn to developing alternatives. This is the challenge taken up by the reformers and the 'other developers' discussed in Section 3.9. To take this seriously suggests not that the market should structure the state, but that non-market principles of solidarity and gift relationships between people, if they can be embodied in our public (state as well as non-state) institutions, can enrich society. Here is one of the most famous statements of that point of view, written before the neo-liberal dominance of reform of the state:

> "... the ways in which society organizes and structures its social institutions — and particularly its health and welfare systems — can encourage or discourage the altruistic in man; such systems can foster integration or alienation; they can allow the 'theme of the gift' — of generosity towards strangers — to spread among social groups and generations. This ... is an aspect of freedom in the twentieth century which, compared with the emphasis on consumer choice ... is insufficiently recognized."
>
> (Titmuss, 1970, p.225)

In other words, our social services structure our societies just as, Gita Sen argues in Chapter 10, non-state forms of public action structure our social services. By working from the assumption that individual private interest is all there is, we might lose more than we know, blocking off crucial sources of security and morality.

Summary

1. The 1970s and 1980s saw the development of widespread political and theoretical critiques of the state.
2. The focus of theoretical critique was the 'public interest' view of the state. This conception of the state underpinned the development policy ideas of the 1950s and 1960s.
3. One set of critics rejected the existence of an identifiable 'public interest', in favour of a conception of the state founded in class and power relations.

4 Neo-liberal critics, on the other hand, developed the 'private interest' view of the state, to counter the 'public interest' arguments.

5 'Private interest' ideas have underpinned efforts to reduce the size of the state, through cuts and privatization, and to introduce competition within state structures.

6 In pursuit of divergent objectives, left-wing and right-wing critics of the state have advocated devolution to non-government bodies of both state activities and decision making.

7 Finally, some critics have sought to reform the state, through 'up-ending' and decentralization, believing that the state *can* be made to operate more effectively in the interests of the less powerful.

8 Not all these views can be reconciled. Reform of the state on a market model conflicts with reform which seeks to strengthen the state as a vehicle of social solidarity.

4

PUBLIC OFFICE AND PRIVATE GAIN: AN INTERPRETATION OF THE TANZANIAN EXPERIENCE

JOSHUA DORIYE

4.1 Introduction

This chapter takes up the theme of 'government failure' which, as Chapter 3 explained, has occupied centre stage in development policy debates in the 1980s. For Africa in particular, commentators and aid agencies have used this idea of government failure as their central explanation of economic crises. Critics pointed initially to dismal results from state-organized production, and evidence of poorly executed development programmes. Later they added problems in the traditional fields of state activity (peace keeping and general public administration) to their list of evidence of disorganized government. Problems in public administration have been particularly pronounced in sub-Saharan Africa, and have given rise to the idea of a 'failure of governance' as the central feature of a general crisis in the subcontinent.

This chapter takes a critical look at this progression of ideas, using the example of Tanzania. Its central proposition is that the failings of African state structures, at least in Tanzania, are not, as some commentators seem to imply, the result of some inherent aspects of African society and government. Instead, there has been a marked, and recent, breakdown in public administration in the country.

Furthermore, this breakdown can be explained. It is the outcome of a series of identifiable economic and social pressures, and is exacerbated by a number of the (externally advocated) policies of economic reform. This chapter describes and analyses the disintegration of a 'public interest' attitude and practice in the administration of Tanzanian social services and policing.

The chapter has a sombre message. It concludes that the emergence of private interests as a driving force in the organization of social services and policing is both disastrous, and extremely difficult to reverse. 'Private interest' use of public office has become a survival strategy for state employees. Both the economic costs, and the cultural changes, required to reverse the situation and to recreate a cohesive public administration, seem too large in scale to be achieved. Public sector reform in Africa faces a severe dilemma.

4.2 The African state and the crisis of development

The search for the root cause of economic crisis in sub-Saharan Africa has been going on for over a decade now. Underlying this search is a belief that the causes of the African crisis are different from those of other regions. It was noted that sub-Saharan economies were performing far worse than those of other developing regions during the recession of the 1980s (World Bank, 1981). Furthermore, the fact that, within the continent, performance appeared broadly similar was interpreted to indicate that there could be a problem that was uniquely African (Bienefeld, 1983).

Initial diagnosis attributed the problem to the over-extension of the African state. That is, the state was thought to have assumed too many responsibilities, in relation to its capacity to carry them out efficiently (World Bank, 1981). Although that diagnosis has been hotly contested, it has nevertheless formed the central plank of policy reforms relating to the public sector (Nunberg & Nellis, 1990).

In debate, the diagnosis was contested on two grounds. There is in fact no evidence that state intervention in sub-Saharan Africa was more extensive than in other regions. Indeed,

> " ... both the proportion of GDP spent by African governments and the functional breakdown of these expenditures, represent typical expenditure patterns of other governments."
>
> (Colclough, 1983)

Furthermore, evidence from successful newly industrialized economies suggests a view quite contrary to the orthodoxy: extensive state intervention appears to have benefited national economic efficiency (see Chapters 4 to 6 of *Hewitt* et al., *1992*). Accordingly, it was suggested, the African economic crisis could not be explained by the extent of state intervention. Nor are there grounds for supposing that state intervention is inherently anti-development. Nevertheless, the search for an explanation of the African economic crisis has continued to focus on the role of the state, and some new justifications for this focus have been proposed.

Commentators now point to the absence of a strong local private enterprise sector in much of Africa. They argue that because foreign private investment is not available to most African countries in the desired forms and on good terms, the state is in fact likely to remain the central institution in development initiatives in Africa. However, this implies that an understanding of the potential of the state to fulfil this role becomes a crucial element of policy analysis. Given the importance of the state in development, the widespread dissatisfaction with the performance of the African state therefore requires an explanation and a solution.

After more than a decade of policy reforms, in the context of donor-supported 'structural adjustment' programmes (of the type described in detail in Chapter 7), the World Bank is hard pressed to find a showcase example of a successfully reformed economy. As a result there is a general feeling that economic and political reform is not working in Africa. This, in turn, has led to a renewed search for the underlying cause of the African crisis.

The dominant view at present among observers is that it is the *quality* rather than the *extent* of state intervention which is the issue. A poorly designed and managed state administration has something to do with the continued African economic crisis. It is this view that this chapter particularly discusses.

A crisis of governance?

A number of studies indicate not only a widespread acceptance of inefficiency in the African public services (Wellings, 1983), but also that there is widespread use of public office for pecuniary gain. The mainstream economic literature has dealt with this issue through the concept of 'rent seeking' (see Chapter 3, Box 3.7) This concept involves the diversion of resources (time and money) in order to profit from public sector regulation and control. Originally developed to analyse controls on international trade, it has been widened to include the general use of public office for private advantage.

The orthodox economic literature sees such behaviour as a normal, not a pathological, response to state regulation and control. This implies that it can only be lessened by liberalization; that is, fewer regulations. However, it is more difficult to take this neutral attitude to diversion of public funds when the behaviour occurs within traditional areas of state activity such as law and order and general public administration. The observation that 'rent seeking' has penetrated police and civil service has given rise to the view that the problem lies in a general 'crisis of governance'. Conable, President of the World Bank, wrote:

> "A root cause of weak economic performance in the past has been the failure of public institutions. Private sector initiative and market mechanisms are important, but they must go hand-in-hand with good governance — a public service that is efficient, a juridical system that is reliable and an administration that is accountable to the public ... "
>
> (Conable, in World Bank, 1989)

Following Haggard & Kaufman (in Sachs, 1989) and Sandbrook (1986), the World Bank (1988a) explains the crisis of governance in terms of the nature of states based on personal rule, under which,

> " ... ethnic and/or factional affiliation tended to replace technical competence in hiring and promotion, and nepotism and corruption to replace impartiality in the exercise of authority ... "
>
> (Sandbrook, 1986)

This seems to suggest that such factional affiliations and resultant nepotism are inherent characteristics of African society. Furthermore, these tendencies are thought to be reinforced by an absence of countervailing power, that is, of effective civil organization to pressure the state. Because of this,

> " ... the leadership assumes broad discretionary authority and loses its legitimacy. Information is controlled and voluntary associations are co-opted or disbanded. This environment cannot support a dynamic economy. At worst the state becomes coercive and arbitrary ... "
>
> (World Bank, 1989)

Whether or not one accepts the above explanation of continuing African economic crisis, there is rightly a growing concern about the performance of public institutions. In fact, in the majority of cases the deterioration of the quality of state action does not date from independence, but rather was associated with the economic crises of the 1970s and 1980s. Accordingly, whatever we believe to be the nature of the state, much of Africa now seems to be in a condition in which processes which undermine state action come into play. This chapter outlines some of the processes which have led to a deterioration of state action in Tanzania. It seeks to explain why the use of public office for private gain has become an increasingly dominant norm of behaviour in the Tanzanian public service. This approach is essential, once it is realised that in Tanzania 'public service continued to function competently' until the late 1970s (Sandbrook, 1986). This would then indicate that there are factors other than 'personal rule' that account for the subsequent deterioration of the quality of state interventions.

4.3 The growth and objectives of the developmental state

Between Independence in 1961 and 1967, the Tanzanian government confined its role to the maintenance of law and order, on the one hand, and the creation of economic and social infrastructure (such as roads and schools) on the other. Investment in the directly productive activities — that is, goods and services to be sold on the market — was left to private enterprise, including foreign firms. Thus within the First Five Year Plan, which was launched in July 1967, the Government was to expand

> " ... public investment in infrastructure and the rural economy, with a major reliance on the private sector, including foreign investment, for other directly productive investment ... "
>
> (United Republic of Tanzania, 1969)

The Government was involved in the provision of social services, but it did not have the monopoly. Both religious and other non-governmental organizations participated in the delivery of social services. The Government's investment in the rural economy, meanwhile, was concentrated in irrigation and large-scale resettlement schemes. Up to 1966, therefore, the public sector was relatively small and consisted mainly of the central government. State enterprises accounted in nominal terms for only 8% of total investment as against 28% for the central government (United Republic of Tanzania, 1975).

However, by 1971, public sector investment had increased to about 65% of total investment and the proportion of investment by state enterprises had increased from 8% in 1966 to about 42% at

Figure 4.1
(Above) Tanganyka meat packers: the company belonging to Brooke Bond was nationalized following the Arusha Declaration. (Right) The Friendship Textile Mill, Dar-es-Salaam 1968. Financed by Chinese capital, this factory was designed to be labour-intensive.

the later date. This growth of the public sector dates from 1967 and was associated with a change in the government's overall development strategy. The change in the latter was announced by President Nyerere in a policy blueprint, the Arusha Declaration, adopted by the ruling party in February 1967 (Nyerere, 1970).

Two factors combined to bring about a change in strategy. The first was the ruling party's constitutional commitments to put the country on a socialist path of development. Underlying this commitment was a strong conception of the state as a developmental state, capable of articulating the public interest and organizing economy and society accordingly. The second, which also determined the timing of the change, was the failure to meet the First Five Year Plan investment targets. To begin with, anticipated private investment in industry was not forthcoming, a failure believed to be the cause of industrial stagnation. On the other hand, only about 24% of the external finance earmarked for the public sector investment programme was actually raised. At the same time the Government was able to raise about 60% more resources domestically than had been anticipated. This was taken as an indication that with a more systematic effort a lot more resources could have been mobilized domestically than was previously thought possible.

As Nyerere declared,

> " ... we discovered at the end of the Plan that we had been pursuing a policy of self-reliance long before we officially adopted it! For, the extent to which we had financed the First Plan from within our own resources was out of all proportion to our expectations when we drew up the Plan. Over 70% of the development expenditure under that Plan had been found from within Tanzania, as against about 22% which we had expected to provide ... "
>
> (Nyerere, 1970)

Apart from the ideological orientation of the ruling party, the failure of private enterprise-based development strategy and the unreliability of external financial resources would appear to have influenced the adoption of the policy of socialism and self-reliance. The latter was defined as the mobilization of local resources for development, while avoiding dependence on foreign resources. It also implied the creation of public enterprises to act as sources of local investment initiatives, in place of foreign investment, which had been singled out as 'a root cause of industrial stagnation' (United Republic of Tanzania, 1969).

Policies also changed outside the industrial sector. A growing concern with rural incomes, and with wealth differentiation in rural areas, led to reconsideration of rural development strategy. According to the then President Nyerere:

> "In the first Plan we talked only in terms of increased production and our efforts were directed at encouraging greater output, even when this meant helping individual peasants to become employers of labour: "
>
> (Nyerere, 1970)

Figure 4.2 President Julius Nyerere, the architect of the Tanzanian developmental state.

Now the objective of rural development policy was modified, in order to achieve increases in output on the basis of cooperative and collective production.

The mode of delivery of social services was also changed. Private delivery of social services on a religious or ethnic basis was viewed as undesirable, on national unity grounds. Furthermore, if users were charged for the services, there was concern that access could be limited to the well-to-do. Accordingly, the objective of social policy under the new strategy became to ensure equitable and free provision of social services by the state.

This strategy was implemented through a mixture of nationalization and the creation of new state institutions. Privately owned commercial, industrial and financial enterprises were nationalized in 1967. This was followed by the acquisition of real estate property in 1971. Privately owned social service institutions were also brought under government control. This resulted in the initial growth of investment both by central government and by parastatals (autonomous state-owned agencies) over the period 1967–71. In addition new institutions were created in both directly productive and social services sectors.

Here, we need not go into detailed description of how the various institutions in the public sector evolved. The essential point to note is that state intervention in various activities was driven by, among other things, a perceived failure of a market-based development strategy. In a farewell speech to Parliament pending his retirement from the Presidency, Nyerere described the nature of the state that evolved since 1967 in the following terms:

> "... at Independence, Government was something which administered and kept the peace. It was not organized for the development of the people or the economy ... The situation is very different now, Government leads and guides the development of Tanzania, this is its primary goal and the purpose of administration and peace-keeping is to promote that objective ..."
>
> (Nyerere, 1985)

Figure 4.3 (a) The school curriculum was also intended to promote self-sufficiency. Learning cultivation techniques in a school garden in Dudoma, Tanzania.

Figure 4.3 (b) Access to clean water was part of the public health campaign in the late 1960s.

Among the achievements of this developmental state, he cited the creation of an industrial base for the economy and a remarkable progress made in the creation of social services infrastructure. With regard to the industrial sector, achievements were described as follows:

> "At Independence ... in the whole of the mainland Tanzania ... what we would really call factories could almost be counted on the fingers of one hand ... by February 1967, there were still only about seven industrial enterprises which were so strategic to our development that it was necessary for us to take them into public ownership or control. The real progress was made between then and the late 1970s ... I can no longer give a meaningful list of our industries ... it would take too long ..."
>
> (Nyerere, 1985)

It is striking that very little industrialization took place between independence in 1961 and the change in the development strategy in 1967.

Box 4.1 The Tanzanian developmental state

A developmental state is one which takes on the function of leading and guiding development. In Tanzania, the three main aims established for governmental action after 1967 were:

- to increase industrial growth through direct state investment in industry
- to promote widespread and egalitarian access to state-provided social services
- to promote collective and cooperative agricultural development.

It was not the nationalizations, but new investments in industry, that brought about an increase in the number of industrial establishments.

The government's efforts in the social services sector were, however, directed at the creation of facilities for the provision of basic services. Investment in education placed emphasis on primary education and adult literacy. Expansion in health facilities was concentrated on the construction of rural health centres and dispensaries. By 1985 the number of rural health centres had increased tenfold as against a 50% increase in the number of hospitals. At the same time, primary school enrolment increased eightfold and the number of people with access to safe water facilities to about half of the total population. It is clear that the aim was to provide for the basic needs of the people.

4.4 Private interests and the quality of public services

The increase in the rate of investment, and the growth in health and social services provision, were undisputed achievements of the Tanzanian developmental state. Criticism has however, focused on the *quality* of what was provided, and the changing *terms* under which it was provided. This raises the questions:

Q How do you measure the quality of state provision?

Q What can cause state-provided social services to deteriorate?

Measuring quality of output is always difficult. For enterprises producing tangible goods, however, it is relatively easy. The output of water and power can, for example, be directly measured, while the quality of health care is much harder to assess. For some state enterprises, therefore, performance measures can be produced. Relative efficiency can be measured by comparing the unit cost of output against a yardstick of expected costs. Such enterprises can also be judged by their effectiveness in meeting specific goals, such as the delivery of specified quantities of goods to specified areas.

Under the new strategy the state enterprises were expected, on the one hand, to become increasingly self-reliant financially, and on the other, to exercise restraint 'in the use of imports', (United Republic of Tanzania, 1975). However, contrary to these objectives, not only did they become a drain on the government budget, but also, the import intensity of production actually increased, from 15% at Independence to 70% in 1984 (World Bank, 1987).

In addition, under the new strategy, the enterprises were given objectives relating to the efficiency and effectiveness of output distribution to the consumers. This meant 'to ensure that things required are available at the places where they are needed' (Nyerere, 1969) and at affordable prices as defined by the government price control agency. Failure of state action in this respect would consist in a diversion of goods from public availability at controlled prices, into 'parallel market' distribution channels at higher prices.

By the early 1980s, these parallel (that is, illegal) markets were developing in earnest for many commodities produced by state enterprises. Such goods were vanishing from official marketing channels. It was also clear that state employees were deeply involved in the distribution of scarce commodities, undermining state distribution systems by diverting the products out of official markets to earn parallel market premia (McGaffey, 1983).

Figure 4.4 'They should have a separate finance minister for the other economy!'

However, it is much more difficult to identify and measure this kind of problem for public (i.e. state) services and for other state activities whose output is not quantifiable. In this case, efficiency, as measured in terms of cost of providing the services, does not necessarily imply that the services are delivered effectively. Furthermore, the breakdown of delivery of social services is not discernible in the same way as the disappearance of official supplies of goods. There are few data. Yet the underlying processes and outcomes are probably much the same.

This section therefore concentrates on assessing the problems of quality and availability of supposedly free social services in the 1980s. In the absence of direct measures of efficiency and effectiveness, the following account relies on the views of the users of the output of such state activities. A careful examination of these views tends to show that the processes under way in social services are very similar to those which give rise to parallel markets of scarce goods. This conclusion is supported by a number of examples.

Health Care

The first example relates to the delivery of health services. The following passage demonstrates what goes on in that sector. It is drawn from a typical letter to editors of daily national newspapers about what the users of public health services think of their quality and the causes of the problems. Thus:

> "Babati district hospital services are unsatisfactory. What is most disturbing is that two days after receiving supplies, one is told by the staff that the hospital has run out of medicines and be told to see them at home in the evening where there are underground dispensaries known to those who can afford to pay. In such dispensaries there are all kinds of medicines taken illegally from the hospital stocks. Even then the cost of treatment in those dispensaries is so high that one would think the proprietors had actually spent foreign exchange to import the medicine. Previously, if one paid the same doctor something he/she will receive treatment there and then at the hospital. But for how long is this state of affairs going to continue? Even then the cries of the poor end up in thin air … "
>
> Signed: Spokesperson of the weak, PO Box 382, Arusha
>
> (*Uhuru*, 4 January 1991)

A number of observations can be made about the above passage. Two factors seem to underlie the poor quality of the services. Medicine is unavailable in the free medical services sector. And at the same time, contrary to official policy, patients who are willing to pay are more likely to receive medical attention than those who are not. It is then clear why public services are poor: because medical supplies have been illegally diverted into private parallel medical practices, and because medical consultancy ser vices have been privatized and commercialized.

Furthermore, for those processes to take place, all or most of the staff involved have to be implicated. This is therefore not an isolated incident produced by a few individuals whose activities are uncoordinated. It follows at once that it is not possible to get hospital authorities to redress the situation.

That the practices described above are increasingly becoming a norm rather than an exception can be seen from the following passage:

> "The bribe epidemic has greatly infested almost all departments and sections in the health sector. Its symptoms start manifesting themselves when one arrives at the reception desk. From then onwards, a shower of artistic gestures and poetic [sic] languages — the contemporary hallmark of the bribe jargon — is let loose. The same staccato of the evil song keeps reverberating against the ears of patients and service seekers in order to see a doctor, get prescribed medicines, obtain a bed in a ward, undergo stool, blood, etc. examination, be accorded services at mortuaries and X-ray departments, etc. In all these departments you can hardly make it through without a whisper of a bribe solicitor from among the personnel. This includes the entire strata — from doctors to cleaners and messengers; the former being the biggest beneficiaries on account of their positions. There are, however, a few who adhere to their professional ethics and laws of the country — never to succumb to the ego of self-aggrandizement by selling of otherwise free public services".
>
> (Maziku, 1990)

Education

A similar practice of privatization and commercialization of the delivery of social services is equally widespread in basic education, particularly in large urban centres. Interestingly, however, it has generated a different response on the part of the users from that in the health sector. While in the matter of health the response is totally hostile, in the case of education the users would appear to encourage the practice.

In primary education, teachers provide private tuition to pupils who are willing to pay for it after normal school hours. Recently the Ministry of Education became concerned because:

> "[its] ... investigations had shown that some teachers were spending most of their time preparing for private tuition instead of teaching in the schools employing them ... "
>
> (*Daily News,* 2 January 1991)

The same edition follows with the statement that on that basis the Ministry had announced a ban on private tuition on the grounds that 'teaching was supposed to impart knowledge and should not be commercialized'.

However, contrary to expectations, it was the Ministry's decision to ban private tuition that was received with hostility by the public. Why was this?

We can first note that, apart from the use of school buildings, the practice in this case did not involve diversion of public resources into private practice, as was the case in the health sector, although it can be argued that reduced efforts on the part of the teachers during official teaching hours affects the quality of education. Nevertheless, the public response seems to have been unaffected by that argument, perhaps considering that the authorities could control that tendency through a regular school inspection system.

More importantly, public opinion is supportive of private tuition partly because it is sympathetic to the view that teachers, like everybody else, need supplementary sources of income. A ban on private tuition would merely force them to seek income supplements elsewhere, outside the teaching practice, further aggravating the quality of education. As one parent commented:

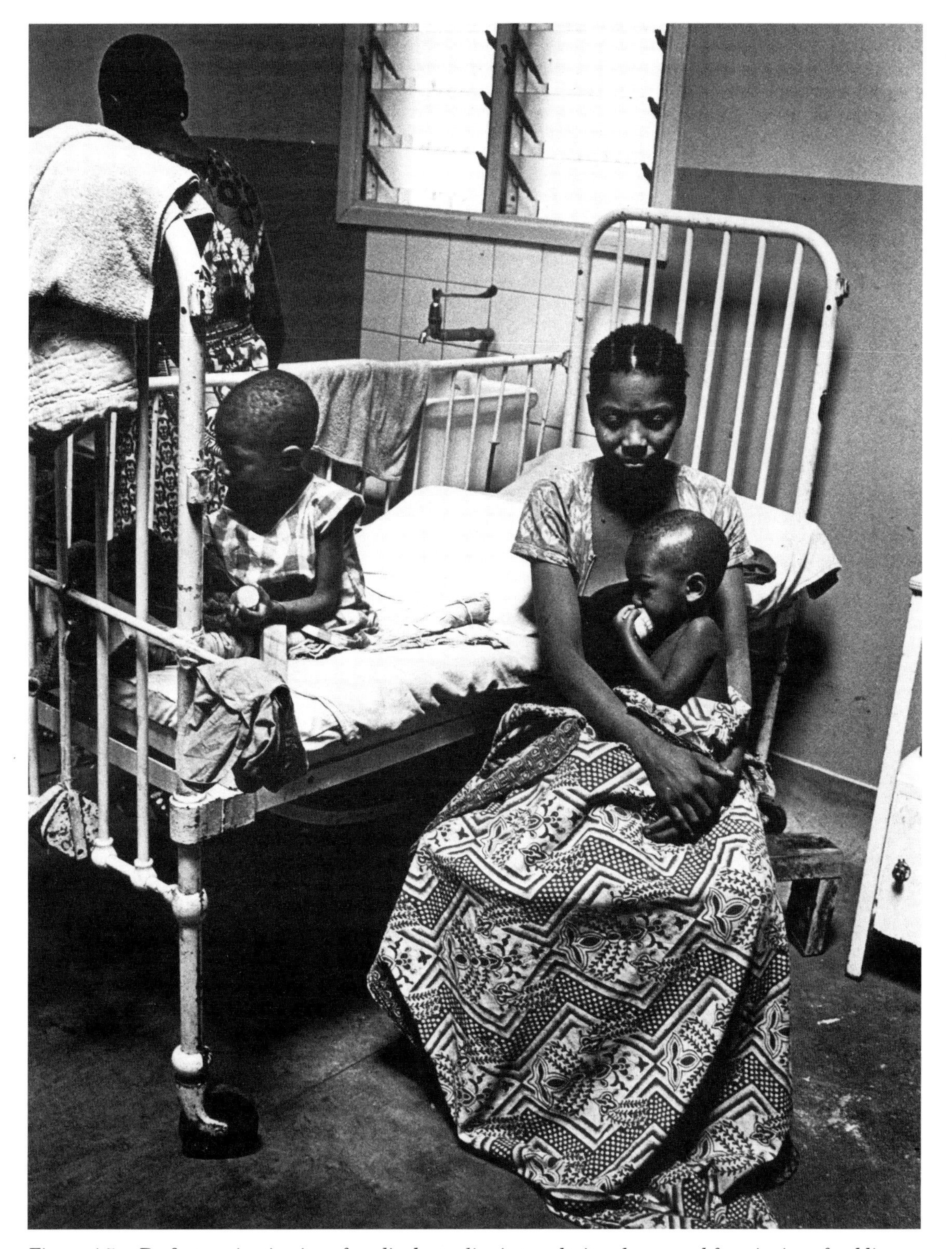

Figure 4.5 De facto *privatization of medical supplies is paralysing the normal functioning of public hospitals and health centres, Mvumi Hospital, Dodoma.*

Figure 4.6 (Left) Contrary to the public hostility against privatization of health provision, the growth of private tuition found greater public acceptance:...

(Right) ... nevertheless, there is evidence of declining quality in public education.

> "The ban could affect the standard of education further. Let the teachers generate (sic) from tuition as they do not have any other source of income ... "
>
> (*Daily News*, 5 January 1991)

The public does not appear to attribute falling quality of education to reduced efforts on the part of the teachers. Rather, it is seen to be due to the inadequacy of resources allocated to education, in the form of textbooks, school desks and other teaching materials, along with higher than average pupil-to-teacher ratios. By implication, whatever teachers did within the public educational system would not have affected its quality one way or the other.

Indeed, for tuition classes, teachers compile additional teaching materials from a variety of sources that are not available in their schools and therefore make up for the inadequacies of the state school system. However, they are assumed to do that only because there is extra pay involved. In support of this system, against those who argued that private tuition is for the well-to-do, a parent, in the same edition of the local press, argued for it on grounds of both quality and value-for-money.

> "The moment I took my daughter to a tuition class she picked up. After all it is we parents who ask the teachers to conduct the tuition class. Tuition is not for the well

off because such people send their children abroad ... "

(*Daily News*, 5 January 1991)

A further reason for the support of private tuition is that parents regard primary education as a preparatory stage for entry into secondary education rather than as an end in itself. Since good preparation at that stage increases a child's chances of being selected for the few places in relatively cheap and better equipped state secondary schools, parents weigh cost of private tuition against that of sending a child to the more expensive, less adequately staffed and equipped private secondary schools.

Nevertheless, public support notwithstanding, the point remains that the quality of state education has suffered, among other things, because of an emergence of a system of private delivery, which has diverted teaching efforts away from public schools to the parallel practice.

This general assessment of the processes that undermine state delivery of social services, though qualitative and imprecise, is widely acknowledged, even officially. On this subject the Second Vice President of the Republic is on record as saying that:

"Unscrupulous public servants have formulated reasons to ... swindle the public for services supposed to be free ... "

(*Daily News*, 27 April 1990)

The police and the courts

There are two other areas in which the presence of similar processes causes great anxiety, even among overseas aid agencies, in that they undermine the very foundations of state legitimacy. The emergence of private interest activity, in the institutions which exercise state power, constitutes what has been referred to above as the 'failure of governance'. This crisis takes a variety of forms, of which two typical examples can be presented here.

A letter to the press has described one such activity in the following terms. The setting is a district in which murders and livestock thefts are a cause of common crimes. There are in all such cases culprits and victims. The forces of law and order are supposed to apprehend the culprits and to restore justice to the victims. The institutions involved are the police and the courts. The police should apprehend the culprits and frame charges, on the basis of which the courts pass judgements. A breakdown can occur at either level or at both.

In this particular example, it is the police who charge an individual for a crime he/she did not commit and they know it. They also know that he/she has the resources and will want to buy his/her freedom back. According to this account this practice is said to be widespread, and the author refers to it as a form of business:

" ... Here at our district police headquarters in Mugumu there has developed a business which is being conducted by some senior police officers who harass people ... The senior officers use junior police to arrest a person who will then be remanded until the following day when a charge will be read out accusing the individual of a murder or livestock theft. He/she will be roughed up and showered with insults before he/she is told to pay at least shillings 40 000 if he/she is to be freed ... "

(Borega, 1990)

There are several other forms of this 'business'. One such form is described by the author quoted above and relates to livestock. The normal role of the police in this case would be to recover stolen livestock and, where they can, to arrest the culprits. A breakdown consists in police organizing a livestock theft. This they do by issuing a certificate of loss of livestock to a band of people with whom they are in collusion. The bearers will then go into a village, identify heads of cattle and claim them to be their stolen stocks. The police then drive the stock to the station where either the owners pay for their release or their best animals will be selected and handed to the holders of certificates of loss, ostensibly as the rightful owners. The account concludes with:

> "But, the cattle are actually then smuggled into a neighbouring country where they are sold and proceeds shared with the senior police."
>
> (Borega, 1990)

Another form is a direct one, where police services are privatized and commercialized. In the latter case, for a burglar to be arrested, even when his identity is known, the victim has to pay the police for the service. Alternatively, an argument with a wealthy person could land one in police custody and regaining freedom could be very costly:

> "... there are some wealthy people in our township of Turiani who can falsely accuse one of crime and bribe the police to arrest them, forcing one's relatives to pay for release. Now I wonder is this police station here for all residents or for the wealthy few?"
>
> (*Uhuru*, 20 June 1990)

Public administration

The final area in which state action is undermined is in public administration. Analysis of this will be confined to the collection of taxes and administration of state expenditure. This element of state action is critical to its ability to finance its activities in the areas described above.

In the case of tax collection, public revenue is being diverted to become private income. This occurs when a tax assessment is made, say for a businessman whose interest is to minimize tax liability. A tax assessor would agree to revise tax assessments to below the appropriate figure, provided the businessman agrees to pay him a certain sum of money. The businessman, on the other hand, would agree to the deal provided his total payment is less than his proper tax liability. In that case a part of potential tax revenue is split between the tax collector and tax payer.

Because this arrangement is mutually advantageous, the payer does not complain. Nevertheless, it is widely believed that large sections of the business community do not pay income and sales taxes as they should. This was admitted at a general meeting of tax officers and collectors reported in the *Daily News* of April 1990.

However, a different system is applied with small business people such as street vendors or kiosk owners. An article in *Uhuru* in April

Figure 4.7 Small businesses such as these street vendors in Dudoma face steep payments for trading permits.

1990 reported that to obtain a permit for trade they would have to pay a minimum sum. Failure to do that could result in not obtaining the permit or in being assessed to pay a fee higher than was warranted by the volume of the business.

The equivalent process on the expenditure side of the government budget would consist in diversion of state expenditure into private income in a number of ways. State employees who administer payrolls can divert state expenditure on this account by introducing fictitious names into the payroll, resulting in an expenditure on 'ghost workers'. A government purchasing clerk will over-invoice purchases, and if a sale of public property is made, an opportunity may arise to under-invoice and to appropriate the difference.

There are numerous mechanisms for diverting state expenditure into private income, some of which are subtle, while others are less so. Amongst the less subtle, Wellings (1983) includes 'a disregard of security precautions and regulations to allow for "armed robbery" to take place'. In disgust, the author of a letter printed in the *Daily News* under the headline "Let's be logical in fighting corruption" listed some of the practices in this category as:

> "... thefts of medicines from medical stores, cases of stores being burnt down; machinery, spare parts being stolen from governments stores, misuse or misappropriation of funds or thefts of fuel from public motor vehicles, inflated prices of official purchases, direct defalcations of money from banks ... "
>
> (*Daily News*, 23 May, 1990)

Proposed solutions

These examples can be multiplied. The essential point, however, is that they illustrate processes which undermine state intervention in practically all fields. An understanding of their nature and causes will assist in explaining what development literature refers to as 'government failure', and will better inform public policy debate than reference to abstract notions of 'personal rule'.

The account presented above has shown a relationship between the inefficiency of public service delivery, and its privatization and commercialization by state sector employees involved in their delivery. But what conclusions are we to draw from this relationship? What solutions are possible? Observers have proposed a variety of responses to the situation described, but many popular solutions seem merely to lead to new problems.

The first response of many commentators, especially in the context of 'structural adjustment' programmes in Africa (programmes which are discussed in detail in Chapter 7), is to propose that *de facto* commercialization and privatization should be legalized.

Fees, they argue, should be charged for state services previously officially provided free of charge. These user fees may allow some state services to be privatized. Private competitors with state services should also be encouraged. The user fees are proposed, we should note, despite a recognition that those fees, like the taxes, are likely to be partly diverted into private hands. Furthermore, the fact that people are observed at present to be prepared to pay for parallel, privately delivered services, is used to justify privatization, even though the original grounds on which services were brought under state control remain valid.

But there is also a more serious problem with these proposals to legalize the pursuit of private interest in the public services. The problem is that this process has clear limits. There are areas of state activity where market solutions simply cannot apply. The fact that a victim of burglary is prepared to pay for police service to recover his property has not been interpreted to imply the need for its privatization. Instead, this and other related practices in the area of law and order have now come to be seen to constitute a 'failure of governance' which requires quite a different remedy. The suggested remedy is the creation of some intermediary institution which would act as a 'countervailing force' against the abuse of state power under a regime based on personal rule. In this way, according to the World Bank, the tendency towards 'coercive and arbitrary' rule can be resisted.

> "Intermediaries have an important role to play: they can create links both upward and downward in society and voice local concerns more effectively than grassroots institutions ... they can also exert pressure on public officials for better performance and greater accountability."
>
> (World Bank, 1989b)

In presenting this argument, the World Bank refers to the role that the National Christian Council of Kenya has played in opposing the Kenyan government. This suggests that the Bank envisages non-governmental organizations (NGOs) acting as intermediaries which will serve as a 'countervailing force' against state power.

But again, there are problems with this approach. It will bring non-governmental organizations directly into conflict with governments, with unpredictable results. Even worse, it does not really address the breakdown of state services.

'Grassroots institutions' which truly reflect the views of those communities that are subject to breakdown of law and order, are not likely to act as the World Bank expects: that is, as a type of opposition party pressing for public probity. Instead, local communities are more likely to take the law into their own hands. Where policing has broken down, local communities have created local security groups, which have initially administered punishments to the offenders. In other words, local communities respond to specific failures of the state in very specific terms. They do not create 'opposition parties', not only because of the political difficulties, but more importantly because that would not resolve their particular problem of state breakdown. Once we admit that law and order cannot be privatized, no development of general 'intermediaries' will resolve the breakdown of the state.

Both popular approaches to the disintegration of public probity in the state are therefore flawed. In order to begin to construct a more effective response, we must first understand the causes of the problem more clearly.

4.5 Real wages and the behaviour of state employees

Q How can we explain the behaviour of state employees which has just been described?

This section gives an account of some of the economic and social conditions in Tanzania, which may contribute to our understanding of the activities of public sector employees as described above. It is not claimed that the discussion which follows fully explains the widespread emergence of these practices. But there is clearly an association between the deterioration of employment conditions, described below, and the declining effectiveness of public services that we have traced.

The decline in real wages

State employees in Tanzania have faced falling real wages. Figure 4.8 shows this for the central government sector. Between 1967 and 1984, employment in the public service more than trebled, while the total wages bill in real terms

Box 4.2 Real wages in Tanzania

We can distinguish:

- *nominal, or money wages* — the cash an employee receives, in Tanzanian shillings
- *real wages* — the goods and services that can be bought with the wage.

Hence, if money wages do not rise, but prices double, real wages have halved. The money wage will buy only half the goods it bought before.

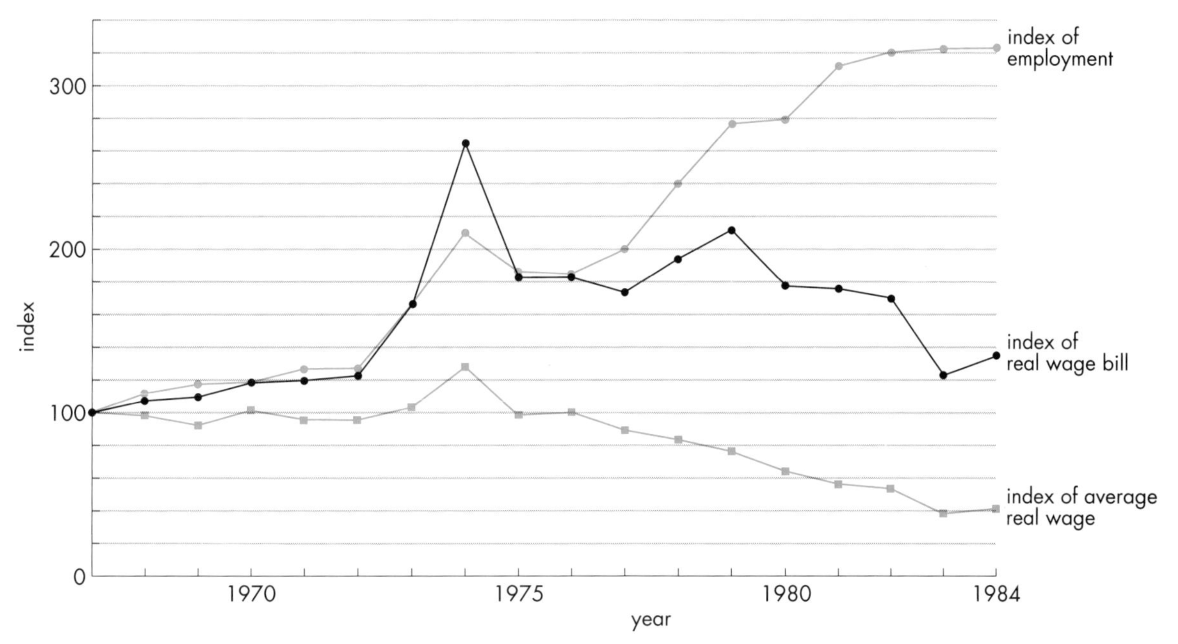

Figure 4.8 Average real wages in central government public service sector, 1967–84 (1967 = 100).

increased by a mere 40%. This means that expansion in public service employment was financed by a drop in the average real wage of over a half. At the same time (although this is not shown in Figure 4.8), state expenditure on other items expanded faster than on wages. So the share of wages in total current expenditure fell from about 39% in 1972 to 20% in 1985. While it is the conventional wisdom that wages have tended to expand to take up most of

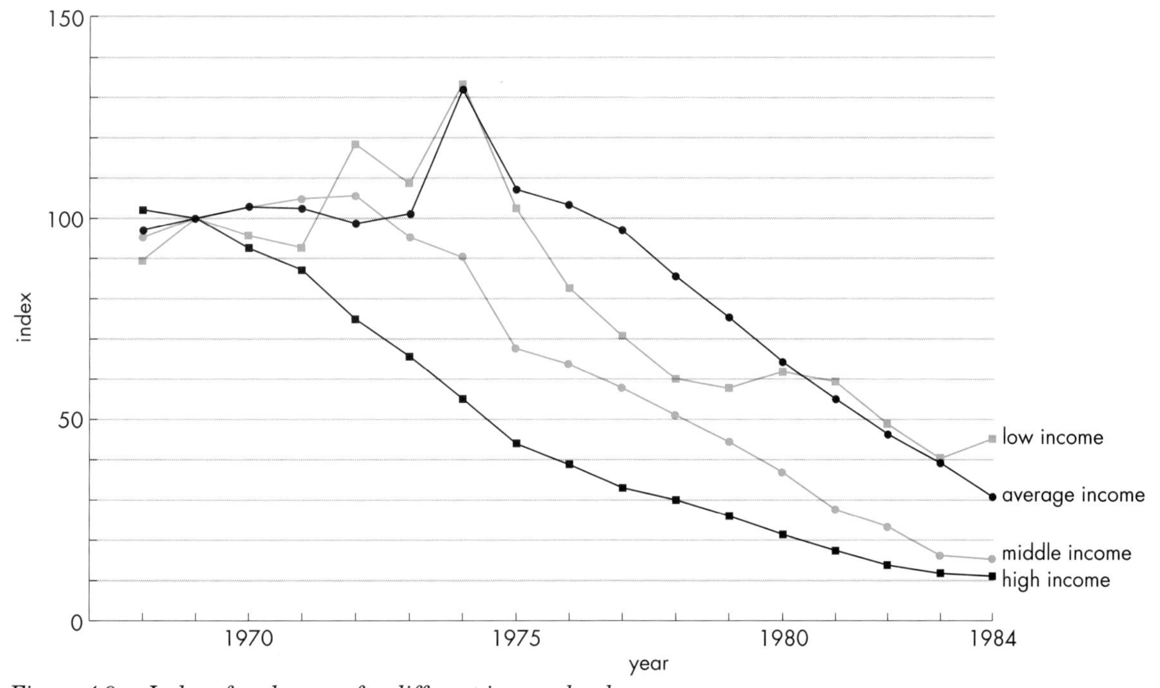

Figure 4.9 Index of real wages for different income levels.

government spending, here, on the contrary, it is wages and salaries that have been crowded out of government expenditure.

Among state employees, the top and middle wage earners suffered most. Figure 4.9 shows the erosion of the real wages of different groups of public service employees. It can be seen that the decline in real wages of the high-income group was well under way as early as 1970 and continued uninterrupted. By 1984, compared with their levels in 1969, the real wages of top level employees were down to just over 10%, and those of middle level employees to about 15%, while the minimum wage stood at about 45%.

Common estimates indicate that by the mid 1980s the monthly real wage of public service employees was sufficient to meet no more than a week's basic requirements. The question, then, is how do public service employees finance the bulk of their basic consumption? A widely held view is that they rely on income generated in the 'informal sector', that is, in small scale employment activities. Thus:

> " ... members of the civil service, in an attempt to make ends meet, had to do other income generating projects which resulted into negligence, absenteeism, reduced efficiency and low productivity."
>
> (Baguma, 1990)

In this view, both the survival of public service employees, and inefficiency in the public administration, are explained in terms of diversion of effort to informal activities. However, a Tanzanian study carried out on this subject (Madihi, 1990) contradicts this hypothesis. The respondents in this survey were drawn from low, middle and top level civil servants. Overall, about 51% of the respondents did not have informal income supplements, and nearly all of these were in the low and middle income categories.

However nearly all the employees in the top income category indicated that they had informal sector sources of income supplements and that over half of their monthly earnings were derived from these. The majority of the low and

Figure 4.10 'A new concept of civil service'.

middle income categories therefore depend on incomes from formal employment. On average, two thirds of earnings of minimum wage earners and nearly 80% of those of middle income employees are from formal employment. A significant proportion of this consists of various allowances and, in the case of middle income employees, consultancy fees.

Nevertheless, all these sources of income put together are said to finance between 50% and 60% of total requirements. Even this seems to be an overestimate, since other confidential estimates seem to indicate that total monthly requirements for various categories of public employees are far larger than those indicated in the study.

There are grounds, however, for doubting the findings of the study with respect to the sources of livelihood for high level wage earners. The value of allowances and fringe benefits is certainly grossly underestimated. Practically all high level civil servants have free fully furnished housing, access to official car, and electricity, water and telephone bills paid, along with wages of two domestic servants. The money value of all these benefits would be several times larger than the earnings actually indicated in the study.

Public money and private income

Perhaps more important, however, the informal sector earnings of the high income category of employees are highly dependent on the facilities and resources drawn from public service employment. As the study points out, activities such as livestock and poultry farming require 'enough space around the operators' premises and access to transport facilities'. In addition, such operations require capital, in the sense of start-up and operating funds. These do not seem to be generated within the activities themselves. Some such operators, when interviewed for a confidential study, did not know how much money they earned from their informal sector activities. Others were frank enough to say they only provided the financing to what are essentially their wives' operations.

Clearly, funding for such activities comes from outside the operations themselves. It is therefore not surprising, that according to the study, many high level civil servants were '[reluctant] to disclose their income generating activities'. This would imply that the informal sector activity explanation is offered as a smokescreen, while the true basis of survival actually lies elsewhere, more specifically in a 'generalized practice of finding avenues to convert public expenditures into private incomes' (Samoff & Wuyts, 1989). And, one might add, public revenue into private incomes.

Figure 4.11 Supplementary income from informal sector activities is probably not the main means by which public servants compensate for falling real salaries.

Some of the avenues through which that process takes place are created *officially* in response to the real wage crisis. For example, an elaborate system of non-wage benefits can be built into an overall pay package. These allowances are normally concentrated in the top wages category, and affect no more than 10% of the public service employees.

There are a number of officially offered justifications for this practice. Since its application is selective and limited to a small proportion of state employees, its effects on the budget are marginal. Furthermore, there are good reasons why allowances are not consolidated into formal salaries. The beneficiaries are also the decision makers, so there is a desire to show that the government's policy of 'wage compression,' that is, progressive narrowing of the top to minimum wage gap, is not being reversed. In addition, given the income tax laws, a consolidation of allowances into wages and salaries structure would mean higher taxes, hence a lower after-tax income than is now the case. Accordingly, the effect of the system of allowances has, in practice, been to increase the net take-home pay of top level employees, while superficially retaining the progressivity of the income tax system, along with a degree of income compression that is politically acceptable.

In recent years, however, allowances and fringe benefits have become a subject of organized employees' group pressures on the government. Recently, medical doctors threatened a work stoppage if they were not granted various allowances and benefits. The report on the Government's response was carried by the *Daily News* in 1990 under the caption 'Mwinyi unveils incentives for docs ... more in July'. The effect of such pressures is clearly to produce an unplanned pay structure, which in turn is an outcome of ad hoc response to the real wage crisis.

Foreign aid donors also contribute to the conversion of public expenditure into private incomes. Some donor programmes are administered outside the framework of general public administration. Donors believe that programmes are more efficiently run if they are administered as autonomous units outside political control. And, in response to widespread inefficiency and perceived lack of accountability on the part of state officials, donors tend to keep close control over projects which receive their funding.

In order to win the commitment of public employees working in their projects, donors then make supplementary payments. Civil servants are given salary supplements on top of their civil service salary, and since such employees remain on government rolls, this practice tends to lead to problems. To begin with, it creates a conflict of loyalty on the part of the individuals so contracted. Furthermore, ' ... the demonstration effect for those government employees lacking access to topping up benefits can be devastating ... ' (Nunberg & Nellis, 1990).

An alternative arrangement is for some civil servants to relinquish their civil service jobs to take employment in donor-funded projects. The effect is to cause 'brain drain' from the public sector into donor projects. It also leads to pressures for perpetuation of donor projects which provide such opportunities. Many donors recognise the 'corrosive and distorting effect' which salary supplements or employment opportunities in donor projects have on civil service morale and management. Nevertheless, the continuation of such practices is justified by the same authors on the grounds that this is:

> "the only way of obtaining good staff to run projects and programmes which ... will enhance the functioning of the civil service and restart the process of economic growth ... "
>
> (Nunberg & Nellis, 1990)

By implication, donor projects are in this case portrayed as centres of efficiency within the public service, which will have a demonstration effect on the rest of the system. However, given the underlying causes of inefficiency, the ultimate effect is that 'a significant group of higher level professionals within the public service' become 'dependent on donor agencies for their livelihood'. This in turn is bound to fragment the civil service, and erode 'the cohesiveness of state action' (Samoff & Wuyts, 1989). Consultancy projects funded by donors have similar effects, and are increasingly being singled out as undermining pedagogy and basic research in institutions of higher learning.

Insecurity and survival

To summarize the argument so far, the erosion of the cohesiveness of public action has apparently been associated with a decline in real wages. So sharp has been the fall in measured real wages that it is widely believed that no more than a week's requirements are financed out of monthly wages, on average. This situation was brought about by:

- a rapid expansion in state employment;
- a reduction in the proportion of government spending going to wages and salaries; and
- nominal wage increments that were a mere fraction of the inflation rate, especially for top and middle level employees.

Contrary to common belief, few state employees earn income supplements in informal sector activities. Few employees can afford such activities, and those who do rely on state facilities and resources. Even in the latter case, this does not appear to be their primary source of income supplements.

Top level employees have been able to get an increase in their net take home pay through an elaborate system of allowances and fringe benefits. This, in turn, has introduced distortions in the pay structure and undermined labour harmony in the public sector. Donor activities have also served as a source of supplementary incomes for a group of higher and middle level professionals. For an overwhelming majority of public employees, these opportunities are not open. Nevertheless, along with their colleagues in the top income categories, survival for them lies in the public sector itself.

How do they survive? For the low-level and middle-level employees, the sale of state goods and services described in Section 4.4 seems well explained by the falling real wage. In order to meet basic daily requirements, these public employees are forced to seek to sell services for extra cash. But this does not explain why top-level employees do it, given the extensive range of allowances and fringe benefits accorded them. A different insecurity seems to lead them to undermine state action.

Many top level public employees have in their background the precarious existence of a peasant family household. Their ascent to public office is through educational achievement. This enables them to attain a standard of living and a way of life that are distinctly different from what they had in their background. The material base of the new-found standard of living is the benefits associated with the public office, free fully-furnished housing, an official car, and a whole range of allowances and fringe benefits.

However, beyond public office there is no material base to support that standard of living. There are conditions peculiar to Tanzania that make the situation more critical than perhaps in other developing countries. Under the leadership code introduced with the new independent strategy, public officials were barred from engaging in business or drawing income other than formal salary.

The restriction was imposed at a time when the real wage was falling. For top level employees, the fall was well under way by 1970. It is not surprising, therefore, that in spite of a generally well functioning public service, incidents of corruption among this group of employees were frequently encountered, leading to the enactment of an anti-corruption law in 1973.

It is also plausible to assume that public employees are also much concerned with being able, after leaving public office, to sustain the standard of living so far achieved. However, since the resources needed for this do not exist except through the use of public office, officials will inevitably exploit their positions for material gain. The greater the insecurity about the term of office, the less subtle this process of accumulation through the use of public office becomes.

This insecurity is affected by the length of period left before compulsory retirement, and by the extent to which progress has been made in reproducing the standard of living so far achieved, independently of public office. The nearer the date of retirement and the smaller the progress made in providing for independent material base, the more frantically public office is used to that end.

Uncertainty is further influenced by insecurity of employment, particularly for political appointees. For the latter the tenure of office depends on the discretion of the appointing authority. The higher the turnover of political appointees, the greater is the uncertainty about the tenure of office and the greater urgency there is to establish an independent economic base.

In addition, political instability and violence, as distinct from insecurity of tenure of office, means that one may be forced to live outside the country. This, in turn, creates the necessity to hold external assets. This effect will be greater the higher is the frequency of political upheavals.

This analysis has a number of implications. The first is the need for a real wage which would finance an average basket of daily necessities. In Tanzania the current wage is far too low, and the balance, for the majority of public employees, is provided by the use of public office for private gain. The inefficiency of public institutions is then a reflection of this process.

Secondly, the real wage should enable an employee, after leaving public office, to sustain the kind of standard of living so far attained. To the extent that the real wage falls short of that requirement, public officials can be expected to use public office to raise the necessary resources.

4.6 Conclusion: the reform dilemma

The analysis in this chapter has attempted to show that the present crisis of the public sector in Tanzania cannot be explained by a general reference to 'governance failure'. The problem is to show how a relatively uncorrupt and reasonably competent public sector

(Sandbrook, 1986) deteriorated into a state where the pursuit of private gain undermined the cohesiveness and effectiveness of state action. Neither poverty as such, nor the specificity of African conditions, necessarily imply corruption. There is no such straightforward link between low levels of development and a high incidence of corruption. On the contrary, corruption is found as a result of specific circumstances, in this case the nature of the economic crisis of the early 1980s and the character of economic reforms thereafter. This led to a situation where private interests emerged as a driving force in public sector provision and administration.

The arguments in Section 4.5 have some depressing implications for those who would reform the state. The survival activities of state employees are undermining the cohesiveness and probity of state action, and those employees are now locked into a pattern of behaviour which will be very hard to break. The question that now arises is:

Q Is reform possible?

Higher real wages at lower and middle levels appear essential. But the government budget has shrunk, and state employment has grown. Hence the restoration of a living wage will entail retrenchment of large numbers of public employees. But since for the majority survival is dependent on public office, the loss of the state job implies a cumulative impoverishment far beyond that measured by the wage lost.

The dilemma that faces public sector reform is then clear. There is, on the one hand, the need to restore cohesion of public action through a restoration of real wages to an appropriate level. However, given budgetary constraints, this can only be achieved through reduction in the number of employees. On the other hand, if mass impoverishment is to be avoided, public resources will have to be used to assist those who have been made redundant to generate income. Hence, given resource constraints, it is unlikely that cutting em-ployment will release sufficient funds for the authorities to be able to raise the real wages of those who remain to an adequate level.

In other words, once the state services have (apparently) expanded, but actually deteriorated drastically in quality, the scale of funding required to restore probity and effectiveness appears beyond the resources of the state. Even worse, the aid funding which is coming into the public services, far from providing the necessary resources, is in some ways contributing to the disintegration of public administration.

Summary

1 After nearly a decade of experimenting with a private enterprise-centred development strategy, the government of Tanzania set out to guide development through state investment and the provision of equitable social services.

2 Although industrial investment and the infrastructure for basic social services expanded, the quality of output of these facilities deteriorated very severely.

3 In social services, as in commodity production, state funds were diverted into private incomes. Parallel commercial markets were created by state employees for supposedly free state services.

4 This privatization and commercialization has extended to law and order and general administration, hence the diagnosis of a 'failure of governance'.

5 The growth of this use of public office for private gain is associated with declining real wages in public services. Insecurity and a search for survival — now, and after public office — leads to 'private interest' behaviour by all levels of employees, and hence the disintegration of a cohesive public administration.

6 The public sector faces an acute dilemma. The resources are not available to restore real wages and to prevent impoverishment through large scale redundancy. Public sector reform with a 'human face' requires additional funding. However, aid funding at present available for social services is tending to make this problem worse. There are no easy answers.

DEVELOPMENT FROM BELOW: NON-GOVERNMENTAL ACTION

5

NON-GOVERNMENTAL ORGANIZATIONS AND THE LIMITS TO EMPOWERMENT

ALAN THOMAS

5.1 Introduction

Up to the 1970s one might have been forgiven for thinking that public action for development could only mean state action. As we have seen, since then, with the dominance of neo-liberalism and its emphasis on private individuals and firms acting in the market, the idea of the state as universal development agent has increasingly come under attack.

However, there is no reason to suppose either that the market can provide other than uneven development or that it can reliably improve the lot of the poorest regions and peoples. This leads one back to look for other forms of public action. The state on the one hand and private individuals and firms acting in their own interests on the other are not the only agents in the development arena. In fact, throughout the 1980s, the 'third sector' of non-state and non-profit organizations has become increasingly important in both industrialized and developing countries. This chapter examines the extent to which the so-called 'non-governmental organizations' that make up this sector can be looked to as a source of independent public action and as a distinctive type of development agent.

Non-governmental organizations (NGOs) form elements of civil society in any region or country. Their importance, in terms of numerical strength, membership, size of operations in relation to the economy as a whole or to particular sectors such as health or welfare services, or acceptance of ways of working, varies enormously from country to country. For example, within the UK, where the term 'voluntary organization' is widely used rather than 'NGO', recent estimates indicate a total of over half a million voluntary organizations of all types. They employ at least 250 000 full-time equivalent employees, and maybe as many as 500 000 or more, representing 1–2% of the employed workforce, as well as managing the equivalent in voluntary workers of over 1 million staff. Among them are over 170 000 registered charities. These alone have a total annual income of over £17 billion, equivalent to more than 4% of Gross National Product compared with 3.4% of GNP in 1980–81, an approximate doubling in real terms since 1976, of which about half comes from the state in grant aid and contract fees for services provided.

The importance of NGOs in developing countries is increasing too:

> "In many countries dense networks of NGOs have memberships that run into millions. For example, in Sri Lanka the Thrift and Credit Movement had a membership of more than 600 000 by 1990 ... and the Sarvodaya Shramadana Movement operates 9 500 community groups with a membership of the same order of magnitude. The Bangladesh Rural Advance-

ment Committee (BRAC) has similarly impressive figures, whilst in India the government can barely keep up with the registration of newly formed NGOs. In Rio de Janeiro more than 1500 spontaneous associations operate ... The situation in Africa is less clear, and varies from relatively well-developed networks of voluntary organizations in some countries (e.g. Kenya) to others where [the third sector] has been actively discouraged and is negligible (e.g. Somalia)."

(Hulme, 1991, p.4)

Here we are particularly interested in the international NGOs that are engaged in development activities. These are generally based in the industrialized countries. Originally drawing on voluntary labour and donations, they have channelled resources to specific relief, welfare and development programmes in developing countries. Recently, increasing quantities of official aid are being channelled via such private voluntary organizations. According to van der Heijden (1987), by 1985 voluntary channels took more than 13% of total aid flows from OECD countries to developing countries, a total exceeding US$4 billion, and this proportion has certainly increased since then. Hulme also notes:

"Not only is the third sector operating on a more significant scale, but it is also extending its role in some countries to substitute for government in what might be regarded as 'classic' state functions. In Northern Uganda and Southern Sudan voluntary aid agencies have virtually operated as local administrations, co-ordinating and planning operations across entire districts at times of turmoil. In Bangladesh the government has ceded responsibility for the allocation of newly formed land in some parts of the Bay of Bengal to OXFAM."

(Hulme, 1991, pp.4–5)

A distinctive 'NGO approach' based on empowerment?

Whereas some have seen these various types of NGO in a 'residual' role, filling in gaps in human needs not adequately covered by state or market, for others they have potentially much more importance as the promoters of a distinctive form of development.

This latter argument emphasizes notions such as 'grassroots development' and 'empowerment'. 'Grassroots' development means direct improvements in living standards by groups of the poorest people in local communities. Promoting development at this level implies working directly with such groups on projects designed and run collaboratively with them. Chambers' (1983) phrase 'Putting the last first' encapsulates this way of thinking. The idea of 'empowerment' takes this notion of collaboration a stage further by indicating that development should be undertaken with the direct aim of increasing the power and control of groups of intended beneficiaries over the circumstances of their own lives, so that they are in a position to become their own development agents in the future.

Many of those who favour NGOs suggest that they are the type of development agent best suited to this way of working. NGOs have themselves been espousing the idea of 'empowerment' and this has been taken up by the World Bank and others (Cernea, 1985; World Bank, 1989) and even portrayed as a kind of 'new orthodoxy' (Poulton & Harris, 1988).

Q Is it realistic to expect non-governmental organizations to provide a new model for development via ideas of grassroots development and empowerment?

A good example of grassroots development brought about directly through local action is provided by the Association of Sarva Seva Farms (ASSEFA) in India. This is one of the movements by which followers of Mahatma Gandhi have attempted to give practical expression to his vision of a non-violent revolution leading to a new kind of society based on the concepts of *Gram Swaraj* (village self-rule) and *Sarvodaya* (the welfare of all).

ASSEFA was created in 1969 by a small group of *Sarvodaya* workers, with the aim of developing the lands donated for the use of the landless through the *Bhoodan* (land-gift) movement and making them into self-reliant communities. Over 4 million acres has been donated since the 1950s by individual landowners, and there are

Figure 5.1 (Above) Vinoba Bhave, a disciple of Mahatma Gandhi and founder of the Bhoodan *(land-gift) movement. (Right) Members of the village council meet in Kaushali, a newly formed village of the* Bhoodan *movement in Bihar.*

also over 15 000 *Gramdan* villages, where land has been donated collectively by a majority of the landowners. But this land is often barren or else under-utilized because those to whom it is allotted lack self-confidence, skills and capital.

Here is an account of ASSEFA and its methods:

"From a first farm of 70 acres in 1969, ASSEFA grew steadily to encompass nearly 100 projects and almost 10 000 acres in seven states by 1985. From an initial emphasis on settlement of allottees and land development on *Bhoodan* plots, it has broadened its approach to include rural industries and the provision of education and health facilities in a more integrated concept of community development The basic elements of the ASSEFA approach however, remain the following.

(1) Initial contact: it is often *Sarvodaya* workers in the field who suggest a particular *Gramdan* village or group of *Bhoodan* allottees, for a future ASSEFA project. An ASSEFA *sevak*, or community worker, may spend 2–3 years in a village, listening to concerns and appraising the level of commitment to development of the area. But the initiative must come from the villagers themselves; the risks posed by change are greatest for the poor, and unless they take the decision to change, no progress is possible. Their participation includes contributing labour, in the case of development of *Bhoodan* land, and in *Gramdan* villages participating in or establishing an active local council or *Gram Sabha*.

(2) Eventually the decision is taken to start a pilot project to test the extent of co-operation and to enable ASSEFA and the participants to better understand each other's capacities and to build a basis of trust for larger and longer-term collaboration. This phase is usually two years.

(3) During this period, plans are prepared for more comprehensive development. Technical assistance and support are introduced, local resources inventoried, and collective self-help activity launched. Typically, a project will involve the introduction of irrigation, the levelling and clearing of land for cultivation, and the building of necessary infrastructure such as schools and storehouses. Most importantly, training and practical experience is provided so that the project is seen as the product of collective endeavour, not the result of external largesse. A group of farmers will manage collectively the land irrigated by a single well or borehole, with advice on a suitable mix of crops for consumption and sale.

(4) Finally, there is a phase of complementary activities, usually including agro-industries, health and education, and the promotion of spin-offs in neighbouring communities. During this phase, which may last 3–4 years, reduced ASSEFA advice and support and continued training in self-management is destined to ensure that the entire project will be run by democratic local institutions, and that the ideal of *Gram Swaraj* will in fact be realised. With self-sufficiency the initial ASSEFA investment is paid back, to be re-cycled to start up new projects.

Support for this programme of activities comes from a range of donor organizations. Early funding was provided by Movimento Sviluppo e Pace (MSP) of Italy, Oxfam, CCFD France and others. More recently substantial funding is being provided by Action Aid from the UK, ICCO from Holland, Inter Pares Canada and Foster Parents Plan International. And, increasingly, ASSEFA is utilizing local resources from the state and central governments and from the nationalized banks.

So far only a handful of the early farms have been handed over totally to the allottees. Utchapatty, near Madurai, is an example of one such farm. It consists of 33.5 acres allotted to 13 landless harijan families, and was funded in 1970. Visiting the farm today one is struck by the contrast between the startling green of this small area rendered productive through irrigation, amid the surrounding barrenness. ASSEFA invested Rs 97 000 (or Rs 2 900 per acre) and the allottees contributed their labour to the task of digging wells and levelling the land; at present 2–3 crops a year are produced, both food crops for consumption and cash crops for sale. In 1978 ASSEFA was able to hand the farm over to the allottees, having recovered Rs 74 000 of its investment from sales of produce, with the balance paid off within two years. Utchapatty today is a thriving self-reliant community, with an active *Gram Sabha*, which handles local affairs and disputes and plans other schemes to improve the village, including small agro-based industries to provide employment and increase incomes. But more than a higher standard of living has been achieved. As some of the allottees remarked to Professor Gupta of the Indian Institute of Management (Ahmedabad) when he visited in 1984: 'We have gained recognition in the village. Other castes, who were our masters earlier, now not only listen but pay attention to what we say'."

(Brodhead, in Poulton & Harris, 1988, pp.48–51)

Both the way of working, with NGO activists (like the ASSEFA *sevaks*) working closely and intensely with groups of the poorest, such as landless harijans, and the ideal of such groups developing through their own self-reliant organizations, are central to this supposed new model for development. However, calling it a model implies more than occasional local successes for local NGOs. It suggests the possibility of a broader impact through regional, national or international NGOs linking directly with many local community groups and helping empower them. Brodhead, in the above article on ASSEFA, is himself doubtful about this even in that case:

> "Allowing for ASSEFA's achievements we may still ask how adequate it is as a response to the vast problems of rural poverty and landlessness in India. It has been estimated that at the present level of activity 200–300 years will be needed to develop just the 1.3 million acres of *Bhoodan* land which have already been distributed — a Sisyphean task given the rate of population growth alone!"
>
> (Brodhead, in Poulton & Harris, 1988, pp.51–52)

This chapter looks critically at the idea that NGOs can be the source of a distinctive new development model in this way. While empowerment is an important analytical concept, the idea that grassroots empowerment through the action of NGOs could be the basis of a development model is extremely limited and could even be potentially dangerous. This is not just because of the difficulties in replicating local successes on a large scale. Empowerment through NGOs is limited because development, and development policy or strategy, needs to be viewed in terms of *processes*, rather than looking for new and better *prescriptions* to replace the faulty state-led development models. We saw in the previous two chapters that the state cannot be regarded as a benevolent neutral agency able to implement whatever is required to solve problems of poverty. So we should equally be wary of assuming that NGOs can implement 'the NGO approach' without problems.

Empowerment at the grassroots through an NGO should be seen as a process that relates to other wider processes. In the example of ASSEFA it is apparent that any success is actually dependent on positive processes outside the sphere of direct local action.

1 Landowners individually and collectively were foregoing their property rights.

2 Resources were made available by state and central government. Also, there must have been a process of legitimizing certain democratic forms of local government allowing the Gram Sabha to be recognized as representing local interests.

3 Gandhian activist workers were recruited to work in ASSEFA. Resources were also acquired from international NGOs which support ASSEFA.

These in fact represent three important general types of process:

1 local community-level processes of social differentiation and changing power relations

2 wider processes relating that community to the state and society at large, including representation of interests, broader social differentiation and democratization

3 processes occurring within the NGO itself or between NGOs, relating the local action to the 'parent' NGO and others, including both mechanisms of accountability and the means by which material and staff resources for such local actions are made available and maintained.

These processes go on within three intersecting arenas, as shown in Figure 5.2. When an example appears to be successful in its own terms, it is likely to be because of favourable circumstances in these arenas, as we have noted in the ASSEFA case. Even then, it is hard to see such examples as more than just individually successful projects. In order to think through the implications of trying to formulate a general new development model in such a way, it will be necessary to consider what limitations may lie in one or other of these wider arenas.

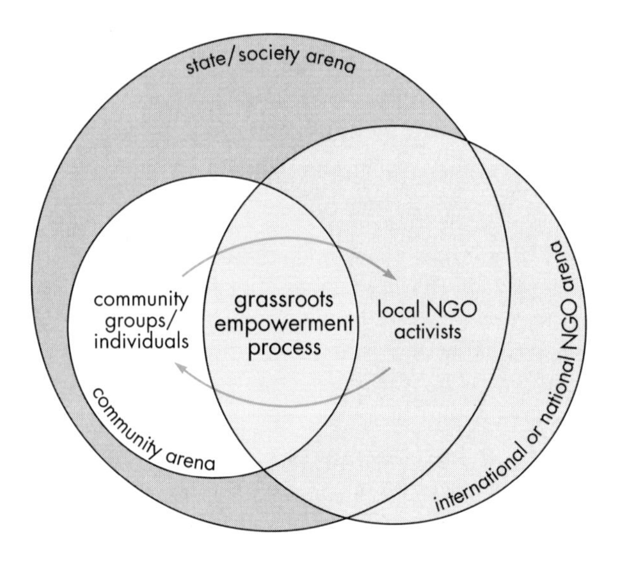

Figure 5.2 Grassroots empowerment through NGO activity as a process within three intersecting arenas.

In the next section we look at what types of organization are covered by this label 'non-governmental organization', and the wide range of what NGOs actually do. The following section explores the idea of an 'NGO approach' based on empowerment, by looking first at what unites NGOs and then at the conceptual origins of the ideology of empowerment. Then we come back to the question posed above and examine the limits to empowerment, concluding by restating the view that the 'new orthodoxy' cannot provide the basis for a distinctive and workable model of development without consideration of wider processes.

5.2 What are 'non-governmental organizations'?

Different ways of defining non-governmental organizations (NGOs) all take it for granted that the organizations under discussion are *non-profit* as well as *non-governmental*. The broadest definition simply comprises *all* such organizations, including all kinds of clubs, work teams, associations, co-operatives, charities, campaigning groups and so on (but usually excluding informal networks and political movements).

However, within this broad definition there is a great variety of types of NGO. One important distinction is between NGOs set up to benefit their own members (clubs, unions, co-operatives, and so on) and those set up for others or for public benefit (charities, campaigning organizations, etc.). Another is in terms of the scope and scale of activity: NGOs at local, regional or national, and international levels can be very different from each other.

Figure 5.3 shows these two ways of distinguishing types of NGO together as a 2 × 3 matrix, with examples in each of the six cells.

In the context of development studies, a narrower usage of the term 'non-governmental organization' is also common. In this sense, 'NGOs' mean private non-profit agencies devoted to international aid and development assistance or to national or regional development.

	Mutual benefit	**Public benefit**
International	International federations of trades unions International Co-operative Alliance	Oxfam Care Save the Children WWF Amnesty International
National	ASSEFA Samata (Figure 5.9)	BRAC Maasai Health Services (Figure 5.6)
Local	'grassroots' or 'people's' organizations village cereal banks Sarva Seva Farm at Utchapatty	local charities 'Passage House' in Recife

Figure 5.3 Main types of non-governmental organization with examples.

This narrower definition would imply that only organizations falling within the top two cells on the right of Figure 5.3 should be called 'NGOs' (and then only if they are specifically oriented towards development). In this chapter, the narrower usage will be indicated by 'international NGOs' or 'national development NGOs' as appropriate.

Let us now look more closely at some examples at each of the three levels in order to understand the variety covered by the term 'non-governmental organization'.

Local NGOs

The local level is where member or mutual benefit organizations are most obviously important. Local member organizations may also be known as 'people's organizations' or 'grassroots organizations'. The situation amongst grassroots organizations in Latin America has been described as follows:

> "If anything is certain about the nature of 'grassroots organizations', it is that — whatever they are — they are *not* homogeneous units. Church groups, labor organizations, Boy Scouts, political action committees, self-help organizations, village potable water associations, communal labor arrangements, squatter associations, worker-owned businesses, ethnic burial societies, transportation collectives, peasant leagues, Catholic 'reflection' groups, tribal federations, microentrepreneur credit associations, all differ from each other and from their equivalents in different countries. ...
>
> Today, there is an elaborate coral reef-like maze of organizations among the poor. With seemingly limitless social energy, they die, are resurrected, and spawn new hybrids of old ideas."
>
> (Annis, 1987, p.131)

Traditional forms of social organization within a village or other community include age-sets, work-teams, different kinds of cultural groups, burial societies and so on. When such organizational forms have a degree of formality as organizations then they are a kind of local member organization, though it is hard to specify just how formal a grouping must be before it is considered to be an NGO.

In one of the other books in this series (Chapters 8 and 9 in *Bernstein* et al., *1992*) there are examples of traditional forms of member organization:

- Woodhouse (Chapter 8) also explains how greater differentiation leads to the formation of economic organizations of particular groups. He gives the examples of 'the transformation of [Senegalese] youth organizations, or *foyers,* from cultural groups into lobbying agencies to gain access to irrigated farming for young people' and of Kenyan women's self-help groups, or *mwethia*, which may be savings clubs or co-operative business ventures but are best known for the voluntary hand-digging of terraces on steep sloping farmland.

Mamdani (Chapter 9) describes different kinds of communal work-teams in northern Uganda, with the older *Wang Tic* form with membership taken from all households in a given area declining as economic differentiation between households increased, and other forms (*awak* and *akiba*) coming to prominence as labouring associations amongst those of a common class position as poor or lower-middle peasants.

Indigenous member organizations are also found in towns and cities. In Addis Ababa, *edirs* (originally burial societies but now providing a kind of indigenous insurance to both rural and urban poor) and *ikubs* (rotating credit societies) are prevalent.

> "People from all walks of life seem to be members of these institutions: university lecturers, civil servants, merchants, salaried employees, domestic servants, and even street children, peasants, artisans, craftsmen, village leaders, and, perhaps especially, women. ... People who are poorer and who have no access to banks or credit learn to rely on *Ikubs* and *Edirs* for prac-tically all levels of support: the bulk purchase of *teff* (a staple food) when prices are high; the cost of materials for the re-pair of eroding houses; the equipping of school children; for information and contacts; and lastly for solace and companionship in difficult circumstances."
>
> (Salole, 1991, p.10)

Such indigenous self-help associations have a role in maintaining the links of urban migrants with their village origins, and this has been criticized as a source of ethnic division in urban communities. However, Salole argues that the importance of such organizations is deeper and that disadvantaged groups tend to copy democratic traditional forms if they are well known and use them to set up their own organizations in response to new survival problems. Thus even street children have set up their own *ikubs*:

> "... how is the necessary credit obtained to purchase the first tin of shoe polish or the first brush? Living from hand to mouth, as many of these children do, they have had to develop strong interdependent links with shopkeepers and suppliers, and they have done this mostly by forming *voluntary associations* amongst themselves. These associations do not stop at merely providing credit and a network, but actually play a role in arbitrating turf battles and ... disputes between different groups of children engaged in rival activities."
>
> (Salole, 1991, p.13)

Salole complains that international development NGOs too often overlook traditional self-help associations when they look to form partnerships with local NGOs. They may prefer to try to help set up modern, formal, democratic and representative forms of 'people's organization' such as co-operatives and unions. These modern organizations incorporate principles such as open membership and 'one person, one vote' and thus differ from groupings based on traditional ways of organizing in village communities or urban neighbourhoods, where who is a member and who holds power may be determined by pre-existing local social and power relations which may or may not include democratic elements.

In other cases, international or national development NGOs have attempted to build on traditional communal forms to develop local co-operative and grassroots organizations appropriate to modern conditions. Perhaps the most important proponent of this approach was Mahatma Gandhi. Any one of the Sarva Seva farms, that at Utchapatty, for example (see above), is a local member organization of the type we are discussing.

A similar case from Africa is provided by the idea of village cereal banks directly controlled by the community, which were promoted by international NGOs, particularly Oxfam, in Upper Volta (now Burkina Faso) from around 1975. A comparable system had existed in the precolonial period. The modern version is described as follows:

> '... a village cooperative group raises funds from an NGO with which it buys grain at harvest-time, when the price is low. If the harvest has been good that year, they may buy the grain within their own village; if the harvest has been poor, they will look elsewhere in the country. The cooperative builds a small grain store and keeps the food for up to six or eight months until members of the group have begun to run out of their personal food stocks and market prices have begun to rise. The group then decides at what price the grain should be sold, bearing in mind inflation and the need to repay the initial loan. Some groups buy and sell just once in the year; others carry out the exercise repeatedly (like the grain merchants), each time providing a service to their community while slowly building up their own working capital fund."
>
> (Twose, in Poulton & Harris, 1988, p.142)

Cereal banks help to stabilize grain prices over the year while assisting villagers to challenge the often monopolistic position of traders. However, different NGOs tended to promote cereal banks separately in different villages. Twose points out that 'aid cannot be at its most effective when hundreds of village-level projects are operating autonomously'. He also comments that the success of the cereal banks varied widely with not much learning from experience being passed from one to another.

> "Research on one village cereal bank in 1982 showed that the poor households were not even members of the co-operative. In the main, grain was bought from the cereal bank by comparatively rich households, perhaps to resell at a profit to poorer members of the community. Four of the six members of the cereal bank committee came from these richer households, and they determined who was able to buy the controlled-price grain. Furthermore, the grain was only sold in sacks of one hundred kilos, a quantity which the poor can rarely afford to buy."
>
> (Twose, in Poulton & Harris, 1988, p.143)

Figure 5.4 A farmers' co-operative in Burkina Faso. Despite the international co-operative principle of 'open membership', this co-operative is predominantly male, which may reflect 'traditional' community organization or may indicate that this is a particular interest group within the community.

Twose also notes that, while initially there was little state involvement in the promotion of village cereal banks, the state tried to take advantage of what was happening:

> "By 1981, two-thirds of the state grain agency's purchases were being made on its behalf by village groups and then resold by the groups — but the resultant profits were recuperated by the state, not left with the community. Many village groups became unwilling to participate further on these terms."
>
> (Twose, in Poulton & Harris, 1988, p.143)

Once again we can see constraints on the spread of village cereal banks as a model, which can be related to the three intersecting arenas of Figure 5.2. For the moment, however, the point is that each village-based co-operative running a cereal bank is another example of a local member organization.

Figure 5.3 also envisages the possibility of local NGOs aimed at public rather than member benefit. Local charities would be typical cases. We have already met some examples of these, among Brazilian street children, in Chapter 2. For example, the Passage House in Recife was set up *for* children and in order to do something about the public problem of 'abandonment' rather than being an organization of the children themselves. To the extent that the founders and organizers of such projects try to involve the local community and the 'clients', in this case children, then the dividing line between local charities and 'people's organizations' becomes a little fuzzy. However, in principle there is a clear difference between those cases where people, including children, organize on their own behalf and other cases where Church groups or other associations form to provide for others and meet public needs.

International NGOs

At the other end of the spectrum from local grassroots organizations are those which operate in several countries. These international NGOs may raise support and resources in one country, generally in the industrialized world, in order to help poor people or do something about development problems in other, developing, countries. Note that these are quite different from inter-governmental organizations such as United Nations agencies, though sometimes they work in similar fields and co-operate closely together.

In principle there could be examples in the top left cell of Figure 5.3: international member organizations. International federations of trade unions or of peasant or women's organizations, or the International Co-operative Alliance might fit the bill; although as federations they would seem to be rather different from the 'people's organizations' discussed above. We will meet other examples of such 'intermediary' organizations in the discussion below on national and regional NGOs.

In principle also the category of 'international public benefit NGOs' (top right on Figure 5.3) includes organizations which try to act directly on general public concerns such as the environment or human rights through lobbying, educational programmes and other forms of campaigning. The WorldWide Fund for Nature (WWF) and Amnesty International are two well known examples.

However, for the purposes of this chapter we are more concerned with international development *charities* or *private service organizations*. They attempt to relieve poverty by providing services to particular groups of people in developing countries. Even when the services are provided for very specific groups of people the NGO may still be regarded as acting in the *public* interest. Indeed, British charity law regards the relief of poverty for specific beneficiaries as one of a number of public benefits that a charity may be set up to provide. This relates to the way the individual needs of the poor can turn into a public need to reduce or contain the problem or threat posed by poverty (Chapter 1).

International NGOs differ enormously in respect of the interests they represent. Some, such as World Vision, are based on Christian churches. Others, such as Care, based in the United States, the French Médecins sans

Frontières and most of the the large Dutch and Scandinavian NGOs, are heavily dependent on funding from the governments of the industrialized countries in which they are based. Others, such as Save the Children and Oxfam, might be regarded as independent development agencies; raising the question of who they are actually accountable to.

International development NGOs engage in an enormous range of different activities the balance of which has tended to change over time. One influential view is that of Korten (1987) who suggested that there are typically three 'generations' of NGO strategies:

- relief and welfare
- small-scale self-reliant development
- sustainable systems development.

Relief and welfare

Korten points out that many of the large international development NGOs began as charitable relief organizations focused on refugee and natural disaster situations. These organizations also brought their expertise to bear in the delivery of welfare services in non-disaster situations.

According to Korten, emergency situations will always occur that demand 'immediate and effective humanitarian action'. Welfare provision, in health, education, shelter etc., will also be required so long as there are individuals and groups within a society who suffer deprivation. Voluntary agencies have a continuing role here as well, in industrialized as well as developing countries, alongside state agencies supplying similar services.

However, there are cases where NGOs compete with statutory services and even set up parallel administrations, thus undermining the basis for improved state services in the future. Hanlon (1991) documents this in the case of Mozambique. For example:

> "Mozambicans are setting up NGOs, but foreign NGOs are moving very much faster. They are flaunting their new wealth and power, and in a very public way displacing the Mozambican government. ...
>
> The issue is perhaps clearest within the health sector. ... Mozambique's health service still has the commitment and administrative capacity to operate good rural facilities, if funds were made available. Instead, donors give the money to NGOs.
>
> In the emergency appeal in April 1989, the government asked for $4.2 mn (£2.5 mn) to rehabilitate health posts, provide vital medicines and transport, and help children. Donors provided only three-quarters of this, leading to severe drug shortages in many areas. But $16 mn (£9.6 mn) was allocated to NGO projects that Mozambique did not ask for ...
>
> As a result, at least six different teams of flying doctors were working with displaced people. Mozambique has three objections to this. First is use of resources. Airplanes and foreign doctors are very expensive, and for a similar amount of money Mozambican health workers could be put in more places providing health care to more displaced people...
>
> Second is that these NGOs are providing a short term emergency service rather than building up a Mozambican one; when they think the emergency is over they will go away, leaving no health provision behind...
>
> Third, some of these flying doctors refuse to even liaise with the government. They fly to towns without advance notice, bring in medicines not on the national restricted drugs list, and do nutrition surveys without giving the government the results...
>
> In areas where they are working, this new breed of NGO doctors are displacing the government health service, and setting up an independent one."
>
> (Hanlon, 1991, pp.216–217)

Nevertheless, so long as they work with the state, NGOs can have a distinctive role in this area through their value basis, the motivations of their staff, and their relations with particular groups of those served. Still, whether services come from state agencies or from NGOs, relief and welfare approaches constitute a very limited strategy for development. They 'offer little more than temporary alleviation of the symptoms of underdevelopment' (Korten, 1987).

Small-scale self-reliant development

The 'second generation' of NGO activities became prevalent from the late 1970s. They included community development projects in areas such as preventive health, improved farming practices, and so on. These projects stressed local self-reliance and the idea that benefits could be sustained beyond the period of NGO assistance. The promotion of co-operative village cereal banks by Oxfam and others (see above) is an example of a second generation strategy. Like the first generation, these were also paralleled by state programmes and in fact a similar shift in the avowed style of all local development occurred in this period.

Sustainable systems development

Korten calls his 'third generation' strategies 'sustainable systems development'. These strategies have evolved as some of the larger and more experienced NGOs, in particular, attempt to find ways of overcoming the limitations of approaches confined too much either to specific services or to the level of local communities. According to Korten, this means less 'direct involvement at village level' and more:

> " ... working in a catalytic, foundation-like role rather than an operational service-delivery role ... facilitating development by other organizations, both public and private, of the capacities, linkages, and commitments required to address designated needs on a sustained basis."
>
> (Korten, 1987, p.149)

Hanlon in his book on Mozambique quoted above was clearly arguing that it is important to work with the state and to help improve state services rather than setting up an alternative in competition with them. This would be a type of third generation strategy, and he does give examples of NGOs acting in such a way as well as the kind of counter-productive activity described in the previous quote:

> "In one province an [international] NGO arrived and wanted to support health. It asked what was needed, and was told the most urgent need was for training — particularly upgrading and refresher courses for existing staff in the districts. So the NGO built a training centre, assigned two staff from the NGO, and provided all the material needs. The centre does two- or three-week courses as requested by the provincial health directorate. These involve normal Ministry programmes as well as special ones that often cater to the needs of refugees. In the first year, 200 primary and community health workers attended courses. The two NGO staff also go out to the districts to follow up and support people who attended the courses.
>
> The centre is popular and successful, both with officials and with the health workers themselves. It satisfies a felt need in the province; the NGO provides the material conditions such as food and lodging, notebooks and manuals, and transport from rural areas; but the centre remains under government control and trains government staff in government-backed programmes. Thus the NGO is seen to be helping the government, not competing with it, and the health workers feel that they are being trained by the Ministry and not by an NGO."
>
> (Hanlon, 1991, p.205)

The case of the village cereal banks leads to an interesting example of a third generation strategy developing from a second generation one partly as a result of processes in the state/society arena (Figure 5.2). In 1983 a new, radical, government came to power which began to limit the freedom of action of international NGOs and made clear its aim of integrating the village cereal banks into state grain marketing strategy. Oxfam reacted by using its experience of cereal banks to promote a particular version of a state strategy, and this apparently led to a fruitful relationship with the state, at least until 1987 when that government in turn was overthrown.

> "Two months after the new government of Burkina Faso came to power [in 1983], an Oxfam paper laid out the way in which

independent cereal banks could become involved in transferring cereals from surplus to deficit areas of the country, based on a programme of grain sales between co-operatives which would be co-ordinated by the state grain agency and could perhaps use seasonal credit facilities from national banks. Such proposals would not only make the state system more efficient; they would also take maximum advantage of the existing village cereal banks by linking them together at a level that would be inappropriate for NGO intervention. It would also show NGOs that they themselves could benefit from collaborating with the state, once it shares their concern to give to the poor greater control over their lives. When local communities were asked to formulate development priorities for their own area in 1985, cereal banks were requested throughout the country and the new regime seized upon the idea, looking to incorporate it into state grain marketing strategy."

(Twose, in Poulton & Harris, 1988, p.144)

This third generation approach means NGOs are trying to use relatively small resources to influence the policies of other agencies with much larger resources. To work in such a role means collaborating with other organizations, forming coalitions and lobbying, as well as working with government.

Although the three approaches are portrayed as leading on from each other, it is not the case that NGOs have moved sequentially from one to the other. Nor can one can identify 'old-fashioned' NGOs stuck in the first generation welfare and relief mode, a second set of NGOs promoting community level self-reliance, and a third, more 'progressive', group that has realized the importance of collaborating and working with governments on broad systems development.

Figure 5.5 Working with government: water quality testing on newly protected springs in Wollo province, Ethiopia. Oxfam's preferred policy in many cases is to work through local NGO partners, but here Oxfam was working directly with the Ethiopian government to build up the capacity for technical services.

But some NGOs do remain specialized in first-generation relief and welfare work. These include some of the church-based international NGOs, and NGOs funded largely by the governments of industrialized countries to carry out specific work on their behalf, which Korten (1989) calls 'Public Service Contractors'. These also tend to be the cases where there is a danger of duplicating and undermining state services in the way Hanlon so graphically deplores.

In general, however, it is programmes rather than organizations that can be labelled first, second or third generation, and many NGOs run a combination of programmes of all three types. In fact, the logic of the third generation type of strategy is to support the other two by overcoming some of the limitations and thus allow policies aimed at providing welfare or local empowerment to work better. Thus the three strategic orientations represented by Korten's first, second and third generation strategies co-exist, both in different NGOs that work alongside each other and in the same NGO.

National and regional NGOs

At the national and regional level it is easy to find examples both of member benefit or 'people's organizations' and of public benefit organizations or charities, as well as of NGOs which combine both.

Annis describes how grassroots initiatives can broaden to the national level and how national NGOs can in turn spin off further local organizations:

> "... poor people organize not only in response to their needs, but in response to incentives. The growth of organizations is driven from within, and from without.
>
> In the rare instances in which needs, incentives, resources, luck, leadership, and hard work all converge, then small organizations tend to grow. But ... they soon confront obstacles that are beyond their control or require resources that are outside their reach. The farmers who found that they could successfully grow vegetables together in their village find they cannot get their vegetables to market without a truck, without bribes, without a road, or without access to markets in the capital. A group of squatters ... find that they cannot get municipal services and build substantial homes until they also have title to the land. A rural community that has just built a schoolroom ... finds that it cannot get a second teacher, textbooks from the Ministry of Education, or force the primary teacher not to take four-day week-ends in the capital.
>
> ... So, small organizations ... link forces with ... similar organizations that have greater lobbying and technical power. These higher-order, intermediary groups prod externally for sources of funding; they form ephemeral political alliances of convenience; they technify and professionalize; and they spin-off lower-order groups."
>
> (Annis, 1987, p.131)

In ASSEFA we have seen an example of a national NGO which grew out of a particular grassroots movement and now acts to promote the local self-reliant organizations which form that movement. Other national NGOs began, like the international NGOs described above, as emergency or relief agencies. Some of these have also changed orientation towards self-reliant community development or towards a broader policy orientation, as in Korten's second and third generation NGO strategies.

One well known national NGO even changed its name, from the original Bangladesh Relief and Assistance Committee to the Bangladesh Rural Advancement Committee (still 'BRAC'), to reflect this shift in emphasis. BRAC is perhaps best known for its research programme in collaboration with local landless people which resulted in the report, *The Net: power structure in ten villages*, which BRAC published in 1980. This was BRAC's response on finding its efforts at organizing groups of landless people to gain access to land and start collective agricultural activities frustrated by a 'net' of powerful individuals well connected with government.

Another example of the work of this type of national NGO is given by the Grass Roots Integrated Development (GRID) Project in a very poor region of North-East Thailand. This is a programme begun in 1984 jointly by three national Thai NGOs and a Thai university institute, using 'Village development animateurs' (VDAs):

"The work of a VDA appears very unstructured and amorphous to the villagers, especially at first. The VDA must get to know

Figure 5.6 National NGOs working with local groups. (Above) A community worker from Maasai Health Services, Tanzanian NHO, helping a woman from the village of Naisinyai as she practises signing her name. The women of this village have established a successful cattle dip separately from the 'traditional' village leadership, and need to open their own bank accounts. (Left) One of many groups of landless organized by GSS, a well established Bangladeshi NGO with almost 100 staff, which runs mainly functional literacy and social action projects. The group pictured is unusual in being mixed-sex.

the people, collect data, develop an understanding of village problems and try to identify the informal leaders of the community and those with leadership potential. Many of the villagers are confused by the 'foreigner' who has come to live among them; the villagers are of Laotian origin and retain many of their distinctive customs and traditions, so that a young educated urban-born Thai is almost as much of a novelty and a mystery as a Canadian would be. Also unfamiliar is the emphasis on discussions, seminars and study tours to visit projects being carried out in other villages. In the past, most government development schemes have revolved around the construction of physical infrastructure such as dams, roads or wells; or they have been welfare-oriented, distributing blankets in the cold season or eucalyptus trees to plant for fuel. The villagers have come to associate 'development' with gifts and wonder why GRID does not also distribute such bounty. Slowly, however, some of the people are beginning to understand the concept of self-reliance, to realize that they themselves can take action to better their lives. Groups are being formed and the success of their small projects is encouraging further ventures. ...

Gifts to the *wat* [village temple] bring merit to the givers. But nowadays ... some of these gifts are being returned to the ... community. Part of the money donated to the temple last year by villagers working in Bangkok was used by the village to dig a communal fishpond. Rice that would ordinarily be given to the *wat* is being used to fill a rice bank set up by the community. ...

Projects such as the fish pond and the rice bank increase community self-reliance and confidence. The men who administer the projects develop leadership and management skills, and the ability to locate and contact outside sources of assistance. ... "

(Durno, in Holloway, 1989, pp.100–105)

Although, as in this case, national NGOs may try to involve local people, they remain essentially public benefit or 'charitable' organizations formed outside the communities they attempt to work with. One of the main questions of this chapter, as you are probably realising already, is to what extent activists from national NGOs like this, as well as those from international NGOs, are necessarily limited in their attempts to work closely with individuals and groups of poor people in local communities.

5.3 What's the big idea?

Despite all their differences, 'empowering the poor' has become an almost universal slogan uniting international and national development NGOs. 'Empowerment' is a desired process by which individuals, typically including 'the poorest of the poor', are to take direct control over their lives. Once 'empowered' to do so, poor people will then be able to be the agents of their own development. Along with the related ideas of 'participation', 'community', 'working at the grassroots' and the promotion of 'people's organizations', 'empowerment' forms the basis for what is suggested to be a distinctive 'NGO approach' to development.

A statement from BRAC puts it like this:

> "[NGOs'] development strategy may well be termed as group-based Empowerment Strategy. The central focus of the Empowerment Strategy is building power at the grassroots and the basic instruments used are (i) function[al] conscientization, (ii) participatory organization based on self-organizing system and social action and reflection, (iii) collective values consisting of value based decisions and alternative social techniques, (iv) building economic power with a focus on local resources control and alternative technology, (v) expansion and institutionalization for exploring new possibilities and overcoming stagnation."
>
> (Bangladesh Rural Advancement Committee)

The same concepts form the basis of agreed policy statements by international NGOs. For example, the final statement of the 1987 UN/NGO workshop on 'Debt, Adjustment and the Needs of the Poor' criticised stabilization policies for exacerbating the enormous social and human problems faced in the Third World, and suggested:

> "An ... alternative policy that responds to the needs of the poor and promotes development requires significant will and the organization and participation of the poor themselves in the decision-making process. NGOs must support the efforts of the poor to empower themselves."
>
> (*World Development*, 15 (Supplement), 1987, pp.256–257)

The international financial institutions have also begun to endorse such ideas. For example, in a 'long-term perspective study' carried out in 1987, the World Bank includes the following remarks:

> "Many basic services ... are best managed at the local level — even the village level — with the central agencies providing only technical advice and specialized inputs. The aims should be to empower ordinary people to take charge of their lives, to make communities more responsible for their development, and to make governments listen to their people. Fostering a more pluralistic institutional structure — including non-governmental organizations ... — is a means to these ends."
>
> (World Bank, 1989, pp.54–55)

One might argue that there is too much of a variety of types of NGO for talk of a single distinctive model. However, many of those with long experience of working in international NGOs have no such doubts. For example, one of the editors of a recent book of essays by NGO workers on their experience of working in the field of poverty and rural development, Robin Poulton, writes with some certainty:

> "The NGOs are building up the structures which will allow the peasants not only to claim, but to gain a share of power. ...
>
> Against 'top-down' urban-designed projects, NGOs are using 'bottom-up' methods which pass decision-making progressively to the people. Against the centralized models of bureaucracy, NGOs decentralize responsibility to local groups and community associations. Against a short-term project approach, the NGO methodology is evolutionary and long-term. Building self-reliant village organizations is a long, slow business: but they are and will be organizations of the people, which can work for developing people."
>
> (Poulton & Harris, 1988, pp.31–32)

There are two parts to the idea of a distinctive 'NGO approach' based on empowerment: the notion that NGOs can be considered as a single united category; and then the idea of empowerment by NGOs. Let us look at each of these in turn.

What unites NGOs?

Many writers have claimed that despite their differences NGOs are based on certain distinctive principles. Brett, for example, describes them as distinctively 'value-driven' organizations:

> "Membership is not based on narrow self-interest, but on the commitment to some normative goal or purpose, while their organizational structures are usually democratic rather than hierarchical. Levels of effort and honesty will be superior to those where people work only for a wage or for profit, and participants will be 'empowered' through their participation in decision making processes. As private agencies their development will not strengthen the hand of entrenched political elites; because they are not maximizing profits but the benefits of members or workers they will not exploit market opportunities by paying or charging exploitative prices or wages. Because they involve all members of the relevant community they will reduce levels of economic differentiation and class conflict, thus playing an integrative role in society as a whole."
>
> (Brett, 1990, p.2)

Fowler (1988) lists twelve areas in which 'NGOs believe that in comparison with governments they have a better ability', and then attributes these areas of 'comparative advantage' to two distinctive features of NGOs:

1 Their 'relations with the intended beneficiaries', which for NGOs is one of 'voluntarism' rather than one of control as with relations with governments.

2 Their 'appropriate organization'. Whereas government typically requires bureaucratic organization with its 'command methods', 'uniformity', 'rigidity', etc., Fowler notes that:

> "The situation for NGOs is very different. They have no particular need to organize along bureaucratic lines; development, change and diversity, rather than maintenance, control and uniformity, can be their guiding image and organizational design. ... Achieving flexibility and responsiveness, correctly coupling process to outcomes, experimentation and risk taking, achieving outcomes at less cost, working with and strengthening local institutions, etc., all depend on the right organizational form being chosen and attained."
>
> (Fowler, 1988, p.11)

Solidarity has been postulated as a third principle of regulation, alongside both the market principle of *price* and the state regulatory principle of *authority* (see, for example, Lindblom, 1977). Thus a charity or a grassroots organization, in dealing with other organizations of such types, will tend to provide a good quality and efficient service not through coercion or the desire to get as much in exchange as possible but simply from a sense of common interests, shared values and an understanding of each others' needs.

This idea of a distinctive form of regulation for NGOs can be extended to the ways in which organizations obtain resources. Korten (1989) points out that whereas state organizations 'command resources through taxation and even expropriation — backing demands with coercion if necessary', and business organizations gain the resources they need by producing goods and services for sale in the market, 'voluntary organizations (VOs) depend primarily on shared values as the basis for mobilizing human and financial resources. People contribute their time, money and other resources to a VO because they believe in what it is doing.'

One could be forgiven for regarding all this as a little too good to be true, or at least, too good to be always true. An alternative view is that NGOs are essentially like other private organizations. That is, they are based on individuals participating in order to maximize their private benefit; except that in this case the output is a public good which is equally available to all irrespective of what they put in. This makes immediate sense for a local grassroots NGO such as one of the village cereal banks discussed above. The benefit to each individual could easily exceed the amount of effort each has to put in. For large international NGOs, one can also think of material benefits to those participating, ranging from salaries and travel perks for staff to tax breaks and public relations benefits for company sponsors and improved career prospects for long-term volunteers.

Hanlon, for example, in the book on the role of aid in Mozambique cited above, has a generally sceptical view of the supposed distinctive features of NGO organization:

> "... NGOs are supposed to be more flexible and responsive, as well as cheaper and more efficient, than traditional large and cumbersome state bureaucracies. NGOs are said to be staffed by younger and more enthusiastic people who work long hours for low salaries, and who are prepared to endure hardships to work at grassroots level. Thus NGOs should be able to respond more quickly and appropriately to the needs on the ground, and resolve bottlenecks more expertly...
>
> ... these assumptions are often false. NGO workers are all too often bumbling amateurs who go to Mozambique for a bit of adventure, to gain experience, and perhaps to further their careers. They frequently totally misunderstand local conditions and are arrogant into the bargain...

> ... of course, neither generalization is always true. ... at their best, NGOs can be spectacularly good."
>
> (Hanlon, 1991, p.204)

Despite his scepticism, Hanlon concedes that the positive features claimed for NGOs are sometimes present. Indeed, it is difficult to sustain the idea that people participate in NGOs *simply* for individual material benefit. First, in the case of local NGOs, as they grow and develop the actual running of the organization becomes too complex to be shared equally. To take the village cereal bank as an example, it will not be worthwhile on a purely material basis to take on any kind of voluntary office in the co-operative, but it will be worthwhile to take one's share in the products while putting in the absolute minimum of effort, a phenomenon known as 'free riding'. To the extent that such NGOs often work extremely well despite a certain amount of dispute over what is a fair distribution of effort, it is clear that many members are motivated by other considerations in addition to material self-interest. Equally, in the case of international NGOs, while there are clear material benefits to be gained for many as a result of participation in their activities, it is perhaps over cynical to suggest that these are the sole or even the main motivations. Other factors, ranging from genuine humanitarian concern to political considerations of various kinds, also play a part.

Thus we may conclude that, while there is some substance in the suggested distinctive features of NGOs, they cannot be said to unite all NGOs.

Empowerment and its conceptual origins

The other part of the 'NGO approach' was the idea of empowering the poor through direct action at the grassroots. Grassroots development as an idea is associated with the name of Robert Chambers, whose work was discussed in Chapter 3. However, Chambers does not necessarily see NGOs as having a special role in grassroots development, and he does not emphasize the idea of empowerment as such. In fact three particular earlier figures are often cited as the source of such ideas: Mahatma Gandhi, the inspiration of rural movements for self-reliance throughout India; the radical Brazilian educator Paulo Freire; and the unorthodox British economist E.F. Schumacher.

We have seen an example from the Gandhian tradition in the Association of Sarva Seva Farms (ASSEFA) which typifies a version of the 'NGO approach'. However, the ASSEFA story also brings out clearly some of the problems with generalizing from a small (or even not so small) number of success stories to a model for development in a broader sense. We will begin to see the basis for two rather different visions of a model for development based on empowerment as we look at the ideas of Freire and of Schumacher.

Paulo Freire and conscientization

Freire worked among poor adults in the North East of Brazil and developed a method of teaching literacy which combined learning to read and write with looking critically at one's social situation. The expectation was that the realization of the injustices of their situation would lead such students to go beyond simply learning to read and to take initiatives towards transforming the society which previously denied them social and educational opportunity.

Freire's writings, particularly *The Pedagogy of the Oppressed* (1972), set out clearly his revolutionary ideas on the liberating potential of education, so long as it is a 'dialogical and problem-posing' education. He constantly places the human search for freedom and justice in the context of *oppression*. He argues that to be fully human implies being active and reflective; the fact that people may be passive and unthinkingly accepting of their situation is the result of their being oppressed.

Freire cites writers such as Fanon and Memmi on the cultural effects of being dominated as under colonialism. The oppressed have 'internalized the image of the oppressor and adopted his guidelines' (Freire, 1972, p.23). This implies being 'fearful of freedom', expressing 'fatalistic attitudes' and 'self-depreciation'. It also tends to mean that instead of working to overcome the contradictions that gives rise to relationships of oppression, the only possibility for improvement is seen as becoming 'sub-oppressors'.

> "Their vision of the new man is individualistic; because of their identification with the oppressor, they have no consciousness of themselves as persons or as members of an oppressed class. It is not to become free men that they want agrarian reform, but in order to acquire land and thus become landowners — or, more precisely, bosses over other workers."
>
> (Freire, 1972, p.23)

Freire identifies education as a source of liberation, but only education of a revolutionary type. He uses the term 'banking' to typify the traditional type of education where 'instead of communicating, the teacher issues communiqués and "makes deposits" which the students patiently receive, memorize and repeat'. By maintaining control over how the student perceives his or her reality, such 'banking' education, for Freire, mirrors the oppressive relations of that reality itself. He proposes as an alternative a type of 'dialogical' and 'problem-posing' education:

> "Problem-posing education bases itself on creativity and stimulates true reflection and action upon reality, thereby responding to the vocation of men [sic] as beings who are authentic only when engaged in inquiry and creative transformation."
>
> (Freire, 1972, p.56)

> "The correct method lies in dialogue. The conviction of the oppressed that they must fight for their liberation is not a gift bestowed by the revolutionary leadership, but the result of their own conscientization."
>
> (Freire, 1972, p.42)

Freire propounds specific techniques for dialogical education. Taken together they amount to a methodology for 'conscientization': a process of learning to perceive social, political and economic contradictions, and to take action against the oppressive elements of reality.

Two comments are worth making immediately on the relevance of Freire's thinking to modern NGOs' ideology of empowerment. First, Freire's language and whole view of the world is clearly much more radical than that implied by the workings of most international NGOs in practice. Nevertheless, the idea of working in dialogue with the poor to analyse their actual experiences and to do this in terms of oppression is clearly very powerful in many circumstances. Often it is the state that appears as a source of oppression, either if it is maintaining order on behalf of dominant interests within the country or if it is obliged, say as part of a structural adjustment programme, to cut back on public services and clear the way for increased export-oriented production. Thus it may be that NGOs, by taking a distanced and somewhat critical view of the role of the state, put themselves in a position where they can adopt Freirean ideas with less contradiction than state educators. In fact many NGOs have adopted Freirean ideas in their adult literacy and educational work for some time, though they may not identify their field workers as 'revolutionary leadership'.

Second, Freire did not use the word 'empowerment' at all in *Pedagogy of the Oppressed*. He began using and commenting on the use of the term some years later. When he did so he was quite critical of the way the term has been taken up by North Americans in particular in the individualistic sense of 'self-improvement' along the lines of manuals on 'How to make friends' or 'How to get a good job'. Used in this way 'empowerment' appears to imply the possibility of individuals thinking their way into being able to take control over their lives, when in fact the reasons for their powerlessness lie in structural inequalities that go far beyond the reach of one individual. Freire insisted that his own comprehension of 'empowerment' was '*Not* individual, *not* community, *not* merely social empowerment, but a concept of social class empowerment' which is 'much more than an individual or psychological event. It points to a process by the dominated classes who seek their own freedom from domination ... ' (Freire & Shor, 1987).

E.F. Schumacher and gifts of knowledge

Schumacher is best known for his book *Small is Beautiful* (1973). He did not in fact advocate breaking everything down into small units; recognising that 'large-scale organization is here to stay', Schumacher argued that 'the fundamental task is to achieve smallness *within* large organization'. He looked for the creative synthesis of opposites: order *and* freedom; entrepreneurs *and* bureaucrats.

When considering what is needed for development, Schumacher, somewhat like Freire, discusses what prevents people, 'the primary and ultimate source of any wealth whatsoever', from simply acting on their own behalf:

> "Everything sounds very difficult and in a sense it is very difficult if it is done *for* the people, instead of *by* the people. But let us not think that development or employment is anything but the most natural thing in the world. It occurs in every healthy person's life. There comes a point when he [sic] simply sets to work. ... The really helpful things will not be done from the centre; they cannot be done by big organizations; but they can be done by the people themselves. ... it is the most natural thing for every person born into this world to use his hands in a productive way and ... it is not beyond the wit of man to make this possible ..."
>
> (Schumacher, 1973, pp.203–206.)

Again somewhat like Freire, Schumacher points to the situation of the poor, which 'degrades and stultifies the human person'. However, rather than seeing misery and powerlessness in political or class terms, Schumacher regards 'certain deficiencies in education, organization and discipline' as the causes of extreme poverty. Participation is required for development to work, but for Schumacher this means involvement in practical work of some kind rather than political participation in terms of reflection and dialogue.

> "Economic development ... can succeed only if it is carried forward as a broad, popular 'movement of reconstruction' with primary emphasis on the full utilization of the drive, enthusiasm, intelligence and labour power of everyone."
>
> (Schumacher, 1973, pp.190–191)

Although sharing in the identification of education as a key to enabling people to become their own development agents, Schumacher fundamentally differs from Freire in suggesting that *gifts* of knowledge are possible. Not only that: 'The gift of material goods makes men dependent, but the gift of knowledge makes them free'. Schumacher's is essentially an extension of the paternalistic notion of helping people to help themselves, without paying attention to whatever structural inequalities led them to appear to need help in the first place. He extends the old saying about teaching a man to fish rather than giving him a fish to assuage his hunger. At a higher level, better than supplying fishing tackle would be 'to teach him to make his own fishing tackle and you have helped him to become not only self-supporting but also self-reliant'. He proposes that aid programmes should aim 'to make men self-reliant and independent by the generous supply of the appropriate gifts, gifts of relevant knowledge on the methods of self-help'.

However, Schumacher realises he is proposing some quite radical changes, and, noting that 'development effort is mainly carried on by government officials, ... in other words, by administrators, [who] are not, by training and experience, either entrepreneurs or innovators', Schumacher argues for a special place for *non*-governmental organizations.

> "It is not enough merely to have a new policy: new methods of organization are required, because the policy is in the implementation. ... action groups need to be formed not only in the donor countries but also, and this is most important, in the developing countries themselves. These action groups ... should ideally be outside the government machine, in other words, they should be non-governmental voluntary agencies."
>
> (Schumacher, 1973, p.188)

Schumacher goes on to argue for an international 'intellectual infrastructure', in which international NGOs would play a big part, though the support of governments would also be required. There would be four main functions to be fulfilled: communication, information brokerage, feedback from the grassroots, and creating and coordinating substructures in the developing countries themselves.

You should note that Schumacher never explicitly analyses in terms of class relations, interests or exploitation. He tends to assume goodwill on all sides and to reduce problems to a lack of understanding or an unwillingness to innovate. He asks 'Why is it so difficult for the rich to help the poor?', assuming apparently that they would want to do so if only they knew how.

Two different versions of empowerment?
It seems that Freire and Schumacher in fact represent two rather different approaches to development, with contrasting versions of what is meant by 'empowerment'. While both emphasize the importance of working with groups of local people to help them pursue their own development, the analyses and thus the methods proposed are quite different.

There are two rather different traditions of grassroots action by NGOs, corresponding respectively to the ideas of Freire and Schumacher. One uses the method of '*participatory action research*' (PAR) and has developed mostly in Latin America and South East Asia. The methods used by Village Development Animateurs in the GRID project described above are an example of this. The outcomes of a particular intervention by a VDA are not known in advance and depend on groups of villagers analysing their own situation and coming up with proposals. However, although the Freirean influence is clear, as described this version of PAR does not seem to be so revolutionary in intent or outcome as the methodology of conscientization espoused by Freire himself.

The other tradition might be characterized as promoting 'tools for self-reliance', though the 'tools' concerned can be organizational innovations such as marketing co-operatives or credit unions as well as (appropriate) technical and training solutions to local problems. Village cereal banks would be an example of such a 'tool'.

Both these versions of empowerment have their limitations, and these are to be found in processes in the three intersecting 'arenas' of Figure 5.2. The next section looks at the limitations to the ideology of empowerment in just this way.

5.4 The limits to empowerment

As we can see from the examples given above, there are cases where NGOs appear to succeed in their aim of empowering the poor. However, there are also cases of failure, and there are no examples of widespread social transformation brought about directly through such methods.

We have seen that there are two main usages of the term: empowerment as the provision of tools for self-reliance; and empowerment as the aim of participatory action research leading to challenge to existing power relations. Different limitations apply in these two general cases, but in both cases they stem from processes in the three 'arenas' of Figure 5.2 (local community, state and society, and the NGO itself), so in this section we examine each of the three in turn as sources of limitations on empowerment.

Limitations in the local community arena

Tools for self-reliance
The 'tools for self-reliance' approach tends to adopt a naive view of community. An international or national development NGO works with a local 'community' as though this were a well-defined and unproblematic concept. This was apparently the case with the village cereal banks in Burkina Faso. In practice, of course, the idea of a group of equals who can be 'empowered' *together* is as flawed as the notion of a household consisting of men and women, young and old, all contributing to and drawing from a 'common pot'.

On the one hand it may not be clear that everyone living in a village or urban location is included as a community member. On the other hand, those who are included will certainly be living in complex and unequal relations with each other. Thus it is probable that in practice community organizations represent the interests of better-off and more powerful individuals and groups better than they represent the interests of the poorest members of the community. The example was given of the cereal bank in which the majority of committee members came from richer households who controlled access to controlled-price grain and were even able to resell grain at a profit to poorer households who 'were not even members of the co-operative'.

PAR

Participatory action research, for its part, meets different limitations at the local community level. To be successful, this approach implies working with groups with similar experiences, in similar positions of powerlessness. This could mean working specifically with poorer sections of a 'community' even though they will tend to be under-represented in community organizations, and it could entail disrupting unequal social relations. The most obvious example is the case of gender relations, but the argument could apply also to ethnic or caste relations.

In the article quoted above on the Thai GRID project, you may have noticed that it was *men* who were said to have gained in self-confidence and organizational skills through running village projects such as the communal fishpond. In fact the author goes on to discuss efforts by the VDA to promote women's groups, and, through analysis of their own situation by women, to help establish women's projects. She notes how 'in the rural areas of Thailand women have not taken a public role in the administration of village affairs', and how, being the 'hind legs of the elephant', 'they work just as hard, but they nevertheless walk behind.' In this particular case, the author appears hopeful that, as villages in this area are 'in fact, largely homogeneous', it will prove possible to introduce participatory action research in the future with women's groups as presently with village groups that largely exclude women, while remaining 'largely non-confrontational'. This hope may or may not prove warranted in this case, but surely in many locations work to empower women would in practice either be very limited or else provoke considerable disruption — so much so that such work may often be impossible.

Figure 5.7 Participatory action research. A women's group organized by Nijera Kari *('We do it ourselves'), another Bangladeshi NGO, that brings together groups with common interests to discuss the causes of poverty, starting with their own position and work out way of acting together. The main problems brought out in this particular group were reported as: the dowry system, lack of education, and women having to work in secret.*

The ASSEFA example showed NGO work with 'communities' of particular castes, and its success in empowering them is underlined by the quote 'We have gained recognition in the village. Other castes, who were our masters earlier, now not only listen but pay attention to what we say.' However, the same comment must be made: despite occasional successes, on the whole caste divisions constitute social relations that would be so disrupted by empowerment that they will prove to be a forceful limitation on what can be done in the name of empowerment.

One can find glowing accounts of the apparent success of PAR even against all the odds. For example, Wikramaratchi in an article (in Holloway, 1989) entitled 'Overthrowing the Moneylenders', tells of how in a particular village he as an NGO action researcher succeeded in empowering the majority of 'small and marginal farmers, landless workers, small fisher folk, rural artisans and others' against the powerful minority of 'traders, money-lenders, the bigger landowners, the elite in general and ... village-level bureaucrats.' He begins his article by noting that 'a 'neutral' or technocratic intervention ... in village development often results in only the dominant interest in the village benefiting' (something which we have noted can easily result, even in the name of empowerment, from a naive view of what is meant by 'community') and ends by telling the reader that 'the middlemen and money-lenders were eliminated'.

In this particular case, whatever the reaction of the powerful minority was, it was apparently not enough to save them from 'elimination'. However, there are certainly many other cases where those whose interests are threatened by the methodology of PAR actually use their power against those trying to 'eliminate' them. Indeed, such programmes can lead to activities which are 'frequently of a confrontational kind — strikes, rallies and the organization of militant unions.' (Durno, in Holloway, 1989). In general, the Freirean notion of conscientization and reflection leading to action does not seem to hold any guarantee of that action being successful, especially when it is directed against powerful and dominant local interests.

Limitations in the arena of state and society

We have already begun to see the importance of how NGOs relate to other development agencies, including other NGOs and particularly the state. It has been suggested (Clarke, 1991) that international NGOs (and well-established national NGOs) are likely to relate to the state in one of three ways:

1 *complementing* the state; by filling in gaps, by providing services they are better equipped to provide than the state, or by working with the state to provide jointly a variety suited to the variety of needs among the population

2 *opposing* the state; either directly by lobbying at government level or in international arenas, or indirectly by supporting local and national groups that are adversely affected by government policies

3 *reforming* the state; by representing the interests of groups they work with at grassroots level to government and working with government to improve policies.

We can try to relate the two versions of empowerment to these possible strategies. How far does the model of providing tools for self-reliance correspond to either complementing or reforming the state? Does participative action research imply a more oppositional, or even revolutionary, stance? If we look at each in turn we will find that things are not quite as simple as this.

Tools for self-reliance

The version of empowerment as 'tools for self-reliance' certainly gives the term a rather weak meaning in a political sense. There is a minimum requirement for a benevolent, if perhaps relatively inactive, state. Schumacher's example of showing a person how to make their own fishing tackle might increase that person's capabilities in a technical sense, but would do nothing to help them gain access to fishing rights, for example. If these were all held by the state and/or powerful local interests represented in the state, little could be done through 'gifts of knowledge'.

In the village cereal banks example, individual NGOs were able to promote projects. But as the projects became numerically important the state became involved. At one stage, the author

Figure 5.8 Tools for self-reliance. Oxfam staff in Burkina Faso discussing a project for the construction of 'diguettes', a technique for preventing water running off the land and allowing it to seep into the ground to aid cultivation.

claims, the state tried to 'take advantage' of what was happening and took so much of the profit from the schemes that might have remained with the villages themselves that many groups became unwilling to continue participating.

While it is clear that such projects must remain isolated mini-successes until and unless the state takes up the idea and formulates policy around them, it is also clear that the participative basis of such projects is at risk unless the state is supportive of the idea of self-help at the grassroots. After the change of government, Oxfam was able to influence the state's strategy towards village cereal banks. But this depended on a basically positive attitude on the part of the state towards this type of development. Certainly there is no question of challenge to local or national power structures on the basis of such projects.

This poses a particularly acute dilemma for international NGOs if the state is, for example, interested in increasing village level production but not in ensuring the participation of the poorest groups in each village.

PAR

The other version of empowerment, participative action research, is indeed potentially revolutionary. It may lead to the articulation of particular local group interests against other powerful interests, but this does not necessarily imply opposition to the state as such. There are a number of different possible cases, depending on the interests represented in the state and how they relate to the grassroots interests that an NGO may be trying to represent. For example, if an NGO is working in a context where there are strong local power differentials and the local elites are well represented in the state apparatus, that NGO may well be led into opposition to that state. However, in another context a state may respond positively to the organization of the poor against local oppression.

One can (idealistically) conceive of an NGO strategy that combines participative action research at a local level with representing the interests of groups of the poor in dealings with other agencies and the state. In assessing how realistic such a strategy would be one can utilize the concept of *political space*. In an extreme case of rural exploitation, such as that portrayed in BRAC's study *The Net*, there may be almost no room for positive action. (Nevertheless, BRAC has been making heroic attempts to prove this assessment wrong and they stress the aim of *group*-based empowerment. See the quote at the beginning of Section 5.3.)

In other cases there may be more political space: more room for NGOs to work with governments in building up people's organizations. Holloway (1989), for example, argues that

> "[a] common problem is government having the desire to help, but being unable to do so well. In such cases, if approached diplomatically, government officials can be

Figure 5.9 Bangladesh has many NGOs but their political space is heavily constrained. Here is a local Samata *(landless people's organization) meeting held after the local* Samata *office was burnt down by landlords and all the records disappeared.*

persuaded of the pragmatic value of involving people's organizations, participatory modes of planning, and all the other activities in which NGOs have a lot of experience."

(Holloway, 1989, p.213)

Opposing or working with government?

The question of whether NGOs are in opposition to government or can work with government is not as clear-cut as it might appear. For example, Annis discusses whether small-scale development can be 'scaled up' and points out that:

> "... grassroots groups in Latin America usually perceive themselves as being in adversarial relation with the state. Inevitably, they emphasize their separateness, their political independence, and their *non*-governmental character. Yet in practice, non-governmental organizations tend to be most numerous and most important where the state is strong; and generally, the larger, more democratic, better organized, and more prosperous the public sector, the greater are the incentives for the poor to barter for concessions or to 'co-produce' with the state."

(Annis, 1987, p.131)

Annis goes on to note that it is in small, poor countries and under politically repressive regimes that the NGO and grassroots organization sector tends to be smallest and least rich in its interconnections. (An exception to this appears to be the case of women's organizations in Chile under Pinochet, discussed in Chapter 6.)

An oppositional stance may help to maintain an NGO's identity and value base, whereas working more directly with government might appear to offer more scope for putting ambitious projects into practice. Fowler points out the dangers:

> "Co-operation with government in the task of national development is generally worthy of consideration ... However, ... NGOs should look critically at how this might affect their two key areas of comparative advantage: relations with the intended beneficiaries and appropriateness of their organization ... the creativity, diversity of learning and pluralism in civic society which NGOs can give may be lost."

(Fowler, 1988, p.22)

So, whether actually oppositional or not, NGO relations with government are not straightforward. We need to ask, if NGOs adopt 'third generation strategies' of 'sustainable systems development', what are the processes relating the activities of those NGOs to those of government that may either sustain or constrain those activities?

NGOs may be able to lobby governments to adopt policies and practices that in turn help build up NGOs' capacities. A simple example would be to persuade government departments to work via NGO intermediaries such as national federations of grassroots organizations, rather than employing their own local officers.

Another example of the potential policy role of NGOs relates to what Korten (1987) calls 'micro-policy reforms'. These are a type of state action which require institutional changes before they can be put into effect. (Korten contrasts these with 'macro-policy reforms' such as a decision to change the level of subsidy on, say, fertilizer imports, assuming that the administrative apparatus for the subsidy is already in place.) An example of a micro-policy reform would be the introduction of a programme of credit for co-operative groups of small farmers. Unless there is an existing process of co-operative development and a network of relevant credit institutions able to respond to their requests, such a 'policy' would remain wishful thinking on the part of some well-meaning minister or official. This type of reform, if achieved, can in the right circumstances back up the work of an NGO at the local level. The NGO's experience may also help to persuade government of the need for such a reform.

On the other hand, processes at the governmental level can have a constraining influence on the work of an NGO aimed at local empowerment. Too many articulate demands from what was previously a quiescent section of the population may lead at worst to repressive state action or at least to expectations raised only to be rudely dashed.

Limitations on empowerment from NGOs themselves

The third of the 'arenas' in which processes are going on that may constrain or limit local empowerment through NGOs is that of the NGOs themselves at a national or international level. At a simple level, if empowerment were to be replicated in enough local projects to constitute a general model for development, there would have to be a process for recruiting, deploying, training and maintaining the commitment of NGO activists to work in all the local projects. Whether or not such a deliberate strategy is attempted, organizational processes continue within NGOs that affect staff attitudes, their 'probity', whether there is a culture that values public service, and so on.

These processes differ among NGOs adopting Korten's first, second and third generation strategies, since these three approaches require very different capacities and skills of the NGO and its staff.

- The first generation requires logistics management and specific professional competences such as those supplied by medical or paramedical personnel. Military analogies may be in order for the type of organization that is appropriate to run such programmes effectively.
- The second generation incorporates examples like the work of an ASSEFA *sevak* over a period in helping articulate local needs and aid the development of a participative project. Project management is required, with community workers capable of extended periods of political and educative work at a person-to-person and group level and thus requiring a good deal of emotional as well as practical support from the wider organization of the NGO they are representing.
- The capacities required to carry out a third generation strategy successfully differ again. What is needed in this case is strategic management rather than logistics or project management, and high level skills both of strategic analysis and in negotiation and collaboration will be required from staff, as well as an ability to coordinate with various programmes of the other two types both within the NGO itself and run by other agencies.

Clearly, questions such as how to retain the high-level staff required and how to maintain the original value base of the organization loom large alongside the operational difficulties of the process of developing and implementing a third generation strategy.

Interestingly, the question of where the motivated activists are to come from seems to be rather a weak point in both Schumacher's and Freire's thinking. Schumacher puts the question like this:

> "You may say, again, [development] is not an economic problem, but basically a political problem. It is basically a problem of compassion with the ordinary people of the world. It is basically a problem, not of conscripting the ordinary people, but of getting a kind of voluntary conscription of the educated."
>
> (Schumacher, 1973, p.204)

It would be easy to dismiss this as wishful thinking. Freire, however, does not even ask who are to be the 'revolutionary leadership' and how *they* are to maintain the required level of conscientization.

Even if, despite everything, there is a good supply of motivated activists, there is also a well documented tendency in value-driven organizations towards 'degeneration'. This can occur both at the level of individuals and their adherence to values and at the level of the organization itself. With regard to the first, we should note that individuals may begin enthusiastically motivated by solidarity and humanitarian values. But not all will be able to sustain their selflessness in the face of temptations offered by positions of power.

There is certainly no easy answer to this. It may be possible to some extent to design organizational incentive structures in such a way as to reinforce a public service ethic. Motivation does not relate only to material and power benefits, but minimum rewards and recognition are certainly required if staff are to remain committed to the positive values of their organization rather than revert to acting purely in their own private interests.

As for organizational degeneration, this can mean a shift in the type of activity undertaken. For example, Korten (1989) notes a tendency for NGOs to move from a base of 'voluntarism' in which work is taken on in fields that relate directly to the values for which the organization was set up to become 'public service contractors' performing services on behalf of states or donor agencies in a variety of fields. Paradoxically it is a combination of the 'comparative advantages' of flexibility, responsiveness, motivation, etc. with specific expertise that may make an NGO a good candidate to be allocated development resources by donors, while the change to large-scale management of donor-sponsored projects may actually reduce the NGO's ability to demonstrate these qualities.

Accountability

This brings us to the last, and in some ways most important, processes: those of *accountability*. Contradictions and conflicts abound here. For example, empowerment theory may emphasize accountability towards the grassroots. But this conflicts with the need to account to donors for how the funds are used. This applies particularly to international NGOs or national NGOs who take money from international donors while supporting local participatory action research projects. More generally perhaps, if an NGO does 'succeed' in maintaining some independence from donors and continues to support local actions that empower local people, to whom is it accountable? What interests does such an NGO represent? Is it no more than a kind of disembodied representation of the interests of NGO managers and activists? If so, how are the interests of those it is working to empower to be safeguarded?

5.5 Conclusion

Let us return to the question posed at the start of this chapter. Does empowerment at the grassroots through NGOs constitute a new model for development? To answer this we can say that while NGOs can conduct grassroots projects aimed at empowerment, they cannot be multiplied to give a simple alternative 'NGO approach' to development. There are two reasons for this. First, NGOs are involved in many types of development activity, not only local empowerment at the grassroots. Second, a general model for development needs to go beyond the actions of NGOs alone to include the place of NGOs in public action in relation to other development agents, particularly the state.

The message from the limitations described in the previous section is that both the idea of providing tools for self-reliance and the method of participative action research are too restricted to form a general model for social transformation.

The so-called 'new orthodoxy' amounts simply to agreement on what is good practice in the kinds of local project that require local participation in order to work well. For some kinds of project, NGOs have a 'comparative advantage' and in these cases their insistence on empowerment and participation may make them better micro-development agents than governments. Such projects include those which are based on local systems of mutual obligation, and those which require community support, local adaptations or significant inputs from the intended beneficiaries (Fowler, 1988). In other types of project (such as the provision of standardized forms of health care or education) NGOs may simply be duplicating government services. The service provided, whether by an NGO or by a government agency, may use participatory methods, but does not depend on such methods and it may be a distraction from the efficient delivery of services to insist on them too much. Similarly, NGOs' emergency and relief work, may well be based on the simple principle of giving help where help is needed. NGOs may attempt not to do so in such a way as to reinforce sources of dependency and inequality, but the ideas of empowerment may effectively be an irrelevance.

While many types of development project do not require a special approach based on empowerment or participation, others require much *more* than participation. Thus if the 'NGO approach' is to be generalized there is a need for supportive regional and national institutions. In other words, one cannot expect to replicate local projects, however successful they may have been in terms of empowering particular groups of local people, simply by working in the same way in other localities. Many of the constraints on local small-scale development are structural and attempting to overcome them implies action at a broader level.

We saw right at the beginning of this chapter that there are several quite different types of NGO. There is a variety of development tasks to be done, and it is arguable that a different type of organization with a different mix of motivations amongst its staff will be best suited for each task. It is equally the case that different economic and political contexts affect what type of development activity is likely to be appropriate.

There is no given set of NGOs, with set characteristics, that can choose what activities to engage in once they see the tasks to be done and the contexts in which they have to operate. In practice, things are much more fluid. NGOs are constantly adapting and changing. Policy is as much a process for NGOs as for governments, and we have to ask what moves this change and adaptation in one direction rather than another.

Summary of Chapter 5

1 In the 1990s, non-government organizations (NGOs) of different types have been increasing in importance in both industrialized and developing countries. This includes grassroots or people's organizations as well as national and international development NGOs.

2 It is suggested in some quarters that NGOs can provide a new model for development based on direct empowerment of groups and individuals by NGO activists working at local level. However, although there are many examples of successful local projects of this type, this chapter has argued that this cannot constitute a development *model*.

3 Viewing development policy in terms of *process* rather than *prescription* applies to NGOs role in development as much as to state-led development.

4 One cannot just prescribe what NGOs should do to achieve development; their actions aimed at local empowerment are constrained by processes going on in the three wider arenas of the local community, state and society, and in the NGOs themselves.

5 The notion of development through empowerment encompasses two rather different approaches: that of *participatory action research* (PAR) which relates to Freire's ideas on 'conscientization'; and the provision of *tools for self-reliance*, which relates to Schumacher's ideas on gifts of knowledge.

6 The two versions of empowerment are limited in different ways, and neither can form the basis of a development model. A general model of development incorporating NGOs would have to go beyond NGOs' own direct actions, whether 'empowering' or not and relate NGOs to other development agents notably the state, looking particularly at how NGOs' own policy is produced and how NGOs influence state policy.

6

WOMEN'S EMPOWERMENT AND PUBLIC ACTION: EXPERIENCES FROM LATIN AMERICA

HAZEL JOHNSON

6.1 Empowering women

> "For the past decade, women in Latin America have spoken out. In different forms, with different voices, shouting and whispering, in what already amounts to a historically significant rebellion, and after having felt confined for too long to private, invisible spaces, women throughout the continent are now invading streets, town squares and other public places, demanding to be heard."
>
> (Vargas, 1990 p.1)

This is one woman's description of a recent groundswell among women in Latin America. Virginia Vargas is a Peruvian academic and activist, long-engaged in feminist and class politics and founder of a centre that carries out research, training and legal aid for women. She says that, in Latin America, recent years have seen a growth in different kinds of women's organizations, from community groups to organizations of intellectuals. This growth has not been equal throughout all countries, which have different histories and social structures, and face different sets of problems. But there has been a general increase in concern among women about: their position in society; the effects of economic and political change on women's lives; the sort of development needed to benefit women; and the roles women can, or should, play in social change. Besides forming their own organizations, women have also become important figures in certain types of community organizations, as we shall see below, and have championed many issue-based struggles against poverty.

Women's involvement in community-based organizations and the existence of different types of women's groups reflect the diversity of women's realities and experiences. One initiative to understand this diversity and discover common ground has resulted in regional meetings called 'Feminist Encounters', which have been attended by growing numbers (230 in 1981, 650 in 1983, 1000 in 1985, 1500 in 1987, and 3000 in 1990 (Vargas, 1991)). This discussion and debate among women from different backgrounds and experiences has allowed a growing understanding of gender relations in the different contexts of Latin America, as well as encouraging debate about action and alternatives open to women.

Although many women's organizations do not openly espouse a feminist position and although feminism is often seen as 'alien' or divisive to more general movements based on class, ethnic or other interests, feminism has been, and is, of growing importance in the development of women's organizations in Latin America. Even groups which may not identify themselves as feminist are often influenced by feminist debate.

Feminism has often been criticized in Third World contexts as a political perspective originating from the concerns of middle-class western women. Some argue that it separates women's subordination from its material conditions — particularly from those of poor women — and from the wider context of poverty and social inequalities, which affects both women and men. But some Third World writers have expressed a different view:

> "feminism was *not* imposed on the Third World by the West ... historical circumstances produced important material and ideological changes that affected women, even though the impact of imperialism and Western thought was admittedly among the significant elements in these historical circumstances. Debates on women's rights and education were held in 18th-century China and there were movements for women's social emancipation in early 19th-century India; ... other country studies show that feminist struggles originated between 60 and 80 years ago in many countries of Asia. In a way, the fact that such movements for emancipation and feminism flourished in several non-European countries during this period has been 'hidden from history'. "
>
> (Jayawardena, 1986, pp.2–3)

As in other parts of the world, women in Latin America have participated in national and class struggles. Early feminism resulted in some autonomous women's organizations which were associated with female suffrage, as happened in Europe; these organizations were predominantly the concern of intellectuals and the small urban middle classes. At present there are 'many feminisms' in Latin America which try to analyse and struggle against women's subordination in ways that are appropriate to local histories and cultural values. As other writers from the Third World have said:

> "heterogeneity gives feminism its dynamism and makes it the most potentially powerful challenge to the status quo. It allows the struggle against subordination to be waged in all arenas — from relations in the home to relations between nations"
>
> (Sen & Grown, 1987, p.19)

The recognition of women's subordination in Latin America and the rest of the Third World, and the search for alternatives to state action (which may often reinforce rather than change existing gender relations), have opened up the issue of women's *empowerment*. Empowerment means attempts by the poor and powerless — or those whose interests are not represented in the state or whose demands are not met by state policies — to meet their needs, either through their own organizations or through pressure on the state to make it act in their interests (see also *Allen & Thomas, 1992,* p. 91; and *Bernstein et al., 1992*, Chapter 12). 'Empowerment' is also part of the discourse of those non-governmental organizations (NGOs) which have an ideology of 'siding with the poor' (see Chapter 5). Such NGOs try to use their relatively superior resources and position to support processes of empowerment.

Empowerment inevitably raises questions about power: What sort of power? For whom and to what end? The poor and powerless are not an undifferentiated mass. Poverty and powerlessness are relative. And the experience of powerlessness is related to social divisions such as gender, age or ethnic group. For those reasons the processes and goals of empowerment will be multiple and various, as we shall see in the case of women in Latin America.

So what is *women's* empowerment about? Vargas's statement at the beginning of this chapter suggests that women's empowerment involves gaining a voice, having mobility and establishing a public presence. Although women can empower themselves by obtaining control over different aspects of their daily lives, empowerment also suggests the need to gain some control over *power structures*, or to change them. Sen and Grown consider empowerment as a means to an end; for them, women's organizations 'must ... strengthen their organizational capacity, [and] ... crystallize visions and perspectives that will move them beyond their present situation .... Developing the political will for the major changes needed in most societies requires organizations that have the strength to push for those changes'.

This chapter examines the extent to which organizing leads to the empowerment of women. It uses case study material from Latin America to address the following questions:

Q Do women's coping strategies in fact lead to organization and empowerment or do they largely remain as individual or household survival mechanisms?

Q How are women empowered and what kind of change can women's organizations bring?

Q Do women's organizations represent something new or different for development?

For Sen and Grown,

> "... the particular perspective of poor women gives centrality to the fulfilment of basic survival needs as *the* priority issue; they are therefore the most committed, militant and energetic actors once avenues for action emerge."
>
> (Sen & Grown, 1987, p.89)

The issues that women organize around may arise from their immediate needs and concerns, or from more fundamental problems of gender relations. And the actions that women take may have short-term or long-term consequences. These distinctions have sometimes been categorized as *practical* or *strategic* needs and interests (see Box 6.1), the former being those of an immediate and short-term nature, and the latter concerning the need for change at a more fundamental level. As we shall see, the experiences of women's empowerment in Latin America involve both practical and strategic issues.

6.2 Coping strategies and empowerment

> "Well, to be honest we did everything, especially the women, because the men went off to work. And we women had to struggle for everything, those of us who were always up at the front. Because, when it came to co-operating on housing, the husband wasn't there. But who was? The wife. And they had to talk to her — it has always been the wife. So that's why I tell you that women have such an important role. Whether they recognize it or not is another matter, because the men are always saying that they own the home. If the men don't do the washing, the cooking or the sweeping and aren't even here, I don't see how they can be head of the household. Owner of the street yes, but not of the house."
>
> (Quoted by Blondet, in Jelin, 1990, p.29)

For most poor women in urban and rural areas, life is primarily directed towards survival

Box 6.1 Practical and strategic gender interests and gender needs

A study of women's mobilization in Nicaragua by Maxine Molyneux (1985) separated practical and strategic gender *interests* as follows:

Practical gender interests arise from women's positions in sexual divisions of labour (for example, their interests as mothers, rural labourers or domestic servants).

Strategic gender interests are those arising from a desire to challenge women's subordination and existing gender relations.

Moser (1989a) uses these categories to analyse the needs which women organize around into practical and strategic gender needs:

Practical gender needs might include sanitation, children's nurseries, higher wages, healthcare programmes for women.

Strategic gender needs might include changes in divorce laws to give women equality with men, or 'affirmative action' to give women political representation.

strategies for themselves and dependants. It is deeply embedded in the material conditions of work, food provision, wages and the daily needs of family members. Survival strategies are partly based on individual actions and behaviour, but they also depend on networks between individuals and households, and on forms of group solidarity. These networks are often particularly important for women because women's earning power is lower than men's, and because of women's responsibilities for children and for household welfare. Can such networks and forms of solidarity transcend their initial purpose and become mechanisms for women's empowerment? Under what circumstances might this occur?

A Peruvian example

An interesting example of a process of changing consciousness has been documented in a study of migrant women in Peru (Blondet, in Jelin, 1990) on which the following account is based. These women went to the capital city, Lima, between 1940 and 1960 to look for work. The study shows their gradual adaptation to city life, and how their coping strategies developed over a long period of time into organizing around community needs and women's needs. The general contours of the women's histories are as follows.

The women come to the city for a variety of reasons: partly economic, partly to get away from oppressive family relations, violence or unwanted marriages. The decisions they make to go to the city are individual and their early lives in Lima are bound up with individual mechanisms for survival. Through migration, the women lose the support networks in their families and among their neighbours in the rural areas, and they try to replace them by obtaining patronage from their employers. The most common form of work for the women is domestic service, which provides them with somewhere to live, food and a small wage, as well as the 'protection' of the household head or his wife. But the quality of survival depends entirely on how individual master/mistress–servant relations work out:

> "I wanted to study and I asked him to show me how to study. 'No,' he said, 'you are not going to college. I got a girl so that she would work.' "
>
> (Blondet, in Jelin, 1990, p.23)

The economic opportunities for migrant women with no skills or contacts are small. So although finding work (and somewhere to live) remain fundamental issues, the social context of the young women's lives (and social expectations) are also important in coping with daily life in the city. Marriage and the formation of liaisons with men result from growing contacts, but also act as a survival mechanism because links with their husbands' families expand women's networks further. Marriage also 'liberates' women from the patronage of domestic service:

> "Sister, get married. What would you rather be — a slave for those shameful people or a slave in your own home?"
>
> (Blondet, in Jelin, 1990, p.25)

As suggested by this question, marriage has its own contradictions and can raise further problems for women: whether husbands or wives can get work; the problem of finding somewhere to live; the problems of intra-marital strife, often arising from conditions of poverty; and the prospect and responsibility of having children in precarious economic conditions.

Such families tend to end up in the 'marginal' areas of Lima; in the squatters' areas or shanty towns (Figure 6.1). Surviving there calls for determination. Hitherto individualistic survival strategies, dependent though they were on more diffuse forms of solidarity and patronage, take on a more collective character, involving the participation of women. Women band together to find land, build shacks, get supplies of water and electricity:

> "In this street we have all been united. Those who aren't, oh well, they have to go without water; so whether you like it or not you have to work together if you don't want to remain without water."
>
> (Blondet, in Jelin, 1990, p.32)

They also establish new forms of individual support through *compadrazgo*. *Compadres* are

Figure 6.1 (Top) Cerro San Cristóbal, a poor shanty town in Lima. (Above) A family goes about its daily affairs in the shanty town of Callao, Lima.

equivalent to god-parents and the relationship with *compadres* is tantamount to a form of kinship (in social analysis, compadres are often called 'fictive kin'; see *Bernstein et al., 1992,* Chapter 6). *Compadrazgo* can involve some sense of hierarchy (the compadres being the superior partners), and the patron–client relations similar to those of employer–employee are often re-established. But women also try to find *compadrazgo* with others of more equal status who can provide moral support as well as material help when there is no money or food, or when children are ill.

These relationships and networks, and the joint actions of women for community and household survival, contribute to a growing awareness of the reasons for material hardship, and this in turn can lead to women's participation in more formal organizations. There are two main areas of participation: community organizations, in which women become centrally involved but are less commonly part of the leadership; and mothers' clubs, which are also community-based but organized and run by women.

Mothers' clubs were initially promoted by the state and the church in many Latin American countries during the 1960s. Reformist governments tended to see them as a way of improving women's capacities as wives and mothers, as well as enhancing their roles in the community. Mothers' clubs were initially 'welfarist' in nature. They provided material support for poor women, but did not actually engage with the conditions of women's subordination or try to change women's position in society. Nevertheless, while the initiatives were often supported by the state, and many clubs at that time were started by middle-class women and served as a base for mobilizing political support for the government, many were also run by members of the local communities and became a means for women to discuss common problems and take joint actions. An interviewee from a poor community in Lima stated:

> "It was my idea to establish a mothers' club in the Sixth Zone so that we could pool our ideas and problems and see how we could help each other out. Besides, we can do a lot for our health and personal defence which benefits the household, and also help one another with the children ... help with everything in the community."
>
> (Blondet, in Jelin, 1990, p.36)

Mothers' clubs in Peru can be seen as performing several different roles at this time. One was to act as a channel for state policies and party politics. Mothers' clubs could also become local power bases for their leaders to establish new forms of patronage with other women. They also acted as focal points for food-for-work programmes run by non-governmental organizations and the government during economic crisis in the late 1970s and early 1980s.

But the continuing existence of mothers' clubs in poor communities in one form or another (Figure 6.2) allowed some poor women to transcend the original welfarist purposes. The clubs could gradually be adapted and moulded to the women's own perceptions of their problems and needs. In particular the clubs have increasingly been adapted to coping with poverty and creating ways of generating informal-sector income compatible with women's reproductive roles in the household (see *Hewitt et al.*, Chapter 8, for discussion of informal-sector employment for women). Some clubs have even been subcontracted to supply marketed goods for export:

> "There's a workshop called Vipol and they get us to knit alpaca jumpers. For example they give us the wool, the alpaca, and we women in the club knit the jumper and they tell us the style they want. Well, that's what we do, with various women here in my neighbourhood."
>
> (Blondet, in Jelin, 1990, p.41)

In addition, the internal composition and dynamic of the clubs has changed over time. Younger women (daughters of the migrants) have joined and have raised issues of internal democracy in the clubs, access to training and the importance of income-earning work:

> "The president did the organizing. She ... did everything. There was no support from the grassroots, no formal work to be done or anyone to teach us. I used to say we had to

Figure 6.2 Mothers' clubs are widespread. This recent photograph is of the Doña Ramitos mothers' club in the coastal town of Trujillo.

> work in something, so that we could produce .... Women can do a job now. They give their children some economic help and can educate them. They are not going to be poor and illiterate like their mothers were."
>
> (Blondet, in Jelin, 1990, p.41)

The clubs' rationale and activities have changed with time, partly in response to political change and the effects of economic crisis, and partly because the consciousness of women in the communities has changed. Although many clubs may still be welfarist in orientation (even when organized by women in the community), many have broadened into a more collective approach to welfare needs, such as setting up communal kitchens and children's dining rooms. Whereas women originally became involved in the clubs to extend their individual networks and engage moral support in the community, they have become increasingly motivated by the collective need to find ways of dealing with economic crisis and poverty (Figure 6.3).

From coping to empowerment

To what extent did the coping strategies pursued by migrant women in Lima lead to some form of empowerment? It may seem that there was little change in the position of the women over time or that the collective actions they undertook were merely part of their daily survival. So when and how does action take on a 'public' character and also become part of an empowerment process?

What becomes apparent from this example is that changes in consciousness and behaviours can take a long time. The processes described took place over a period of up to forty years. What

Figure 6.3 Collective strategies for dealing with poverty and local needs. (Top) A community kitchen in Independencia, Lima. (Above) A meeting of women representatives of the health committee in El Augustino, Lima. The poster top left says 'We children all need basic services. TODAY! Tomorrow is late'.

were the moments of transition? The first one was when the women began to organize around community needs, such as water and electricity, and to become conscious that they were the ones who did most of the work. That is, it was when they organized around *gender-related* issues (Box 6.2). The second was when the women began to realise that they could use the mothers' clubs for their own ends. Although this involvement in the clubs was often still directed to gender-related rather than *gender-specific* goals, it was an important step forward in exerting more control over their daily lives.

However, neither of these moments of transition was particularly straightforward. Women may well have been the ones who worked together for the welfare of the community, but men still

Box 6.2 Gender-related and gender-specific issues

Gender-related issues: Issues that affect practical and strategic gender needs and interests but are not specifically directed to them. Examples might be an improvement in community health services or access to piped water.

Gender-specific issues: Issues which directly concern the practical and strategic gender needs and interests of women. Examples might be improvements in maternity care or equality in wage rates.

maintained their leadership roles. In that sense, women undertook public action but were not empowered by it in an obvious or direct sense because existing hierarchies of gender relations in local power structures were maintained. But their awareness of their own strength and importance grew.

Women's participation in the mothers' clubs could also be a somewhat contradictory experience. Initially, the clubs were often objects of charity from the state, receiving items like cookers for distribution to their members; or they were patronized by those women in the community who had the time and resources to organize clubs and run them. Changes in the orientation of the clubs came from the impact of external events combined with growing awareness among the women of ways to confront their conditions of material hardship. Particularly significant among external events was the reformist period of the Velasco government (1968–75) during which the importance of the mothers' clubs declined while other forms of organization such as the 'National Support System for Social Mobilization' (SINAMOS — a state initiative) offered a different kind of participation in neighbourhood activities. Although this process was also initiated by the state, it gave women experience of different kinds of community action.

Another external 'event' was the growing economic crisis in Peru from the late 1970s. This made life even harder for poor urban families. The crisis created an additional pressure on women to find some means of generating income — but in conditions which would allow them to combine paid activities with the demands of domestic work. But the economic pressures had an ambiguous effect on the relationship between women and the mothers' clubs, as the clubs remained the focal points for donations and food-for-work schemes mentioned above. Thus many women tended to use the clubs opportunistically for individual ends. The growing collective consciousness and the drive to make the clubs into a different kind of organization was not necessarily universal. But for many women, although the mothers' clubs did not 'give them power', the process of involvement in the clubs increased their productive capacities and their consciousness of their interests, demands and needs, and in this sense acted as an empowering (or enabling) process.

Thus, in the processes of:

- establishing themselves in the city
- meeting the gender-related needs of their families
- meeting their own gender-specific needs

women became more aware of their own capacities to organize, generate income, and find new survival strategies. Although these activities were directed to *practical* gender interests and needs rather than having a *strategic* character (at least where this story ends...), they highlighted the integrated nature of women's reproductive and productive roles. This integration of the worlds of production and reproduction is particularly evident in the lives of poor women, and is reflected in the types of organization and action that women engage in, as we shall see.

6.3 Empowerment and change

Women's daily struggles and strategies for survival can lead them to organize themselves in ways that help change or improve some basic conditions of existence. Organizing in this way is an empowering process, even if the organizations created do not change formal power structures. What are the elements of these processes? In the Peruvian example, important issues included:

- Discovering forms of solidarity that were based on common identities with other women in similar circumstances, rather than on hierarchical forms of patronage.
- Realizing the importance (and greater effect) of collective action, compared with action based on individual need.
- Recognizing that skills and new sources of income were needed rather than a reliance on state- or voluntary-sector handouts; in effect, realizing that any serious change in their situation was going to have to come from their own actions.

This section explores two facets of empowerment: collective action and class identity. Women's empowerment presupposes some form of collective identity and action. It does not simply involve individual struggles for survival or change, or for social recognition and status. However the formation of collective identity has individual dimensions: the recognition of one's own class position; one's role in sexual divisions of labour or the family; ethnic or cultural background; age. For example, what mainly gave the Peruvian women their identity, both as individuals and as a group, was their poverty. But they also shared identities as wives and mothers. Some women would also have had a sense of leadership. There would therefore have been a number of cross-cutting identities which affected how they organized and the actions they undertook.

The question of *class* identity can be a particularly important dimension of women's empowerment. The experience and recognition of subordination by class is a powerful motive for action and, when combined with awareness of subordination by gender, can be a unifying force for many poor women. However, class and gender awareness do not automatically go together, and do not necessarily concern the same issues. In the Peruvian case, the women's group identity seemed largely formed by the problems of poverty, although it also involved women's roles in reproductive and domestic work as well as in earning incomes. But women's awareness of these roles, and their organization around them, did not apparently extend to issues concerning intra-household relations between women and men, or 'broader' gender awareness, at least at the point of the study.

So what is the relationship between class-based action and action taken by women based on awareness of gender needs and interests? Can they easily be distinguished? Are they linked? To explore the nature of collective action by women and their class identities, we first look at women's involvement in community-based action during the socialist government of Popular Unity in Chile, which lasted three short years from 1970–3. This example also demonstrates other dimensions that are important in understanding empowerment as a process.

Class and gender under socialist reforms: what kind of empowerment?

The Popular Unity period in Chile offered many opportunities for organization and mobilization: among trade unions, political parties, and community organizations. The state nationalized key economic sectors such as mining, implemented a land reform programme and carried out many social measures designed to redistribute income and protect the livelihoods of the rural and urban poor. There was open debate on the 'democratic road to socialism' and many different organizations struggled to occupy political space and, eventually, to defend standards of living in the face of economic crisis.

What role did women play in this process? Did they organize around their own needs and interests (be they gender-specific, gender-related, practical or strategic)?

Figure 6.4 (Top) Salvador Allende, Chilean president and leader of the Socialist Party (Partido Socialista, PS) and Popular Unity had considerable support in the shanty towns. (Above) Demonstration in support of Allende and Popular Unity. The placard depicting Allende says 'We shall overcome': the poster is in support of the Socialist Party and advocates 'a central trade union federation for socialism'.

In the Chilean context of that period, these questions immediately raise a particular issue, namely the relationship between different types of public action, particularly those of the state, popular organizations and other groups. On one hand, a government was in power that claimed to represent the interests of the rural and urban poor, whom it attempted to mobilize in defence of a 'socialist model' of development, in addition to implementing measures in their favour. On the other hand, there was a diversity of local groups and organizations which supported the state's initiatives, and which also took action on their own account in the atmosphere of radical reform (Figure 6.4). So it is important to see how the state's interests and 'popular' interests could converge or be intertwined, and to appreciate that the division between types of public action is sometimes more complex than might at first be imagined.

With respect to state action, there were policy changes which affected the needs and interests of women. Local health centres were established and free milk was provided for mothers with small children. Childcare centres in factories and businesses above a certain size were made statutory, as were maternity benefits for waged employees. New laws were put in process to give illegitimate children the same status as legitimate children, and to legalize divorce. There were also plans to establish a 'Ministry of the Family' but the military coup in September 1973 prevented their implementation.

With respect to community-based action, the mothers' centres (similar to the mothers' clubs in Peru) were divided between their support for the Popular Unity government and for other political groupings. 'Progressive' centres engaged in political education and literacy training programmes, and in other activities such as savings and loan schemes. They also helped to improve conditions of domestic work and household survival. Other 'grass-roots' organizations involving women were the *Juntas de Abastecimiento y Control de Precios* (JAPs or Associations for the Control of Supply and Prices). Although the JAPs were instigated by the state to mobilize people to help with distributing basic consumer goods in shanty towns and other poor urban areas, the actual organization and control was community based and women were central participants. Other community-based groups such as the *Comandos Comunales* (Communal Associations) were established to counteract actions to paralyse the economy, such as the famous lorry drivers' strike in 1972–3, and had support from mothers' centres.

To what extent did these forms of public action empower women? There were two important characteristics of women's mobilization during this period. One was its class character. The issues around which women organized were crucial in the defence of class interests, and to a large extent they were identified with the state's own agenda (although there were political differences between members and supporters of Popular Unity, which was a coalition of several left-wing parties). The other characteristic of women's mobilization was the absence of women's organizations that were independent of class interests or political parties. Although the prevailing ideology of the time was to encourage women to participate in the public sphere, the predominant 'image' presented of women conformed to existing sexual divisions, and did not, as such, seriously challenge prevailing gender relations.

Paradoxically, the one arena in which women did apparently organize as women was through *Poder Feminino* (Female Power). The existence of *Poder Feminino* aroused controversy at the time (and in later analyses) because of its apparent cross-class support. The controversy concerned who and what *Poder Feminino* represented, and the nature of its political and class links.

Poder Feminino focused on the ideology of 'motherhood' and 'patriotism' to mobilize against Popular Unity. Its first appearance was in the 'march of the empty pots and saucepans' in December 1971, in which women protested about food shortages. The 'leadership' and main supporters were wealthy upper-class women (the emblem was a gold brooch of a saucepan), but there was also middle- and working-class support (Garrett-Schesch, 1975). Although there were serious problems of food shortage and distribution during Popular Unity (exacerbated by hoarding and other types of economic subversion), which would have been experienced (and responded to) differently by different groups of women, it is revealing how the ideology of the 'patriotic mother' could be appropriated to cut across class divisions.

The extent to which women benefited from reforms and went through an empowering process was nevertheless partly class related. Women participated at community levels in the 'national project', and played important roles in the defence of class interests. But although some gender interests were furthered through state action (mainly of a practical nature, although changes in laws could have had strategic implications), women's voices arguing in support of more strategic gender interests were not audible at this time. In fact, later assessments by Chilean women acknowledge the fact that the politically progressive nature of this period largely set aside the concerns of changing gender relations in the overwhelming drive towards class-based reforms.

Apart from material gains that might have accrued to women, either through government reforms or through their own participation in public action, were there other ways in which women in Chile might have been empowered? To some extent women's participation in political and community action because of their commitment to social change increased their visibility. In addition, by organizing and taking action, women had the opportunity to learn and to show leadership at the local level. But paradoxically, it was in challenging the authoritarian state that came after Popular Unity that women increasingly mobilized 'as women' and became more concerned with gender issues, as well as those of class.

Women and the state: what kind of change?

> "... when women began organizing against the Pinochet regime, it was like after a rain when you see mushrooms appearing everywhere. It is during this brutal repression, with this monetarist economic model and the disastrous economic consequences, that we see the flourishing of the women's movement in Chile."
>
> (Letelier, 1989, p.126)

So far, I have shown that empowerment has the following elements. It involves moving from individual to collective action. It also involves an increase in capabilities; for example, becoming able to understand and articulate needs and interests, learning new skills, or increasing one's ability to earn an income. Empowerment also involves challenging existing structures, including class structures, and social relations at many levels: in households, in the community, in places of employment, and in relation to the state.

The state is critical because of the different interests it may represent, the ideologies it may promote, its role in public policy and its control of forces of law and order. Although the process of empowerment may involve women in trying to gain a voice in the state, women's organizations and actions are often forms of resistance against it.

In Chile there was a stark contrast between the role of the state during Popular Unity and the role of the state after the military coup in 1973, which established a dictatorship that was to last 17 years. During the dictatorship, many women's organizations sprang up and the agenda for women's empowerment grew. The following 'cameos' illustrate some of the dynamics of this process.

Struggling against political authoritarianism

> "It is a tragedy and an irony that it took a military coup and horrendous suffering, hundreds of thousands of widows and orphans, mutilated people, and exiled people for women to start organizing again. Women had been active in the 1930s ... but the history books don't talk about women except as lovers or as good supportive ladies ... or as some incredible hero."
>
> (Letelier, 1989, p.126)

After the military coup, a particularly important area of action for women was the defence of human rights. Women as well as men were imprisoned, were tortured or disappeared after the coup. In the early years of the dictatorship, however, women were relatively less affected by political and physical repression than men. For this reason it has been suggested that they were able to take a prominent role in organizing campaigns of defence, as well being able to do liaison work for banned political organizations (Letelier, 1989; and Chuchryk, in Charlton *et al., 1989*). Two women's organizations were established: *Agrupación de Mujeres Democráticas* (Association of Democratic Women) and *Agrupación de los Familiares de Detenidos Desaparecidos* (Association of the Relatives of the Detained and Disappeared). The first was an organization of middle-class women that provided food for families of persecuted Chileans, liaised with political parties, and worked to bring a return to democracy. The second was similar to other groups of relatives of detainees and disappeared people in Latin America. It comprised women across the class spectrum who had lost relatives. They campaigned through the courts and by other means, such as hunger strikes, to try and bring about the release of detainees (Figure 6.5). It also was part of the movement for a return to democracy.

There were other women's organizations during the dictatorship in Chile that were involved in human rights work of this kind and in the struggle for democracy. What relationship did these organizations have to women's empowerment? By participating in democratic and anti-authoritarian struggles, women were brought together in new forms of organization that had not existed previously. Their consciousness about how authoritarianism affected women's lives in a more

Figure 6.5 Mothers or relatives of the disappeared at their office in Santiago, Chile. The placards ask 'Where are they?'

general sense was also changed. Many women became aware that gender relations had been largely ignored by progressive forces under Popular Unity and that authoritarian structures that controlled and oppressed women could permeate other areas of life, such as political organizations, the workplace and the family. Thus for many women the struggle for democracy in the state became linked to the struggle for democracy in the home.

Women's struggles against authoritarianism made them realize their own sense of purpose and importance. Writing from exile, the Chilean Isabel Letelier states:

> "Why women, I don't know. I do not want to pass judgement on the Chilean male ... But it was the women who went to the military places looking for them .... The men were not around. The political field was empty. A new voice emerged, a voice that had never been heard in that political arena ... women could do something."
>
> (Letelier, 1989, p.126)

Economic struggles: empowerment or retreat?

> "They [the union leaders] never understood the work we wanted to do with the ordinary woman, the ordinary housewife ... we wanted to get them out of the house ... have them take that first step, and then develop an awareness of themselves as people, as women and pobladoras [residents of a shanty town or poor community], of the working class. After that, to work with them to develop a political and social consciousness. We were working towards developing a space where the pobladora could achieve an awareness of her identity, not just class awareness, but also a gender awareness, through a process of personal and social development, and through looking at her own daily life."
>
> (Quoted by Bronstein, in Wallace & March, 1991, p.257)

This statement is made by a leader of an organization called MOMUPO (*Movimiento de Mujeres Populares de la Zona Norte*, Movement of Poor Women from the Northern Zone [of the Chilean capital, Santiago]). The organization she represents was set up in 1982. It emerged from attempts to form a group directed to women's issues in a workshop organized by the church and trade unions. The women had received little sympathy for their initiatives and eventually set up their own group to work with poor women from the shanty towns. By 1982 MOMUPO was a federation of 12–15 shanty town groups which ran communal kitchens and projects for generating income. The federation also met to discuss political and legal rights issues, health issues, education in relation to women's identity and feminism. It also had yearly campaigns on similar themes (Bronstein, in Wallace & March, 1991).

That such an organization managed to establish itself and survive under the dictatorship relates to another aspect of the authoritarian state in Chile: the effects of the neoliberal economic model set in motion after the coup. Economic policies typical of the model included: deregulation of prices and liberalizing of markets, free trade (elimination of customs barriers), incentives to foreign investment, state intervention to favour a free-market business climate rather than direct intervention in companies, rescinding protective labour legislation (Bravo & Todaro, 1985; Lago, in Deere & Leon, 1987; and see Chapter 7 for further elaboration of these policies). There were policies which particularly affected women: a decline in government spending on social services; education and health services were subject to privatization; childcare for waged employees was no longer statutory, and pregnant women could be fired (Chuchryk, in Jaquette, 1989).

Twenty years later, Chile is often held up as a model of neoliberal success. But neoliberal policies can cause grave hardship for some sections of society. The urban poor were particularly squeezed: there was growth in unemployment and a redistribution of employment from the formal to informal sectors as companies laid off workers (Gálvez & Todaro, in Jelin, 1990). In agriculture, while capitalist production and some types of peasant production remained viable and wage rates improved, there was a large sector of impoverished peasantry and growing proletarianization.

What were the effects on women? How did they respond? In the rural areas, the effects were varied. In general, because of economic pressures, women had to produce more of their own food or intensify existing production on family farms (Lago, in Deere & Leon 1987). They also sought work in export crops such as fruit-picking in the capitalist expansion zones. In urban areas, poverty also forced many women to look for paid work, particularly in the informal sector and in domestic service (Chuchryk, in Charlton *et al.*, 1989).

Poor women also developed new survival strategies in what have been termed *organizaciones económicas populares* (popular economic organizations, OEPs). These were community-based groups which ran projects to generate income, such as knitting, creating the now famous tapestries (*arpillerías*, which often carried protests against the military regime — see Figure 6.6), collecting and selling old clothing, and maintaining communal gardens. They also set up shopping collectives to try to regulate prices, and established communal kitchens and dining rooms. Estimates indicate that by 1985 there were over a thousand such OEPs alone in the inner city of the capital, Santiago (Chuchryk, 1989b). MOMUPO's work illustrates what such organizations can do.

The growth of OEPs shows how poverty, class, gender interests and gender needs can become combined. On one hand, the OEPs were forms of resistance against the government's economic and social policies in poor areas — where women had lost many of the material improvements and the political representation they had gained under Popular Unity. In this sense, OEPs can be seen as a form of 'retreat' or defensive action. Their existence also meant that the state did not need to assume the burden of sustaining the poor. On the other hand, although the activities of OEPs to provide food and money were responses to economic hardship, the existence of OEPs had another dimension. Under a repres-

Figure 6.6 A woman making a patchwork tapestry at a community centre in a shanty town in Santiago. She is embroidering 'Silence in the means of communication (the media)', referring to political censorship.

Box 6.3 Solidarity as a practical necessity for survival

One lesson that has emerged, especially from the examples of the Peruvian women and some of the organizations under the Chilean dictatorship, is a different conception of solidarity. Solidarity is often seen as something purely political. It is also often idealized and romanticized. The solidarity shown between the women in Chile and in Peru was based on practical as well as collective responses to problems of daily survival. It was also a form of resistance in both cases, although in Chile this resistance often had an overtly political character too.

sive regime, they were one way of maintaining solidarity between those who suffered the consequences of state action (Box 6.3). Within the limited space for action, they offered a legitimate means of assembling and talking, and they became a focus of self-education and political organizing.

Hence as well as protecting people from the adverse effects of social and economic policies, many OEPs engaged in action that increased women's awareness. They extended their concerns from action on practical gender needs and daily survival to more strategic questions and alternative possibilities for the future. MOMUPO, for example, continues to exist after the demise of the military regime because many women continue to face material hardship, and MOMUPO still has a role to play in political education for poor women (Bronstein, in March & Wallace, 1991).

Motherhood as a weapon

> "Woman, from the moment she becomes a mother, expects nothing more in terms of material things: she seeks and finds the purpose of her life in her child, her only treasure, and the object of all her dreams."
>
> (General Pinochet, head of the military junta and President of Chile, 1973–1990 quoted in *El Mercurio*, 19 October 1979)

Pinochet's view of Chilean women was an integral part of the ideological controls enforced by the military junta. For the military regime, women's role in the future development of Chilean society was to be directed to the home and to socializing children to be good citizens. This ideological stance was encapsulated in the following objectives of the newly created and state directed National Women's Secretariat:

> "To create in women a national consciousness and an adequate understanding of the dignity and importance of their mission within the family and society.
>
> To train women, and through them, the family in order that they may be fully incorporated into the social and cultural development of the nation.
>
> To promote women's support of government plans and to encourage women to voluntarily collaborate in the development programs promoted by the government.
>
> To qualify women volunteers by means of specific programs.
>
> To honour the importance of women and to cooperate in preparing women to better carry out their roles as mothers, wives and housekeepers."
>
> (Larrain, 1982, p.50)

To reinforce this conception of women's roles in Chilean society and to promote associated policies, mothers' centres, which had been closed down after the coup, were reopened and appropriated as mechanisms of political and ideological control.

Chilean women, organizing in defence of their own and their class interests, turned this particular view of motherhood on its head. Women learned to use 'motherhood' to fight for gender-specific as well as gender-related needs and interests. Poor women used it as a legitimate way of organizing in the shanty towns, and in the process raised their own awareness of their

Figure 6.7 A health education workshop for women in the shanty town of La Victoria, Santiago. The speaker is talking about justice.

positions as women in their communities and in Chilean society more generally (Figure 6.7). Community groups socialized tasks normally carried out by individual women in their homes, by providing food for children through communal dining rooms. Many women also began to discuss feminist ideas as well as taking action in defence of their interests as poor women.

The use of motherhood as an organizing tool and a justification for action raises a number of points about women's empowerment. Being mothers was one obvious and important shared identity among many women. They were united for instance by such experiences as: their loss of sons and daughters in the repression; the need to provide subsistence for the family; and their general responsibilities for family welfare. By organizing, women combined this identity with others derived from shared poverty, class position, and their roles as income earners and political activists. Hence the concerns of family life were brought into the political arena. This may not have meant that existing sexual divisions of labour in the household were challenged or changed — in many contexts women's activism is often an additional commitment on top of existing responsibilities (and often meets with resistance from husbands and fathers). But the *public recognition* of women's roles in reproductive and domestic work, and the recognition of their close integration in other social and economic activities, were important steps in the process of empowerment.

That motherhood can be used as a controlling as well as a politicizing mechanism shows how important the role of ideology is in women's empowerment. In empowering themselves, women often have to transgress prevailing norms and values related to their roles in society and in the home. Although these norms often act as forms of social control, women can also use them to challenge subordination.

What kind of empowerment? What kind of change?

The different examples of organization under the military regime in Chile that I have considered do not cover the whole spectrum of groups and movements during that period. Many of the movements were urban-based because Chile is a predominantly urban society (and about 30% of the population lives in Santiago, the capital). What other organizations existed?

The writer Patricia Chuchryk has identified a remarkable, and under the circumstances surprising, growth in autonomous women's organizations under the military dictatorship. The Women's Department of the National Trade Union Coordination tried to organize women workers and improve their conditions of work. In 1987 it succeeded in holding a conference on women's rights in the workplace. Furthermore, although women had not historically participated in *rural* labour unions, when they began to work in export fruit production in the early 1980s they joined unions and campaigned on gender-specific issues such as the double day and childcare (Lago, in Deere & Leon, 1987). The Women's Committee of the Chilean Human Rights Commission, set up in 1979, carried out solidarity work and popular education around women's rights. The Committee for the Defence of Women's Rights also engaged in political education and organized around economic and welfare problems among poor women. Likewise, Women of Chile worked in the shanty towns and participated in the struggle for democracy. There were also intellectual, feminist organizations such as the Centre for Women's Studies, which emerged out of the Women's Study Circle and which carries out research. Other organizations were the Movement of Women for Socialism and the Feminist Movement (Chuchryk, in Charlton *et al.,* 1989).

The existence of an authoritarian regime obviously created rather special conditions for the emergence of these different groups. Political and economic struggles were combined as women sought to defend human rights and livelihoods. But in this process, a new awareness arose among many women about the more general issues and problems they faced, in addition to those resulting from the combined effects of state repression and a neoliberal model of development:

> "A different kind of organization emerged ... Being together created the most extraordinary change in their lives. They were enormously empowered, sometimes with the help of the church, sometimes with the backing of political parties, and sometimes through adult education which taught them the techniques of empowerment. They started listening to each other, not being afraid of each other; they started telling each other the situation in their homes."
>
> (Letelier, 1989, p.127)

Other writers (for example, Alvarez, in Jaquette, 1989) have commented on a similar process in Brazil, where women's organizations grew during periods of authoritarian government in the 1960s and 1970s, especially in urban areas, and which were able to promote policies serving the needs of women in local government in the democratic openings of the 1980s.

So what does this brief look at some of the 'empowering' experiences of Chilean women during the dictatorship show? To summarize:

- It illustrates the potential mobilizing force of issues that are close to women's daily lives and to the needs of family survival.
- It demonstrates how the accepted roles of women as wives and mothers can be challenged as a result of organizing around domestic concerns.
- It shows that women's organizations have provided a focus for popular mobilization that was an alternative to conventional institutions, such as political parties and trade unions. The organizations were more effective in meeting women's needs and interests, and have been able to continue political work and maintain solidarity when the actions of political parties and workers' organizations were repressed.
- Chilean women's organizations linked democracy and gender relations, and showed that democracy was an important part of human development at all levels of social life, from the household to the state.

6.4 Redefining the political

> "Political activity does not only concern the power of the state, institutional organizations, economic organizations and the dialectics of the exercise of power. It also, and equally importantly, seeks to re-organize the daily life of women and men. It questions in order to reject, or at least to sow seeds of doubt regarding the real need for circumscribing two distinct areas of experience sharply divided into public (political) and private (domestic) spheres that stereotype mutually exclusive and rigid areas for the activities of men and women."
>
> (Kirkwood, 1983, p.637)

This quotation is from a Chilean feminist writer and is an observation arising from the experiences discussed in the previous section. The content of Kirkwood's statement has much broader implications than a simple comment on Chilean politics. It questions the nature of political action and the links between such action and daily life.

A number of writers have recently explored the nature of political and public action in order to redefine them on the basis of different types of experience. For example, Scott suggests that the concept of the political should be broadened to encompass all forms of individual and collective struggle and resistance (Scott, in Colburn, 1989; and see *Bernstein et al., 1992,* Chapter 12). Others, such as Slater, have put women's organizations (or 'the women's movement') in the category of 'new social movements' which resist the alienating effects of capitalist development, search for new alternatives, and place a high value on participation and democracy (Slater, 1985).

We have already seen some of the ways that women's organizations and women's contributions to public action have helped to redefine the political. For example, the actions of the migrant women in Lima were based on needs that were related to family survival (access to building plots and materials, access to basic services such as water and electricity). But the needs were also the social needs of the community, and involved struggles over control and distribution of resources. Thus women's survival struggles also had a political content. Women's forms of action and the types of organization they established have also raised new issues for social change. For example, the actions of women's shanty town organizations in Pinochet's Chile raised the issue of communal responsibility for feeding and caring for children, and for earning income. They also showed the communal responsibility for protection against politically motivated physical and economic repression.

How do processes of empowerment alter perspectives on social change and public action? Are women's organizations and empowerment part of wider political processes directed at more fundamental change? Does public action by women give a new meaning to the political? Redefining what is political can happen in several ways. For example, it might happen by virtue of

- the types of organization established
- the issues that are given priority by women's groups (or organizations that women participate in)
- the methods of organizing
- women's own perceptions of their activities

I look briefly at these four aspects, drawing in some comparative material from Brazil.

Types of organization

Sen and Grown, who were cited in Section 6.1, and who have written a monograph on Third World women's perspectives on development, identify six types of organization that involve or mobilize women or are established by them (Sen & Grown, 1987). The six they isolate are:

- *Service* organizations of a welfarist nature. Some types of mothers' centres might be in this category, particularly if they are run by the church or the state. Some NGOs also fall into this category.
- Organizations *affiliated to political parties*. For example, women's departments or women's sections of political parties. None of these have been discussed in this chapter. They existed during the Popular Unity period in Chile but were precisely the forms of organization that were subordinated to (male) party interests that were later criticized by Chilean feminists.
- *Worker-based* organizations, such as women's departments in trade unions. An example is the Women's Department of the National Trade Union Coordination in Chile, mentioned above. This department managed to campaign on gender-specific issues during the military regime.
- Organizations *set up by funding agencies* (NGOs) largely resulting from activities during and after the UN Decade for Women (1975–85). I have not discussed those here.

- *'Grass-roots'* organizations, usually based in poor urban and rural areas, in which women are key participants or which are set up by women for gender-related and gender-specific purposes. The networks and groups in Lima were of this kind, as were the OEPs in Chile. The women's campaigns on human rights often combined a grass-roots elements when women who had lost relatives were involved, but also they also included professional, middle-class women.
- *Research* organizations, such as the Centre for Women's Studies and the Women's Study Circle in Chile, mentioned in Section 6.3.

Thus Sen and Grown suggest that there are many types of organization that women participate in; and the different types have different potentials for empowerment and change. For many women in Latin America and other parts of the Third World, material hardship, the struggles for daily survival, and gender subordination, greatly affect the kinds of organization that they form, and the issues they focus on.

We have seen that community-based action by poor women can develop their skills and awareness, and can give them the confidence to speak out and act on local development needs. These needs are often *gender-related*. A study by Pires (in Jelin, 1990), examining women's involvement in community organizations in the Brazilian city of Sao Paulo in the early 1980s, shows how women become engaged in a wide variety of community-based groups. Some of these organizations are groups linked to the Catholic or other churches, some are residents' organizations and others are issue-centred groups, as well as mothers' clubs and women's groups. Movements and organizations in shanty towns are largely run and constituted by women. The only organizations in which women are a minority are local branches of political parties, in other words, the more conventional types of political grouping. The reasons women give for their involvement are similar to those of the Peruvian women: building networks, getting out of isolation, learning new skills. As with the Peruvian women, the Brazilian women identify with each other mainly through their poverty and through experiencing the common problems they face *as women* in that situation.

These experiences suggest that the process of empowerment involves women's participation in, and learning from, many types of organization; and that a shared identity arising from common problems can have the effect of changing women's perspectives on the nature of their struggles. However, the growth of consciousness about *gender-specific* issues seems to be associated with the establishment of *independent* women's organizations, as happened with a women's group in a Sao Paulo neighbourhood that broke away from its church origins. One of the members states:

> "... I discovered myself as a woman, as a being, as a person. I hadn't known that women ... have always been oppressed. But it never occurred to me that women were oppressed even though they had rights. Women had to obey because they were women."
>
> (Quoted in Sen & Grown, 1987, p.59)

Issues

Women's activities as organizers, housewives, income earners, and mothers, tend to overlap in form and content: women tend to choose forms of organization that fit in with their daily lives and responsibilities, and the issues they organize around demonstrate the interconnectedness of productive and reproductive activities. In fact, one of the most important contributions made by women's organizations and by women's participation in community action is to bring domestic and reproductive issues into the public arena, and to make them collective rather than individual concerns.

This phenomenon was demonstrated in part by the Peruvian women and their activities in the mothers' clubs, but it was also strikingly evident in the communal feeding of children in the Chilean shanty town under Pinochet. This combined a practical solution to a material need with a way of organizing against the regime, and made what was usually

Figure 6.8 A nursery school set up by residents in Favela Rocinha, Rio de Janeiro, Brazil.

considered a private matter (feeding one's children) into a public and political issue. Another example is discussed by Pires. One of the most important campaigns in the city of Sao Paulo in the 1980s was the Movement to Fight for Nurseries which grew out of links between feminists and groups in the marginal areas and city neighbourhoods (Figure 6.8). This movement was so successful in bringing the issue of childcare into the public domain that it was ultimately taken on by the local authorities. However, it also had some adverse political repercussions from the point of view of women's participation because it was taken out of their control (Pires, in Jelin, 1990).

To what extent are such issues about empowerment? To what extent do they relate to longer-term change? Many of the issues women organize around are practical gender needs and relate to immediate, and often family-related, interests. The process of organizing around practical gender needs and interests achieves two important goals. One is that women learn from this experience and are able to take on new roles and identities in the community. The other is that by organizing around immediate needs, and by dealing with all the problems and obstacles that that involves, women often develop a deeper awareness of the need for more fundamental changes in power structures and in relations between women and men.

Methods of working together

It has been suggested women's organizations, or activities where women are the key participants, tend to have a more democratic and less hierarchical character than more conventional forms of political and public action. Although hierarchies may become established in women's groups, women often have informal ways of drawing each other into their organizations and defining new spaces appropriate to the issues that concern them (Figure 6.9). For example, Pires recounts how women in church-related mothers' clubs in Sao Paulo began to meet separately to discuss sexuality and contraception — topics which the church frowned on but which were of direct importance to the women.

Other accounts of women organizing in Sao Paulo talk of how women build on personal relationships (as we saw also in the case of Peru), how women often meet in each others' homes (that is, in places convenient to them and to their working lives), and how they tend to communicate with each other in informal ways:

"... after the meetings had been going on for a while a woman stopped me in the road and said, 'Look, what are those bits of paper you keep throwing over my fence?' It never occurred to us that many people didn't know how to read or write even though most of us are illiterate. The house meetings and house-to-house visits resolved many of those problems."

(Quoted in Corcoran-Nantes, 1990, p.257)

Another example of effective action taken on the basis of informal communication was the sacking of supermarkets in 1983 by members of the Unemployed Movement in Sao Paulo and Rio de Janeiro:

"... I was looking out of my window when I saw a group of women filing past my house with empty shopping bags. I just knew there was going to be a sacking so I grabbed my bag and a few paper sacks and followed them. No-one said a word ... and as we went along more women joined us. By the time we got to the local supermarket there must have been 50 of us or more and that was the first time anyone spoke. One of the women at the front ... said, 'Only take the basics, nobody touch the till', and that's what we did. When we had filled our bags, we all quietly returned to our homes."

(Quoted in Corcoran-Nantes, 1990, p.259)

Spontaneous action of this kind can probably only work when there are specific goals and a relatively short-term perspective. But the different ways that women have learnt to organize reflect a wide range of initiatives and responses appropriate to their particular needs and circumstances. They also show that formal structures and hierarchies are not necessarily the only forms of resistance, nor the only way to bring about change.

Does this mean that women are always non-hierarchical and democratic and that 'power' is not a direct concern? Obviously different types of organization will run their activities differently and have different internal structures or modes of operation. The informality typical of some types of community action is not always characteristic of women's organizations. Nor is solidarity between women always easily achieved. Some Latin American feminists recently tried to dispel some myths about women's 'political practices'. One 'myth' was that

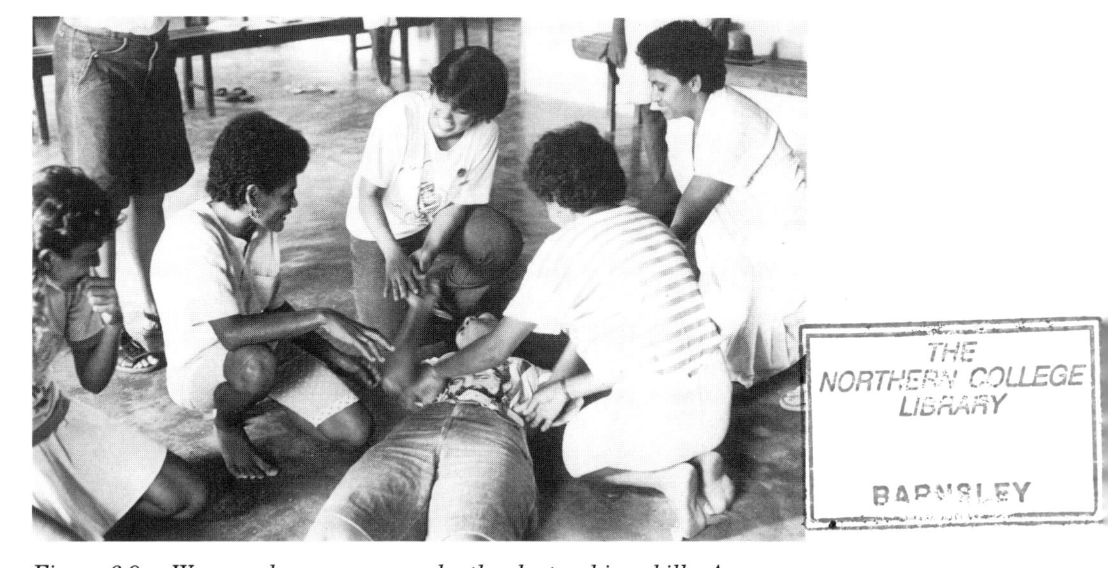

THE NORTHERN COLLEGE LIBRARY BARNSLEY

Figure 6.9 Women also empower each other by teaching skills. A woman's group practises first aid in Peixinhos in Recife, Brazil.

'feminists do politics in a different way'. To this, they commented:

> "Yes, we do politics in backward, arbitrary, victimized, manipulative way. Theoretically we attempt to do it differently, but if we are honest our practice leaves much to be desired, and this has to do with the difficulty of accepting there is unity in diversity and democracy, not only as a necessity but as a prerequisite of our action."
>
> (in *Catalyst*, 1990, p.15)

Even though such myths might be dispelled, this statement still suggests that women ought to try and incorporate a new set of values into the aims and methods of public action.

Women's perceptions

To what extent do women actually perceive themselves as being empowered by their organization and action? To what extent do they see their action as political?

These questions should make us pause and reflect again what empowerment is about, as well as what we mean by 'the political'. We have seen from our examples that there is both a process and an outcome involved: the process of organizing and taking action itself is a process of learning and change; through organizing, women may also achieve changes and improvements in their daily lives (or those of their families). But these changes may also involve challenges to authority and political power structures. This was particularly evident in Chile under the Pinochet dictatorship.

How do women involved in community action see their activities? Those who become involved in organizations directed to gender-related or gender-specific issues tend to express the positive effects of participation in activities outside (but related to) those of their daily survival strategies. And although they take pride in some of their achievements, their statements suggest a rather diffuse satisfaction rather than something overtly goal-centred or of apparent political importance:

> "One gets to know people and find out about problems of others in the neighbourhood."
>
> "... if you go out and go to a meeting, you get a bit more time to yourself ... we can learn like this, by exchanging experiences and learning things about life through practical experience."
>
> (Pires, in Jelin, 1990, pp.56–58)

Although women may achieve particular objectives, such as obtaining services in the community or setting up schemes for generating income, they also articulate quite personal identity changes that take place and which reflect the importance of household gender relations in their lives:

> "When I discovered myself as a woman, that I have the same rights as everyone else, my relationship with my children changed ... My relationship with my husband got better ... He wouldn't accept it at first but in time he began to see that it was a need, my need."
>
> (Pires, in Jelin, 1990, p.66)

However, the process of change is not necessarily easy:

> "My husband doesn't like me to participate ... but even if there's a fuss, I go."
>
> (Pires, in Jelin, 1990, p.65)

> "Women don't need to learn anything, only the men ... women are always left behind."
>
> (Pires, in Jelin, 1990, p.66)

> "When the political parties in Chile were telling people not to be involved in civil disobedience and not to do any public demonstrating because it was too dangerous, women came out."
>
> (Letelier, 1989, p.126)

Thus women's commitment, and their ability to effect change, however limited, is not necessarily identified as political action. Although their daily lives and actions are part of public life, especially at a community level, and although what they actually do in practice is political, they do not identify with this public arena. Pires suggests that many women in the Sao Paulo communities she investigated tend to see 'politics' as a male world:

> "There are more men in the party. Men ... are more interested in personal things ... And women, in the community, do it because they want to."
>
> (Pires, in Jelin, 1990, p.63)

> "... I don't understand anything about politics ... it's something that doesn't affect me."
>
> (Pires, in Jelin, 1990, p.55)

Some community-based groups may, however, be influenced by the actions and discussions of feminist groups, which are all quite strong in Peru, Chile and Brazil, but this does not necessarily mean that they identify their actions or awareness as feminist. They may even be vociferously against feminism.

Do feminists have a different perception of the political and of empowerment? To return to those Latin America feminists who have tried to dispel 'myths', another myth was that 'feminists are not interested in power'. To this, feminists answered:

> "If we start out with the premise that power is fundamental to transforming reality, it is not possible not to be interested in it. ... Yes, we want power. Power to transform social relations, in order to create a democratic society in which the demands of each social sector find space to be resolved. This requires rules of the game that guarantee the presence of a plurality of social actors: in short, we want power in order to construct a democratic and participatory society."
>
> (Pires, in Jelin, 1990, p.14)

This statement suggests that empowerment, politics and methods of organizing are all interconnected.

6.5 Conclusions

> "When I discovered myself it was like maize growing in the field. First you have the leaves, green, tender, so timidly emerging out of the heart of the plant. Then the whole plant is geared towards producing la mazorca [maize cob], strong, bright yellow. It's a process I've gone through."
>
> (Francisca, a rural trade union leader from Brazil, interviewed by Eugenia Piza Lopez, in Wallace & March, 1991, p.261)

Women's empowerment is a *process*. Achieving particular goals may empower women. But the process of engaging in public action is also empowering because women begin to recognize their capacity to bring about change.

The quote above also indicates that empowerment involves *self-discovery* — a change in personal consciousness as well as the development of collective identity with other women and participation in collective action with them. There is no 'correct order' for this process. As we have seen from the examples in this chapter, women become involved in public action for many reasons, and their 'self-discovery' may be a result of their involvement rather than the other way round. Nor are 'self-discovery' and establishing common identities easy or straightforward processes. As the Latin American feminists have suggested, accepting unity as well as diversity have been difficult issues. Vargas, who was quoted at the beginning of the chapter, has written about the different 'streams' of types of women's organizations in Peru and has analysed their contradictions and difficulties as well as achievements. She aptly calls her paper 'Streams, spaces and *knots*' (my emphasis).

Empowerment may be quite gradual. It may involve many kinds of *enabling* processes, as well as the achievement of particular goals. Enabling processes may include confidence-building and skill development. They may also involve deeper lessons about the sources of and reasons for subordination.

Public action (whether by organized women, NGOs, or the state) can have many *unintended effects*. What may start out as organizing around practical goals can have the effect of changing awareness and priorities. By experiencing the problems of practical action, people can become aware of the deeper structural constraints that prevent them satisfying their needs. This is why thinking about women's empowerment has to go beyond the concepts of practical and strategic gender needs and interests.

The processes set in motion by women's empowerment and meeting practical and strategic gender needs and interests may also have much wider repercussions for thinking about *how society is organized* at a more general level, as well as questioning what democracy, equality and participation mean in practice for different sectors and social groups.

Empowerment may entail changing the nature of *power structures*, rather than making gender needs and interests fit into existing ones. Acknowledging diversity as well as unity also means that the questions: 'Whose needs?' and 'Whose interests?' also have to be answered. Practical and strategic gender needs and interests may be different for different groups of women, even though there may be common problems.

Finally, women's empowerment is about *understanding development differently* and thinking about alternatives. Public action by women raises many questions about development priorities, as well as the mechanisms by which development is to be achieved.

Summary of Chapter 6

1 Women's empowerment may start with quite imperceptible changes in consciousness and identity. Women may organize around gender-related survival needs and in the process become aware of more gender-specific concerns.

2 The issues that women organize around frequently arise from their class position and conditions of poverty, but the integration of productive and reproductive activities in women's lives gives a particular character to the form and content of women's organizations, as well as the strategies women adopt for daily survival.

3 The adoption of development strategies to change conditions of poverty and exploitation (such as those under the Popular Unity government in Chile) may bring benefits to women but do not necessarily change the more fundamental aspects of gender relations that may subordinate women.

4 Struggles by women around gender-related questions of human rights and economic hardship under authoritarian regimes have generated new levels of consciousness about women's needs and interests. In particular, women's organizations have brought a new perspective on the nature of democracy and its importance in everyday life.

5 The wide range of organizations and action by women suggests that there is no single way forward for the process of empowerment. Women's empowerment involves self-discovery and enablement as well as challenging structures of economic and political power.

6 The issues women's organizations have raised, and the different ways they have organized, also pose questions about the purpose and direction of development, as well as the means of achieving it.

STRUCTURAL ADJUSTMENT

7

DEPRIVATION AND STRUCTURAL ADJUSTMENT

MAHMOOD MESSKOUB

7.1 Introduction

Previous chapters of this book have referred to 'structural adjustment'. In this chapter, we finally come to grips with this important set of ideas which dominated economic and social reform throughout the Third World in the 1980s. And we link this back to the themes of poverty and deprivation in Chapters 1 and 2 of this book.

In the last decade, deprivation has worsened in many Third World countries. At the same time, many Third World governments have implemented structural adjustment policies. This chapter is centred on one key question.

Q How does structural adjustment affect the poor?

To answer this question, we shall first need to answer two other questions. One sounds straightforward.

Q What is structural adjustment?

The other is more complex. The effects of central government economic policies are determined by the responses of the particular economic, social and political institutions of a country.

Q How are the outcomes of structural adjustment modified and transformed by a country's public and private institutions?

True to the theme of this book, therefore, we are approaching structural adjustment policies as a set of *processes* in the public sphere. More usually, these policies are presented as a set of technical problems in economic management of a national economy. Our attention in that case is centred on central government decisions, and on measuring final outcomes. We shall be interested in both of those elements. But this chapter also looks closely at what happens in between: that is, at the filtering of central government policy decisions through the very varied public and private institutions (markets, households and gender relations) in different countries. This is what we mean here by 'policy as process'

Economic reform policies under structural adjustment are of two kinds. Governments can:

- intervene in markets, in order to change their organization or their pricing behaviour
- act directly to restructure their own taxation and social provision.

Section 7.2 explains the set of such policies characteristic of structural adjustment. The rest of the chapter then examines the processes and the impacts that they create. A final section takes a critical look at proposals to protect the poor during structural adjustment programmes.

Why structural adjustment?

Q Why have so many Third World governments, and only Third World governments, been forced to implement structural adjustment programmes in the last decade?

Many Third World economies went into economic crisis in the late 1970s and early 1980s. Before this, there had been a couple of decades of widespread improvements in both 'wealth' and 'health' (Chapter 1). But by the late 1970s Third World governments were struggling with a number of severe economic pressures. As a legacy of the colonial past, many Third World economies still relied on the export of primary products: products of farming and mining. These exports financed imports of manufactured goods, including inputs to keep their own factories, farms and mines running. As the world economy changed sharply in the 1970s, that colonial legacy turned into a trap.

International markets became more competitive, as the world economy overall slowed down. Governments in developed countries increasingly protected their own producers against Third World competition. The sharp increases in oil prices in 1973 and 1979 compounded (but did not cause) the slow-down (Korner *et al.*, 1986; Loxley, 1986). And most Third World countries, those which were not oil producers, faced huge difficulties in importing enough oil to keep their economies running.

Worse was to come. The international recession reduced the demand for primary products, and by the late 1970s their prices were falling. Many Third World countries resorted to borrowing abroad to pay for imports. The crunch came in the early 1980s, when the United States, faced with problems at home, pushed up interest rates. By 1982, many Third World governments were already unable to repay, or to pay the interest on (that is, 'service'), their debts. In 1982, some really big debtors, such as Brazil, joined this group. Overnight, most Third World countries faced great difficulties in borrowing abroad, whether for investment, trade credit, or imports of food or oil (see Chapter 2 in *Hewitt* et al., *1992*, for a detailed account).

This brought many Third World governments to the door of the International Monetary Fund and the World Bank, seeking finance to continue to trade, and hence to keep their economies functioning. Increasingly over the last decade, these two institutions have worked together. They have provided some finance to Third World

Figure 7.1 Primary commodity exports finance manufactured imports for both productive and non-productive use. (Left) Export of goats from Somalia. (Right) A Mercedes Benz being imported to Angola.

economies in crisis. They have played the role of intermediaries in renegotiating debt obligations with Western banks; and they have exacted a price. The price has been the acceptance by Third World governments of economic reform programmes, aimed to 'readjust' these economies to the lower immediate standard of living enforced by the multiple economic crises just described. The stated longer term aim is to reorganize these economies so that they develop more effectively, that is less vulnerably, in the future. These economic reform programmes have come to be known as 'structural adjustment'.

7.2 What is structural adjustment?

Structural adjustment influences the way people live and work in many parts of the Third World. Yet relatively few people would claim to understand what it is all about. Why is this? The problem is that the language of structural adjustment is impenetrable to many outsiders. This section aims to break with this tradition and to explain, in simple terms, what the World Bank or the International Monetary Fund mean by structural adjustment: what it is and what it aims to achieve.

Obviously, we shall need some concepts to analyse structural adjustment. Fortunately, some of the ideas already introduced in this book allow us to get a good initial grip on the basic issues. Particularly useful are the concepts of *endowment* and *entitlement* (see Box 1.2). But in this case, we apply them to a country or, better still, a *national economy* which we shall treat as if it were a person.

A national economy

The language of structural adjustment concerns policy measures to get a country back into shape. The national economy is seen to be in crisis and urgent remedies are required to cure the disease inflicted upon it. But what is the disease and what are its symptoms? To answer these questions we need some concepts which will lead us to the remedies required to cure the disease.

Box 7.1 The endowments of a country

These include:

1 *people*: women and men, young and old; the present labour force and its future potential (children); people with education and skills; people in various occupations

2 *productive assets*: farms and factories; equipment and livestock; roads, bridges, railways and harbours; stocks of raw materials, food or other commodities; urban and rural infrastructures

3 *financial assets:* cash and bank deposits (not its own money, but dollars, pounds, etc.); debts which other countries owe the country *minus* the debt it owes to others; shares in enterprises of other countries.

Let us start with a country's *endowments*, as listed in Box 7.1.

Given these endowments, a country's entitlements refer to its potential command over goods and services. A country can exert this command over goods and services in various ways.

- It can use its productive resources and its labour power to produce goods and services for use at home.
- It can also *trade* with other countries part of the wealth produced at home. That is, it can sell goods or services abroad (*exports*) or buy them (*imports*) from other countries.
- Imports do not necessarily have to be paid for from export revenues. For example, imports can also be financed by incurring a debt (which involves a transaction in financial assets) or, alternatively, using cash accumulated in previous years.
- Imports can also be backed by *grants* from abroad. For example, food aid from foreign donors increases the availability of goods in the country.

- Finally, the wealth produced in a country does not always accrue to its residents and, conversely, residents in the country may receive income from other countries. For example, a country may receive remittances from migrants working abroad but may also have to pay interest on its foreign debt. That is, it is GNP (the gross national income) which matters in assessing a country's *income* from its endowments, not the GDP (the gross domestic product) which only counts the wealth produced within the country (see Boxes 1.1 and 3.7).

Two lessons follow from this brief analysis of a country's entitlements to goods and services. First, trade allows a country to produce what it does not consume and to consume what it does not produce. Second, in any given period, a country, like an individual, may well *spend* more (or less) than it *earns* in income. Let us look at both aspects in more detail.

Specialization

Trade allows for specialization of a country within a wider division of labour. Orthodox economics tells us that this is a good thing:

Figure 7.2 Zambia's comparative advantage. (Above) Drilling copper ore 1000m underground. (Right) Copper plate ready for export.

Box 7.2 The balance of payments

How does a country pay for its imports? We can illustrate this schematically as follows:

(1) *Imports*
Total imports of goods and services

(2) *Exports*
Total exports of goods and services

(3) *Trade deficit*
Trade deficit (or surplus) = (1) – (2) :

(4) *Deficit on transfers*
Income received from abroad (e.g. migrants remittances) *minus* income payments abroad (e.g. interest payments). The latter normally exceed the former in most Third World countries.

(5) *Surplus on grants*
Grants received from abroad (minus grants made to other countries). This item is normally positive for a developing country.

(6) *Current account deficit*
The overall deficit (or surplus) on trade, income transfers and grants = (3) + (4) + (5)
The current account deficit needs to be paid for. This can be done by:

(7) *Decline in reserves*
Foreign exchange reserves can be run down to pay for imports.

(8) *Capital account transactions*
A country can finance imports through dealing in financial assets, or through direct investment from abroad. This may take the form of loans from abroad, trade credit, or longer term foreign investment.

countries can specialize in producing those goods or services for which they are relatively well endowed and share in the benefits of trade by obtaining goods and services they lack or can only produce less efficiently than others. The argument, as it stands, seems reasonable enough. But, as we have seen in Chapter 1, markets often entail the vulnerability of many people, especially those who own few resources other than their own labour. The same could be said with respect to countries: ownership of resources and the terms on which a country comes to the market matter a great deal. Zambia, for example, depends solely on copper to secure a foothold in the international division of labour, while it needs to import a great variety of goods which it cannot produce domestically (Figure 7.2). This situation is potentially much more vulnerable than, for example, that of South Korea which has a much more diversified economy, both for home production and in trade. We shall return to this question of vulnerability in the next section. At present, we follow orthodox economics in its assertion that countries gain from greater *openness* to world markets.

Spending and earning

A country can spend more than it earns. It can do so, like a person, by using cash accumulated in the past or by incurring a debt. What this implies is that the country imports more than it exports or receives in income from abroad (see Box 7.2). Economists would say that, in this case, the current account of the balance of payments of the country goes into red (becomes negative).

The current account deficit indicates whether or not an economy is 'living above its means'.

Q But how dangerous is it when a current account goes into or stays in the red?

The answer depends on what the country is spending its money on. For this reason, in economics we distinguish between two broad categories of expenditure on goods and services: *consumption* and *investment.*

Consumption and investment

This distinction between consumption and investment appears straightforward but it is not without its problems. Intuitively, consumption concerns all expenditure on the goods and services that secure the day-to-day standard of living of a population. Investment, in contrast, has to do with expenditure made to enhance future economic growth and development. For example, food and clothing are consumer goods, while building a bridge or a factory is an investment.

What about education and health care? If we view them as consumption it follows that wealth must be created before it can be spent on education and health. But would it be more sensible to consider them as investment? Education and health care after all are essential assets for future growth and development, in the same way as a factory is an asset. Health and education, therefore, *create* wealth since they involve investing in people. This explains why, in most countries, the provision of health care and education is, in part, financed through *public expenditure*.

So it may be desirable for a country to spend more than it earns if the additional resources thus acquired are used for investment, and not for consumption. Then the additional spending will contribute to the future growth and development of the national economy. So long as some of this growth is in exports, the country will be able to service its debts out of increased revenues from exports. However, this assumes that the investments are effective in enhancing future growth; otherwise, it would just be money down the drain.

Symptoms and causes of crisis

So much for basic concepts; let us now put them to work. In the view of the International Monetary Fund and the World Bank, the major symptom of economic crisis in a country is its inability to service its foreign debts. That is, a crisis occurs when a country ceases to be able to pay the interest on, or to repay, its debts. Consequently, the country also ceases to be able to borrow abroad, whether for investment, inputs for its industry, food or oil.

Inflation

Symptoms of economic crisis never come singly. Often the country in crisis also suffers from severe inflation; that is, a persistent rise in prices. It is a sign that an economy is under severe strain. Inflation is a complex phenomenon which affects different people in society in different ways. In Chapter 1 we saw several examples of how adverse movements in prices can put people into severe distress. Those in society who cannot protect their income or assets from being eroded by inflation will be hardest hit.

Inflation is not just 'too much money chasing too few goods'. Demand pressures on supply and production obviously matter, but prices rise unevenly, and for a variety of reasons. In some cases, inflation may be provoked by the collapse in imports, which also include inputs to keep factories, farms or mines running. Droughts or floods and the inability to import food may cause steep rises in the price of foodstuffs, followed by further price rises in other commodities. Inflation, therefore, does not hit everyone equally hard nor does it always operate in the same fashion. Much depends on the character of an economy and the security, or lack of it, that the state provides to the people.

Figure 7.3 *'What prices shall I put?'*

Disarray in state finances

Another symptom, one that the World Bank and the International Monetary Fund never fail to point out, is that often the finances of the state are in disarray. Public expenditure by far exceeds revenue from taxation and foreign aid

Figure 7.4 The IMF building in Washington, just across the street from the World Bank. The IMF and the World Bank have worked hand in hand in the complete overhaul of many economies.

grants. But is this a cause or, as some argue, a symptom of the crisis?

Those who argue that it is a symptom point to the fact that economic crisis often necessitates increased public expenditure to deal with distress. Not uncommonly, the crisis also affects income from taxation. Take once more the case of Zambia. A large share of its taxes is derived from the copper mines. So when the price of copper fell sharply, so did incomes from taxation.

Others, including the World Bank and the International Monetary Fund, argue that the *financial indiscipline* of the state is a major *cause* of crisis. The state is 'living above its means' with dire consequences for the economy. By printing money to finance its deficits, the state fuels inflation in the economy and, consequently, increases the demand for imports. As such, it creates an unfavourable climate for investment and private enterprise to thrive. Investors are reluctant to invest and the country's creditors lose confidence in the government's ability to manage its affairs and to service its debts.

But the accusation is not just that the state cannot manage its own affairs. It is also often accused of pursuing misguided policies, in the way it intervenes in the economy at large. Hence, it is argued, rather than promoting the greater *openness* to foreign trade of the country's economy, many states do the opposite. They insulate themselves from the benefits of free trade and export-promoting strategies and, instead, erect artificial barriers behind which they sustain an inefficient economy. As a result, investments are often misdirected and ineffective. They do not propel economic growth and greater opportunities for all, but rather add to the burden of future payments difficulties.

In a nutshell, the argument of the World Bank and the International Monetary Fund is that the crises happen because the state spends too much; giving the wrong signals to the economy, and creating an unfavourable climate for investment and economic growth. What is needed, in their view, is a complete overhaul of economic policy, state action and the general climate for investment and growth in these countries. This is what structural adjustment is all about. Not muddling through or tinkering at the margins, but a complete overhaul of the economy and state action within it.

The policy package

As we have seen, in the view of the World Bank and the International Monetary Fund, the state is often the main *culprit* in matters of economic crisis. The cure is to redress state action in the economy through comprehensive economic reform policies. These are of two kinds:

1 The state should act directly to restructure its own taxation and social provision.

2 The state should alter the way it intervenes in markets to change their organization or their pricing behaviour, in order to create a favourable climate for investment and growth.

The first type of reform has been discussed at length in Chapter 3. The main objective is that the state should get its act together, by cutting its expenditures, raising its revenues and changing its mode of operation. Social provisioning, in particular, must face expenditure cuts, a shift towards targeting 'those in need' and charging user fees to boost revenue and limit 'excessive' use.

But what about economic reforms to change the general climate for investment and growth? Here, the key words are 'getting prices right' and 'deregulation'. What does this mean?

'Getting prices right' means that the economy should be guided by the world market, which is seen as reflecting the real cost of goods and services. Deregulation means that the state should dismantle its artificial barriers which prevent the economy from responding to market pressures in general, and to world market signals in particular. In short, the market should reign supreme.

The main policy measures aimed at getting prices right are

- devaluation (jointly with the abolition of exchange controls)
- raising interest rates
- abolishing subsidies and price controls which distort prices.

What are these measures intended to achieve? Let us first look at devaluation. What are its effects on an economy?

> **Devaluation:** changing the price of the domestic currency (say the shilling) in terms of foreign currency (say the dollar). After a devaluation, the dollar will be more *expensive* in terms of shillings. A devaluation is simple to effect and has widespread consequences for an economy as a whole.

Effects of devaluation

After devaluation, imports will become more expensive. Why? The prices of food, oil, tractors, etc. remain the same in dollars, but will now be higher in shillings. Consumers will have to tighten their belts as the prices of food or other consumer goods rise. Industries relying on imported raw materials and spare parts have to pay more for these imports or, as happens in many developing countries, lay off workers and underutilize their machinery. Agriculture also suffers when imported fertilizers, pesticides and oil become more expensive.

What about exports? Will they become cheaper in world markets? This would be true for Britain or Germany, but not for Tanzania or Bangladesh. Tanzania, for example, cannot influence the prices of coffee, cotton or tea. These are determined on the world market. As a result, the dollar price remains unchanged after a devaluation, but the price in shillings will increase. Farmers will get a better price for their coffee, cotton or tea. This will stimulate exports.

A devaluation, therefore, tightens the belt, stimulates exports and encourages a more prudent use of imports. In short, a devaluation seeks to correct the deficit on the balance of payments and to promote greater openness of the economy.

Raising interest rates

Allowing banks and other financial institutions to charge higher interest rates is seen as a measure to restrain spending in the economy at large. More particularly, it forces investors to be prudent about where they invest their money. That is, it aims to weed out bad investments, such as costly white elephants with shaky returns.

Abolishing subsidies and price controls

The abolition of subsidies and price controls seeks to redress distorted prices. One example is food subsidies which, it is argued, favour the town consumers at the expense of the farmers.

The policy package as a whole, therefore, seeks to open up the economy to world market competition, to promote exports and to restrain domestic spending. If the policy works, it should in theory create a better climate for investment and, consequently, enhance economic growth. Hence, the policy is not just a question of tightening the belt, as some critics would argue, but also of growth and development. The benefits of economic growth, it is argued, will trickle down in the economy and society, and enhance general welfare.

This is the story as told by the World Bank and the International Monetary Fund. Its philosophy is rooted in a belief that increasing reliance on market forces enhances growth and welfare. It is a story of an economy in the abstract, adjusting *flexibly* to world market signals. But does it work? And what about real people in this process of adjustment?

7.3 Structural adjustment and vulnerability

Markets entail vulnerabilities. Strangely enough, as we saw in Chapter 1, orthodox economic theory praises the *flexibility* and *adaptability* of markets, but fails to point out that these desirable features also account for the *vulnerabilities* inherent in the normal working of markets. As the price of a commodity falls in response to changing demand conditions, resources will be withdrawn from its production, which has become less profitable, and supply will adjust to demand. But this also means that wage labourers lose their jobs, peasants or craft workers face hardship or ruin, and enterprises go bankrupt.

Q How, then, does adjustment affect the vulnerabilities of countries and of people?

Vulnerability of a country?

Not all countries enter world markets on equal terms. Some are rich and have diversified

Figure 7.5 Billboards for Japanese products in post-reform China. The market power of the OECD is evident in most parts of the world.

economies with a capacity to trade in a large variety of goods or services. Others are poor, often predominantly agrarian, with a few raw materials to trade. Some like the USA, Japan or the European community exert considerable market power; others have no control over the conditions in which they trade. As such, vulnerabilities and the flexibility to adapt to changing market conditions also differ widely among countries.

Take the case of a country which relies principally on *one* export commodity to finance a wide variety of imports, some of which are needed to keep its factories, farms or mines running, or even to feed its population. In such cases, adverse movements in its *terms of trade* can spell real disaster (Box 7.3).

Box 7.3 The terms of trade

The terms of trade are the ratio of prices of goods or services a country sells (exports) relative to the prices of its purchases (imports).

Take, for example, the case of Zambia. Its terms of trade *worsen* if the price of copper (its export commodity) falls relatively to prices of imported goods. This can happen where (i) the price of copper falls while import prices remain constant, and (ii) import prices rise faster than the price of copper. (See also Chapter 5 in *Hewitt* et al., *1992*.)

A collapse in the price of the export good can provoke a major crisis in the economy. Cuts in imports, in turn, reduce production in various sectors, including these producing exports, which depend on imported inputs. This reliance on a few export commodities (mainly raw materials), coupled with a high dependence on imported goods, can lead to a deeply vulnerable situation. Many of these countries, therefore, have little capacity to adjust, apart from tightening their belts. Merely opening up these economies to world market pressures will do little to enhance future growth or development. The basic problem is one of *entitlement failure* — an inability to adjust — and, as argued in Chapter 1, markets have little ability to deal with such distress.

Vulnerability of the poor

The language and discussion of economic policy often mask the fact that real people will be affected. It is not immediately obvious, for example, that child immunization programmes may be jeopardized because of a 1% rise in international interest rates. So if we are concerned about the welfare and well-being of people in the Third World, and if we want to influence policymaking in the developed world, it is important that we understand what structural adjustment means, however complex the issues involved.

Indeed, the welfare of households is bound up with national economic performance and changes in the international economic climate. These are issues that have been implicit in our discussion so far, but have not been examined explicitly. Economic recession not only means unemployment and loss of livelihood for thousands of people, but also puts pressure on state expenditure. Government revenues decline, and yet more people try to rely on state support because they cannot maintain their living standards.

At the international level, a decline in commodity prices, rising interest rates, trade policies of developed countries and exchange rate fluctuations have had serious consequences for development in the Third World. These international developments are less tangible, but no less important, than what happens at national level in terms of their impact upon the lives of millions. All of these national and international developments come together when the issue of 'adjustment' of Third World economies is discussed. Moreover, given the global nature of economic problems of LDCs and the dominant role of the developed countries in international economic affairs, policy making in the West should also come up for questioning when we address economic problems in the Third World.

Child welfare

Of particular importance is the welfare of children in a context of structural adjustment. Children matter as human beings who need care, but also as the future potential of a society. Investing in children is investing in the wealth of a nation:

> "The wealth of nations has come to be predominantly the acquired abilities of people — their education, experience, skills and health ... The future productivity of the economy is not fore-ordained by space, energy, and cropland. It will be determined by the abilities of human beings. It has been so in the past and there are no compelling reasons why it will not be so in the years to come."
>
> (T.W. Schultz, quoted in UNICEF, 1989, p.10)

Looking after the needs of children also has a major bearing on other aspects of development, such as reducing population growth. It is an accepted fact that falling infant and child mortalities are important contributory factors to the decline of fertility in the Third World. Faced with high infant and child mortalities, parents usually opt for a larger family in order to achieve what they consider an ideal family size.

Most developing countries have made major advances in the area of child and infant mortality and education since the 1950s. But the debt crisis and the recession of the early 1980s has threatened many of the gains made in the earlier periods. The evidence available suggests that in a number of countries in the 1980s, the situation of children deteriorated, or the rate of progress slowed down.

The United Nations Children's Fund (UNICEF) sponsored a study of the situation of children in 10 developing countries in the 1980s. The countries were selected so that they represented different regional, economic, ecological, and social–cultural conditions. The study revealed that, in most of these countries, the decline in child and infant mortality rates slowed down, malnutrition and educational attainment indicators worsened, and disease prevalence increased. (Cornia *et. al.*, 1987). These results do not make the headlines, they are not as dramatic as the picture of famine stricken children of Ethiopia, yet they do have far reaching implications for the future development of LDCs, mainly because of the deterioration in the quality of so-called 'human capital' : the health, energies and skills of people.

Figure 7.6 A school classroom in Uganda — no desks and only one schoolbook to three children.

Decline in education
In sub-Saharan African countries, primary school enrolment declined by about 6% over the 1980–85 period, at the same time as a sharp increase in the incidence of poverty in this region. This decline is even more depressing if we note that in 1980 only 60% of children of primary school age attended school. Child (under five) mortality rates also showed little improvement. Despite being the highest in the world in the 1970s, they showed the least percentage reduction in the 1980s.

UNICEF surveyed a number of countries which experienced economic problems in the early 1980s; in Africa, Latin America and Asia. They found that, where there were no interventions to cushion the poor, school attendance declined. Moreover, the quality of education and educational attainments invariably also declined.

The decline in school enrolment provides an important clue to what happens to children of the poor during recessions. They mostly have to look for work to supplement the meagre and declining income of the family. The incidence of child labour will, therefore, tend to increase at times of economic hardship.

There is evidence, therefore, that structural adjustment has increased the vulnerability of the poor. The next section abandons the conventional approach of looking at a policy and its outcomes at a national level. Instead it develops a framework to look at *policy as a process*. This will enable us to come to terms with the processes through which structural adjustment affects the poor.

7.4 Social institutions and impoverishment

Governments propose, but people and institutions dispose, or at least they influence the outcome. It is now generally agreed that, under structural adjustment policies, the poor and vulnerable suffer more than the better off. But the extent to which inequality grows, or extreme deprivation develops, varies greatly from country to country, despite a remarkable similarity in the policy 'packages' implemented. We can only understand this variation, and can indeed only grasp the extent and implications of the deprivation caused, if we look closely at a country's social institutions. Governments, in other words, take decisions, but it is the responses of people in institutional settings which determine the outcomes.

We must be clear what we mean by institutions. Societies are not the arithmetic sum of the individuals within them. Rather, there are codes and cultures that shape our lives. We even come to market, not only as individuals, but also within collectivities and formal organizations, households and firms. It is not true that, as Margaret Thatcher once averred when she was Prime Minister of Britain, there is no such thing as 'society'. We have argued that public needs and responses are socially constructed. In tracing the impact of structural adjustment, we will focus here on three of the most influential institutions in our lives: households, gender relations, and markets. These institutions are among the most important in our society. Economic policies are filtered and given a specific form by the way these institutions work (Box 7.4).

Box 7.4 Institutions that 'filter' economic policy

The household: the institutional form of domestic life. Within the household, people, especially women, try to manage resources to provide for working adults and dependents, including children.

Gender relations: the institutional form of the relations between men and women. These structure the division of labour in society, including the division between paid and unpaid work.

Markets: the institutional forms of buying and selling. Real markets vary greatly. Some are competitive, some are not. In some, producers respond rapidly to price changes; in others they do not.

All of these institutions vary greatly, within and across countries. This section traces the impact of structural adjustment policies through these institutions, using the examples of health, food policy and child welfare to explain the problems.

Households and child welfare

Sustaining the welfare of children, it was argued above, is an investment in future development. But children's welfare, as Chapter 2 has shown, depends upon the survival strategies of people within households. This insight helps us to trace the impact of structural adjustment on children.

Children's entitlements to food, health and education depend largely on the economic situation of their households. This in turn depends on:

- entitlements to market goods
- public provisioning
- the management of those resources at the household level.

We can think of the welfare of children as 'produced' by the way the household uses its resources. Almost invariably it is the women in the household whose time, education and skill determine how the needs of children are met.

Figure 7.7 illustrates the relationships between policy and household management of resources. At the top of the diagram are some of the typical immediate effects of structural adjustment in the 1980s:

"Adjustment policies affect the poor in

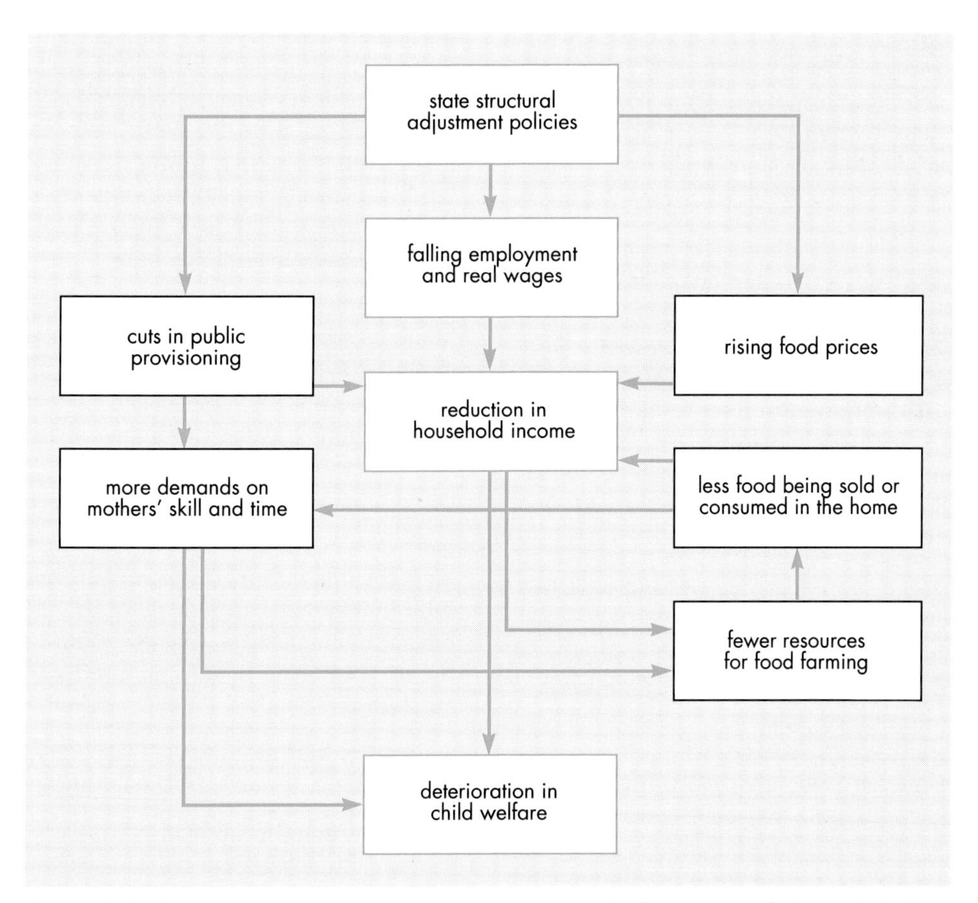

Figure 7.7 The causal relationships between structural adjustment policies and child welfare. [Adapted from Cornia et al.*, 1987]*

three broad ways: first they affect their incomes, either through changes in wages and employment, or through shifts in prices, altering the returns from productive assets; second, they change the prices of their most important purchases; and finally, they shift the level and composition of government expenditures, particularly those in the social sector."

(ODI, 1986, p.1)

At the bottom of the diagram is the 'outcome': child welfare as measured by indicators such as mortality, disease, disability, malnutrition and education levels.

The economic and financial squeeze associated with structural adjustment typically reduces both employment and real wages. There is a downward spiral of investment, output and employment; though the size of the contraction depends on the economic structure and the extent of government involvement in the economy (Khan & Knight, 1985). This recession reduces the household income of those who rely on wage employment and also of those who sell goods to wage earners. The more urbanized the country, and the higher GDP, the more of the poor come into this category. The poor in countries like Brazil, Mexico and Egypt are very vulnerable to falling employment. But some much poorer countries also have a high proportion of their labour force dependent on wages. Sri Lanka, for example, in 1981 had 67% of its population dependent on wage labour, including wage labourers in rural areas (Ginneken, 1988).

General inflation, often worsened by structural adjustment, contributes to falling real incomes. The effect is sharpest when food subsidies are reduced and food prices rise sharply. Poor households, including urban and rural wage workers, as well as many small farmers who have to rely on the market for food, are badly hit. If such households try to maintain their purchases of food, usually the largest item in their budgets, they will have to cut back other essential spending. This is called the 'income effect' of price rises. Often what gives way is household expenditure on health and education. And even this may still not protect food consumption:

> "In the Gambia, for instance, child malnutrition increased when a [IMF] Fund–[World]Bank supported food adjustment programme led to an increase in food prices."
>
> (Cornia *et al.*, 1987, p.67)

Only farming households selling food will benefit from food price rises, though their income will often be hit by rising prices for other essentials, from hoes to soap. Households which farm may be able to eat more of their own produce too as prices rise. But this depends on household assets and incomes. Households often need cash from wages to provide inputs to farming. When employment is lost, the ability to farm can also be undermined.

As poor households try to cope with falling incomes, children share the burden. As incomes fall, the poor try to work harder. Declining school enrolment of children from poor households is associated both with falling incomes (no money for school fees) and with rising child labour, as children are sent to look for work to supplement the meagre household income.

This tendency for everyone in the poor households to work harder as incomes fall partly explains a statistical puzzle: structural adjustment does not always worsen measured income distribution. A study in Mexico showed how this worked.

The study shows a decline in real wages of the poor, lower-middle and upper-middle classes, with the latter two groups registering the largest fall. In the case of the poor the extent of the hardship is, of course, larger. Faced with falling living standards people adopted various coping strategies.

- Individual members of households began to work more hours in the same job.
- They sought additional income-generating activities, perhaps as non-wage earners.
- More members of households joined the work force.

"The hardship imposed by the crises and adjustment should thus be measured not only by decline in income, but also by changes in other indicators of the quality of life such as the length of the working week and the hours available for leisure and rest."

(Lutsig, 1990, p.1337)

If people are working harder at earning income, there is less time for managing resources effectively at home: less time for childcare, education of children, food preparation, nursing the sick. The physical, emotional and psychological costs of trying to cope rebound on the most vulnerable, including children.

In turn, this effect is worsened by the decline in public provisioning caused by structural adjustment (the left-hand side of Figure 7.7). How badly this affects poor households depends on the kinds of cuts implemented in social services, including provision of cheap or free food. In many cases, the changes in public provision have been 'regressive'. That is, they have hit poor households hardest.

For example, in order to cut state expenditure, some governments have cut the health budget indiscriminately. Evidence from Brazil shows that

" ... delays in the implementation of the Expanded Programme of Immunization in São Paulo State ... led to an outbreak of deadly communicable diseases among children."

(Cornia *et al.*, 1987, p.67)

Similarly, cuts in primary health care expenditure in Ghana resulted in sharp deterioration in indicators such as incidence of infectious diseases and disease-specific mortality rates. In Sri Lanka, cuts in food subsidy led to an increase in severe malnutrition among the children of the poorest.

Changes in public finance make this worse. Indirect taxes (such as sales taxes, or VAT in Britain) raise prices for everyone, so they hit the poor hardest. Increasing charges for public services, such as basic health or education, has the same effect. Such indiscriminate and hence regressive policies are common under structural adjustment:

"76% of the programmes supported by IMF between 1980 and 1983 included increases in indirect taxes and 46% in tariffs, fees and charges, as against 13% involving increases in personal, corporate and property taxes."

(Cornia *et al.*, 1987, p.67).

Poor and vulnerable children tend to live in poor and vulnerable households. Structural adjustment tends to make poverty worse. In many countries, especially in Latin America, the income distribution has shifted in favour of the rich. However:

"A massive increase in poverty can occur in the absence of growing income inequality if output declines substantially — as in Costa Rica, where poverty rates increased between 1979 and 1982 from 17 to 29 per cent owing to the sharp GDP decline following the introduction of severe adjustment policies."

(Cornia *et al.*, 1987, p.66)

What we have shown here, is that the impact on the poor, and notably on poor children, is worse even than the data on falling incomes suggest. The declining incomes of poor households have interacted with regressive reform of state taxation and expenditure, to put a cumulative burden on the poorest households. The coping strategies of the poor, specifically trying to survive by working harder, worsen further the impact on the welfare of poor children. The whole country is then 'disinvesting' in a large part of a future generation of its people.

Gender relations and the impact on women

As has just been made clear, it is particularly women who face the responsibility for managing household resources in economic crises. Both sexes have to work harder outside home to make ends meet, but women have to work harder at home as well, to try to maintain living standards and living conditions in the face of rising prices, recession and cuts in state expenditure. Hence the impact of adjustment on women seems

Figure 7.8 Proponents of structural adjustment treat the cost of unpaid labour as zero. In this way, women absorb a significant part of the cost of the adjustment through intensification of their household labour. (Left) Collecting fuelwood in Zambia. (Right) Preparing food in Indonesia.

greater than its impact on men, another fact which threatens children. Here we look at gender relations and the way in which they can push the greatest burden of adjustment onto women. The following account draws heavily on work by the economist Diane Elson (in Onimode, 1989; Elson, 1991).

Those designing adjustment programmes frequently do not realize the burden placed on women, because they ignore the importance of unpaid labour in the economy.

Adjustment programmes are designed on the assumption that what matters is the monetized part of the economy. They assume that labour can be transferred from one type of productive activity to another (e.g. from agriculture to manufacturing, or from handicraft to manufacturing) without any costs to the economy. And policymakers work in macroeconomic concepts, such as consumption and investment, which as Elson (1991) notes, 'appear to be gender neutral'.

None of these assumptions is convincing. All societies use immense amounts of unpaid labour. And all societies display a marked sexual division of labour which allocates the bulk of caring activities to women, and which more generally allocates some tasks typically to women and some to men. As a result, gender relations have major implications for the impact of structural adjustment. For example, adjustment programmes often seek to develop new labour inten-

Box 7.5 Paid and unpaid labour

Paid labour includes all work which earns a (normally cash) income: wage work, self-employment, cash crop farming.

Unpaid labour includes all work not done for a cash return: for example cooking and cleaning for a household, childcare, care of the elderly at home, cultivating food to be eaten at home.

The sexual division of labour in most societies tends to allocate a disproportionate amount of the unpaid labour to women. This unpaid labour is rarely measured in the economic statistics used by policymakers.

sive manufactures for export. Companies establishing such assembly plants (e.g. for clothes or electronic components) tend to prefer to employ women: a Brazilian electronics company for example claims that women have more patience and concentration, and a better sense of touch (Hirata, in Elson & Pearson, 1989).

So women's employment, at low wages and with poor working conditions, expands during the crisis. Under the pressure of poverty, women add this work to their household chores. Women can only cope with pressure by reducing their resting time, at the expense of their physical health. Part of the housework of a mother may be taken up by other female members of the household, daughters and sisters for example. So adjustment requires reallocation of female labour, while making some men unemployed. And the sexual division of labour within the household means that the pressure on women will increase.

All of this happens at a time when state service provision is falling. Some state provision is complementary to women's unpaid work. For example the government may ensure food availability at subsidized prices, while women's work ensures its delivery and the maintenance of nutritional standards within the household. Other state provision substitutes for women's household work. For example the sick may be nursed at home (generally by women) or by paid nurses in hospital. In both cases, cuts in state provision add to women's unpaid work. It is harder to produce nutritious meals as food prices rise. And women take on more of the health care burden, as access to hospital declines.

That there are limits, and costs, to this increasing burden is ignored by policymakers. Structural adjustment treats the cost of unpaid labour as zero. So an increase, substituting for public services, is seen as an unequivocal benefit to the

Figure 7.9 During the 1980s crisis and structural adjustment reforms, women were forced to diversify sources of income in an attempt to keep up with food price rises. These Zambian women now brew maize beer for cash. Each batch takes 7 days labour time away from working on food crops.

economy, freeing monetized resources (that is, people who have to be paid and equipment which costs money) for export industry (Elson, 1991). Such an attitude completely ignores the (economic) costs we have been tracing: for it is not 'consumption' but investment in people for the future which is squeezed when women's work burden reaches its limit.

Research is patchy, but it brings out these hidden costs of adjustment; for women and for investment in people. A study of structural adjustment in Zambia (Evans & Young, 1988) revealed that a 16% cut in real per capita public expenditure between 1983 and 1985 caused great pressure on household and community, and in effect on women, to take up the slack in health care. The cuts meant that rural child immunization programmes, and mother and child health clinics were neglected. Consequently people had to travel longer distances for treatment and medication. Women reported that they 'could not afford' to be ill because of the time it would take away from their work. They also had to spend more time attending to the sick members of the household; both at home and in hospitals which were suffering from shortage of staff and equipment.

Worse, from the point of view of the adjustment programme logic, these pressures on women prevented the expansion of farming output. The extra demand on the time of one woman was such that she missed an entire planting season. This is 'a perfect example of the interdependence between the labour that macroeconomic models do include, and that which they ignore.' (Elson, 1991)

Within the household, work is redistributed from older to younger women, and then to girls. Hence the sexual division of labour in the home increases the girls' school drop-out rate, and damages their future prospects.

Adjustment therefore has brought a working day which is longer, harder and more stressful for women. In a revealing study of 141 households in a low-income urban community in Ecuador, Moser (1989) found that only 30% of women were coping, with 55% just hanging on, by drawing on the labour of their children, especially daughters, and thus having to risk their future in order to survive. The other 15% of women were exhausted, with their families disintegrating and their children dropping out of school, roaming the streets and becoming involved in street gangs and exposed to drugs.

There are echoes here of the story in Chapter 2. When women suffer, children suffer and the future is compromised. The gender relations which structure the sexual division of labour in society worsen the impact of structural adjustment upon the most vulnerable.

Market responses

Finally, let us turn to that institution which is supposed to be central to structural adjustment programmes: the market. As we have seen in

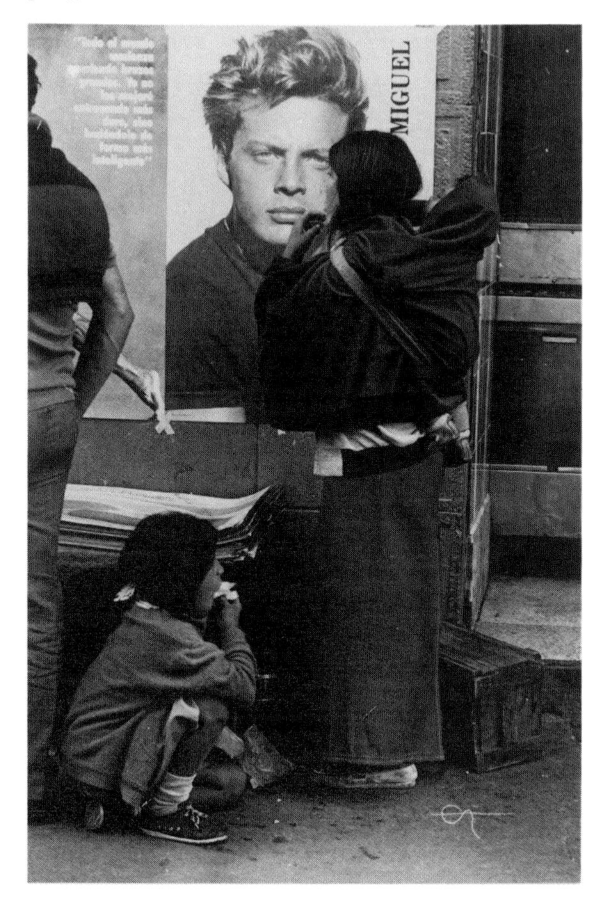

Figure 7.10 A woman with two children, selling posters on the streets of Quito, Ecuador.

Section 7.2, a central aim of adjustment programmes is to change market incentives and hence to *increase supply* of goods and services. If this can be done, then the cuts just discussed need not be so severe.

But will it work? Will producers respond to higher prices by producing more, including more for export? Economists call this the 'supply response' problem. And there are deep-rooted problems in the structure of the actual developing country markets, which suggest that rising food prices, or higher bread prices following devaluation, may not easily call forth greater supply.

The first problem is that the price of a product is only one factor that influences production decisions. Let us take the case of the agricultural sector.

Agricultural prices

Third World governments are often required to deregulate food markets under orthodox adjustment programmes (a policy discussed in detail in the next chapter). The lifting of price controls, the argument goes, will lead to increased production. But to produce more, food farmers, who mainly operate on small and medium sized farms, must either increase the area under crop and/or increase the productivity of their land.

However, the land tenure system often prevents small farmers increasing the areas they cultivate. And to increase productivity of the land, farmers need to improve their production techniques; for example by using high-yielding varieties of seeds. These improvements in turn have costs. They require complementary additional inputs of water, fertilizers and pesticides which are beyond the means of many small farmers. The 'supply response' expected therefore may not materialize.

Need for infrastructure

Similarly, many developing countries are susceptible to floods, pest attacks, drought and other calamities. This is because of poor or non-existent preventive infrastructure, such as flood control systems, water storage and irrigation facilities for management of water resources. In these conditions only additional (public) infrastructure investment will allow the price incentives to bring forth more production. But public investment is being cut back.

The evidence shows that these problems matter. Price increases do tend to increase output. But the effect is often slow. In industry, under-utilized factories could soon be put to work in response to price incentives, provided that inputs such as labour, raw materials and energy are readily available. But to increase industrial capacity takes time, and requires new investment.

In agriculture, the timing of supply response is determined partly by the vegetation and gestation cycles of crops and partly by the availability of inputs. And there is a low supply response of total agricultural output to a rise in farm prices. For example, doubling prices will lead only to a 10% to 30% rise in farm output in the short run, and between a 30% and a 50% rise in the long run (Cornia *et al.*, 1987).

Finally, the hoped-for effects of devaluation suffer similar problems. Devaluation raises the local currency prices, and hence the profits, of those producing for export. But as noted earlier, foreign currency prices of Third World primary exports have been falling, and more production may make that worse. New exports require imported inputs such as new machinery, which devaluation makes more expensive. New production and exports, in a stagnant but inflationary economy, may be very slow to come through. Real markets respond more slowly and less predictably than the simple models suggest.

7.5 Adjustment and deprivation

Adjustment 'with a human face'?

We could sum up the last section as follows. Real institutions modify the expected effects of adjustment policies. The supply response is slower than expected in real markets, prolonging recession. The sexual division of labour constrains the expansion of output on markets, and increases the social costs of adjustment. These immediate costs are much greater, especially for women and children, than was originally expected, and the long-term implications for development of the collapse in social investment in a whole generation of people, are far worse than had been appreciated. So what is to be done?

The best-known response is associated with the UNICEF work already quoted, and is known as 'adjustment with a human face' (Box 7.6). This set of ideas has had some influence in the late 1980s and early 1990s on the World Bank's approach to the 'human aspects' of adjustment:

> "Adjustment with a human face ... [adds] ... a poverty alleviation dimension to adjustment ... and may be thought of as the 'basic needs' approach to adjustment."
>
> (Cornia *et. al.*, 1987, p.7)

This focuses on *the protection of the most vulnerable groups* (the under fives, pregnant women, and nursing mothers) in countries where, as a result of severe economic crisis and adjustment programmes, output is falling, and the burden of adjustment is falling disproportionately on the poor.

Box 7.6 Adjustment 'with a human face'

The four most important elements of the UNICEF approach to adjustment are:

(i) more emphasis on economic growth, hence a less sharp economic squeeze and more investment to restart production

(ii) more development of small scale production to spread employment

(iii) a shift in public spending towards basic needs provision (e.g. from higher to primary education)

(iv) targeting scarce state support on the poor and vulnerable (e.g. by moving from general food subsidies to targeted subsidies or feeding programmes for the very poor).

Some elements of this programme are clearly improvements on the orthodox adjustment approach. It emphasizes *gradualism*: gradual changes in state spending, slower devaluation, and no sharp squeeze. And it makes a case for *more finance* from abroad during adjustment, in order to sustain state spending on infrastructure such as transport and power, and to maintain imports of vital inputs to production (fertilizers, machinery). These points are increasingly widely accepted.

Much more problematic is the emphasis on prioritizing, selectivity and targeting in social policy and state expenditure. The UNICEF approach accepts in principle the privatization of public services, except for those services for the 'poor and vulnerable'. Even for the latter services, UNICEF accepts that community-level provision may be devolved to NGOs and foreign aid donors, in the manner discussed in Chapter 3.

I shall argue in this section that, far from being a sensible piece of economic management, this approach to the public services is destructive, in the long run, both of the public sector and of the interests of the poor and vulnerable.

It may seem perverse to criticize policies that are designed to salvage some public services for the poor in the short run. But this can mean establishing a two-tier system of social service provisioning: one tier of under-funded, under-resourced public health and education services for the poor; another tier of privately or publicly run fee-paying services accessible only to the rich and privileged. So in the medium term, these policies could result in the collapse of effective public provisioning, which would then be very difficult to re-establish. Finally, 'adjustment with a human face' embodies a top–down approach to policy which also undermines effective public action.

Fees and local financing

There is something of a contradiction in the UNICEF proposals between their acceptance of charging fees to part-finance services, and their desire to aid the poor. Cornia (1987) distinguishes basic needs services (primary education, disease control, basic curative medicine) for which fees are not recommended, from other services (university education) which do not serve the poor and which should be fee paying.

In this, UNICEF recognizes the evidence that charges reduce use of services, especially by the poor, and compete with other essential purchases such as food. In Southern Nigeria UNICEF found that the reintroduction of school fees caused primary school enrolment to fall from 90% to 60%. Even in Britain, eye tests fell about 20% following the charges introduced in the late 1980s. However, UNICEF does accept charges for university, high school and most curative medicine. Even if means testing of charges for curative medicine is possible (and lack of information will make it difficult), remaining charges will still restrict access to these services by the poorer population.

Furthermore, the proposed protection of basic services from fees conflicts with another recommendation of 'adjustment with a human face': 'local financing'. This means that health and education projects should be supported where possible by local funds and resources.

Such proposed community financing can take three forms: charging for services or payments for drugs, mobilization and use of free community labour, and fund raising activities. Such projects existed in many countries before the adjustment programmes. Their record, as far as the benefits to the poor and vulnerable are concerned, is a mixed one. Self-financed projects have mainly developed in poorer regions and areas, and have tried to tap the already overstretched resources of the poor. A study of 100 self-financing primary health care projects pointed to the undue pressure brought on people with little ability to pay (Stinton, 1982). UNICEF is aware of this problem, and suggests that self-financed projects should only serve to complement public resources from the state, or other agencies.

That may be feasible when the economy is growing, but under adjustment it is inevitable that community financing will be viewed by policymakers as an alternative to and a substitute for state financing. And as experience has shown, the burden shifts onto the poor, who in most cases can only provide local inputs, such as labour and food, but need outside financial help to run the service. So inevitably local financing worsens the disparity between services in poor areas and those in better-off areas.

This contradiction between a desire to protect free basic services and a recommendation to increase local financing arises because the overriding aim is to reduce the demand on state-provided, tax-financed services. Hence, the pressure for 'self-provisioning' in health, sanitation and nutrition, wherever possible. The UNICEF authors know that this increases the burden on the poor, especially, as Section 7.4 showed, on women and children. But they end up accepting the costs and inequity in order to manage the crisis.

Targeting

The other major policy of 'adjustment with a human face' is the targeting of remaining public services on the poorest. We can ask two questions. Can it be done? When does it work?

Targeting is difficult. A general cut in food subsidies will not hurt the poor, it is said, if the remaining subsidy goes to them. But how do we identify the poor and vulnerable? Minimum income, nutritional level? Information is poor, and very narrow definitions of poverty may miss many people on the margin who then lose out badly when general subsidies are removed.

Identifying the 'targets' and delivering services selectively is also expensive. The high administrative costs are not justified by effective coverage: many of the 'target group' may not come forward through lack of education or information. Finally, the somewhat discretionary nature of these targeting schemes may make them prone to corruption.

An alternative is to target, not people, but geographical areas where the poor live or the type of goods the poor consume. Such schemes are not discretionary, and they may have lower administrative costs compared with the income and needs-based systems. For example, a general cereal subsidy can be replaced by a subsidy to the specific cereals and root crops consumed by the poor. Or in a geographically based scheme, all households in a given area can receive a service or subsidy. Such schemes can be combined with further targeting of the severely malnourished children in a poor neighbourhood (Cornia *et al.*, 1987).

Targeting in other words is complicated and costly; and it demands a well organized and honest public administration. Its results so far in the Third World have been mixed. Only in Third World countries with a long history of social welfare provision, an experienced administration and an active organization of the poor, has targeting provided a degree of protection to the vulnerable; and even then only partially and for short periods.

Chile, as Chapter 6 has already indicated, is an interesting case. In the early 1980s Chile faced a serious balance of payments problem, resulting from the world recession, a 26% drop in its terms of trade, a fourfold rise in interest payment on foreign debt, and an 80% drop in lending. The neo-liberal and authoritarian Pinochet government introduced a drastic structural adjustment programme. Government expenditure fell, including health and education spending. Open unemployment rose from 11% in 1981 to 21% in 1983; poverty increased; prices rose, especially of goods purchased by the poor, while incomes fell.

Within this context, the government social policy was heavily targeted on those defined as vulnerable. Employment-creating public works schemes, and targeted nutritional and health measures together covered large sections of the population. In 1983 public works employed 13% of the total labour force. All children under the age of 6 were covered by a nutritional surveillance and supplementation programmes. A narrower band was also set for giving subsidies: those under the age of 8 and pregnant women in conditions of extreme poverty. Similarly, severely malnourished children under the age of 2 came under a nutritional rehabilitation programme. Social expenditure on these policies increased, despite the general decline in government expenditure.

For a while, this worked. Infant mortality and immunizable diseases continued to decline, as did the incidence of malnutrition among pre-school children. But in 1983, the budget for these programmes was cut back, and the gains were reversed (Cornia *et al.*, 1987).

The experience of Chile is a revealing one. A military regime with a harsh adjustment policy did for a while cushion its effects on some of the most vulnerable sections of Chilean society. In part, Chapter 6 suggested, this was because of active pressure and organization by women. But it also emerged from the historical commitment in Chile to the welfare state, and to the health, nutrition and education of children. An administrative infrastructure had been developed over many decades. And on this the military government relied to deliver its programme, despite the ideological attack on social democracy. The Chilean adjustment programme was, therefore, undermining the very system on which it relied for its social policy.

Targeting, therefore, can only work if it is part of an integrated social programme, and if it is undertaken in the context of a long-term commitment to the public provision of social services. But if targeting is introduced mainly for the financial savings that it can offer, and is coupled with the dismantling of the social and administrative infrastructure, and with an ideological assault on the principle of collective responsibility for collective good, then the small and targeted 'nets' that have replaced the 'big' social welfare 'net' will fail to provide the protection needed by the poor and vulnerable. Then the future of whole sections of the population will be compromised.

A public commitment to support, maintain and develop the population is essential for the future. This is a long-term process. Small-scale, sporadic and 'targeted' efforts will not work, and will not be sustained.

Top–down adjustment

The final criticism of 'adjustment with a human face' concerns the way the policy is to be developed. Like orthodox adjustment, these are top–down policies. Schemes are to be developed and implemented by people in positions of power and influence. The poor and the vulnerable are to be passive observers whose interests are supposedly to be taken into account in the design of adjustment programmes. So the people who will be most affected by the policy still have no direct say in its formulation and implementation.

Policy, in this model, is not the result of discussion and negotiation by the concerned parties. Rather, it is based on what the dominant group or class and its representatives perceive to be the interests and needs of society at large, and of the poor and the needy in particular. However, such a process of arriving at a policy not only has implications for its content, but also means that its implementation will become merely a matter of bureaucratic management, while its monitoring is simply a matter of comparing quantifiable outcomes with a set of predetermined targets.

Not only is this process inequitable. It will not work. Public action to provide security for the vulnerable cannot be sustained by such philanthropy alone. State institutions matter, as does the political process of their reform. The neoliberal agenda of cutting back the state, restructuring and reducing the state's role in the economy (better known in the Third World as structural adjustment), has therefore taken different forms in the developed and developing countries.

In the developed countries, restructuring and institutional reforms have followed a political debate and struggle in which the political parties who favoured a neo-liberal and free-market approach to the running of the economy came to power. In most developing countries on the other hand, the implementation of the market-oriented economic policies, with a strong emphasis on the institutional reform of the state, followed in the great majority of cases from the IMF and World Bank stabilization and structural adjustment programmes.

So structural adjustment is not just a programme developed from economic analysis. Politics and ideology influence it heavily. Orthodox adjustment programmes, and their 'human face' variants, set out to reform all the institutions of state and society: industrialization and trade policy, welfare provision, the organization and role of the state itself. This institutional restructuring affects the lives of millions, and all without any public political debate about the future direction of society.

7.6 Conclusion: adjustment and democracy

So what is our conclusion? What coherent alternative to adjustment can we put forward to tackle the economic and social problems of developing countries? There is no doubt that, like developed countries, developing countries are in need of periodic reform. But how, what, when and by whom?

Long-term investment in people is essential to development: that is our first conclusion. And short-term protection of the poor through patchy welfare 'safety nets' will not do this. Nor will the market mechanism alone achieve it. Building up the human base of an economy requires intervention. It implies co-ordination of health, nutrition and education services at the national, regional and local levels in response to the needs of households and individuals.

What is the role of the state in this? State institutions, which have long played an important role in providing these services in both developed and developing countries, have now been seriously weakened. However, replacing state institutions partly by privately owned, market-oriented and profit-making institutions, and partly by non-governmental, charitable and community-based institutions, will not necessarily improve the availability of and access to social services. Such a strategy can produce inequitable and fragmented services, with deteriorating provision for the poor. Non-governmental services can also, as Chapter 5 has already argued, be just as top–down and unresponsive to local needs as provision by the state.

Nor is simply strengthening the state's role a solution. What is needed is to make state institutions more responsive to public need, and more responsible for their own actions, while maintaining their financial viability. This in turn requires a democratic process that increases the participation of those who are affected by state policy. Public participation should not be limited to national and local elections, but should draw people into policy development. The form of, and the forum for, such discussions has to be determined by reference to national and local traditions and practices. But it is through such democratic mechanisms that national policy to 'adjust' the economy to outside shocks, and reforms to correct past mistakes should emerge. If this were to happen, then public debate on economic reform would have to address investment in people, distribution of income and access to services, the impact of reform on women (half of the population), and the balance between the public and the private sectors.

Economic reform, in other words, is much more than decision-making on the basis of economic models and theories. In reality it involves conflict and struggle between opposing social groups. The resolution of the conflict is determined by the balance of power in the society. If the poor are hit, it is because they are the weakest section of society and have no say in critical decisions which will affect their livelihoods. Protection of the poor through targeting and similar measures can only offer a partial and temporary palliative to their problems. Only empowerment of the poor, through their effective participation in public action on their own behalf, can provide some protection for their interests. And only this empowerment can ensure that economic 'adjustment' takes the form of investment in the well being of the whole population, as the basis of development.

Summary of Chapter 7

1 Structural adjustment is not a technical set of economic policies, though that is how it is often presented. It is a process of institutional reform.

2 'Structural adjustment' refers to a set of policies propelled by the IMF and the World Bank, with the explicit aim of opening up Third World countries to world market competition, to promote their exports, and to restrain domestic expenditure. Such policies seek to create a better climate for private investment and to enhance economic growth.

3 However, some countries, and some groups of people, are acutely vulnerable to impoverishment as a result of structural adjustment.

4 The actions of governments who are implementing structural adjustment packages are filtered and modified by economic and social institutions, including the household, gender relations and the organization of markets. In a number of ways these institutions can worsen the impact on the poor.

5 'Adjustment with a human face' is an ineffective and partial palliative for the destructive and impoverishing effects of orthodox adjustment programmes. It cannot protect the poor for long, or sustain investment in people.

6 Only democratizing economic reform, and empowering the poor can create better reform programmes, which can overcome the development failures and the acute deprivation engendered by orthodox structural adjustment.

8

TWENTIETH CENTURY FREE TRADE REFORM: FOOD MARKET DEREGULATION IN SUB-SAHARAN AFRICA AND SOUTH ASIA

BARBARA HARRISS AND BEN CROW

8.1 Introduction

This chapter continues our exploration of structural adjustment, turning our attention now from social provision to agricultural markets. As Chapter 7 explained, the deregulation of trading, national and local, is a central element of the structural adjustment 'package'. But, as argued in Section 7.4 , real markets do not necessarily respond as the World Bank's models predict. This chapter develops that argument further by looking closely at the implications of this new 'free trade' policy for farmers and for the consumers of the food they grow.

Treating agricultural markets

Agricultural markets are often described using medical images, as though they were sick. In this chapter we shall examine:

- the crises which led to concern
- the ailments identified, and the diagnoses proposed, by the medical practitioners (actually the World Bank and International Monetary Fund)
- the medicine prescribed (deregulation and liberalization)
- some issues which the diagnosis ignored
- some second, more sceptical opinions
- the experience of selected economies undergoing treatment.

Opinions are easier to find in this field than adequately documented case histories. There is some documented evidence of the illnesses of markets and of government roles in them, and some forcefully presented cases about what medicine should be given, but few have been watching what has been happening to the patients as they recuperate or decline after treatment.

We shall be investigating the outcomes of what prove to be varied and experimental therapies, their beneficiaries and casualties, with detailed histories of Somalia, Bangladesh and Malawi.

The argument

Food distribution is a key area for structural adjustment. Deregulation of food markets has been justified on the basis of new ideas about the legitimate roles of government, and enforced amid pressing financial, social and administrative crises.

For ailing agricultural markets and for problems of state food trading in many developing countries, the 'private interest' view of the state described in Chapter 3 provided an orthodox diagnosis (too much government intervention) and recommended the orthodox treatment, that is, expansion of market processes.

This chapter argues that such *quantitative* conceptions of government intervention, that is, too much or too little, combined with abstract models of market processes, have provided inadequate guides to treatment. It demonstrates that limited success has been achieved only where deregulation is combined with *re-regulation*, that is, with carefully considered *new* roles for government. The conclusion reached here, and shared by other researchers, is that a precise consideration of the *nature* and *quality* of necessary government intervention may lead to better treatment than simple prescriptions based on slogans of too much or too little government.

Figure 8.1 Real markets. (Left) Weighing grain in an open-air market in Bangladesh. (Below) Sacks of grain awaiting river transport from the same market.

Food supply and economic crisis

Problems of agricultural production, food supply and food insecurity were central elements of the generalized crisis under which many underdeveloped economies were reeling by the early 1980s.

In the case of sub-Saharan Africa, one influential report described the crises as follows:

> "... for most African countries, and for a majority of the African population, the record is grim and it is not exaggeration to talk of crisis. Slow economic growth, sluggish agricultural performance coupled with rapid rates of population increase, and balance-of-payments and fiscal crises — these are dramatic indicators of trouble."
>
> (World Bank, 1981, p.4)

In Asia, where governments are generally stronger and have a larger revenue base, a comparable though less dramatic crisis also emerged, over a longer period. On both continents, balance of payments problems threatened food imports, food insecurity worsened, and deprivation escalated.

On both continents, countries turned to the IMF and the World Bank, and accepted the medicine of structural adjustment discussed in Chapter 7. The diagnosis of both financial institutions stressed neither the global recession, nor their own past advice, but rather domestic maladministration of policies now considered wrong-headed. Both institutions identified increased local food production as crucial, and problems in agricultural markets as the blockage (Thomson & Smith, 1990). Food deficits, malnutrition, and food aid dependence were all blamed on government regulation of food markets.

8.2 Regulation in agricultural markets: the ailments and the diagnosis

Q Why was market regulation diagnosed as a central cause of problems?

The World Bank's World Development Report 1983 noted that intervention in markets could be justified by market failures (see Chapter 3, Section 3.3). But it went on to argue that *failed* state intervention was the cause of many problems in developing countries.

> "Some market failures are so evident that they cannot be ignored; in addition governments will always have legitimate non-economic objectives that can be pursued only by intervention. The challenge for every government, whatever its political complexion, is to intervene in ways that minimize economic costs to achieve desired goals. Designing mechanisms that alleviate market failures without creating 'bureaucratic failures' has been a difficult task. *All too often the attempted cure has been worse than the disease* [emphasis added]:
>
> - Regulations have created large black markets. Some are associated with the vast traffic in drugs, but in other cases clandestine activities are linked to crops that governments have overtaxed or under priced. Coffee and cocoa in West Africa are examples of how black markets can undermine controls, at great cost in lost revenues and often in outputs as well.
> - Prices set low to benefit consumers — especially for food — have frequently discouraged producers, creating scarcities and greater dependence on imports.
>
> (World Bank, 1983a, pp.52–53)

The key point of this quote is italicized — the attempted cure, state intervention, has been worse than the disease, market failure. This is what we call the Free Trade view. It is closely associated with the 'private interest' view of the state. The Free Trade view identifies a particular set of market ailments as causes of the crisis in food production and distribution. Some of the costs and problems of regulation that it identifies are considered below.

Deficits of state marketing boards

Many governments established state-owned agricultural marketing institutions, either to

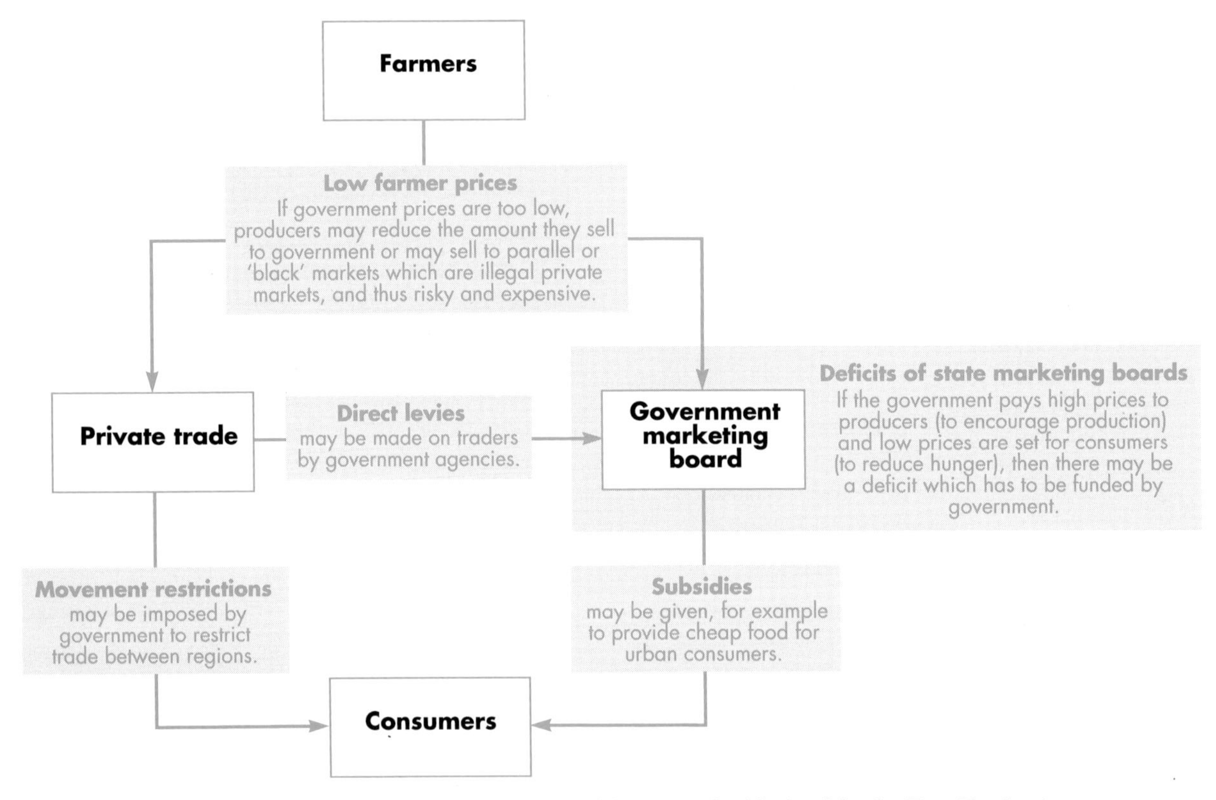

Figure 8.2 A schematic view of grain markets and of the issues highlighted by the Free Trade view.

provide the infrastructure necessary for the spread of markets, or to influence prices for the benefit of producers or consumers. Direct state involvement, particularly to extend markets, is common in sub-Saharan Africa.

Under pressure both to keep consumer prices low for staple foods, and to pay high producer prices to farmers, marketing boards in many economies have run deficits which drained government revenues. In Tanzania, for example, crop trading authorities worked up a combined deficit equal to 15% of GDP, while in Zimbabwe, 3.5% of the government's budget subsidized the agricultural marketing boards. Meanwhile, the prices the marketing boards paid were still too low to prevent producers from switching to crops without controlled prices, or selling crops onto illegal parallel markets.

Low farmer prices and parallel markets

Governments may reduce producer prices, in order either to keep consumer prices low, or to profit from state trading. These controls can reduce prices to the level of the production costs of grain or even below them. In response, farmers reduce production.

Meanwhile, however, the low consumer price may have increased the demand. In this situation, both producers and consumers have good reason to bypass the state-controlled grain market, and establish a parallel (or black) market. This is an illegal market, and to the extent that laws are enforced, it may be a difficult and risky market to operate. It is therefore more expensive than a legal market and may divert resources from more productive uses (Roemer, 1986).

The costliness of parallel markets may be further increased by the need to make payments to government officials whose job it is to regulate markets, so that they 'look the other way'. Low official prices thereby give some regulatory officials, including marketing inspectors, the police, and food and agriculture ministry officers, the opportunity to charge 'rents'. Free traders argue that damaging state policies may be maintained in order to maximize these rents, thus wasting significant resources (see the discussion of 'rent seeking' in Chapter 3).

Direct levies

In another common form of state regulation, state marketing bodies may make a direct levy on traders, legally requiring them to supply grain at a fixed price lower than the market price. Traders will thus make a loss on the quantity of grain provided to the state, and will recoup the loss through an increase in the price of remaining grain sold.

Movement restrictions

State regulation also frequently involves restriction on the movement of grain from one region to another. This is intended to prevent food scarcity, but it can make food insecurity worse. Consider, for example, restrictions on trade between a food deficit region and a surplus region. If there is no restriction, scarcity in the deficit region raises prices and attracts grain from the surplus region. Once trade is restricted between the two, the deficit region price will tend to rise even higher (because there are no longer uncontrolled supplies from the surplus region), threatening hunger for those who have to buy food. In the surplus region there is less demand, hence lower prices and lower production.

Subsidies

In agriculture and food distribution, two types of subsidy are often introduced by government.

Producer subsidies operate through cheap inputs such as fertilizers, credit and irrigation pumps;

Figure 8.3 Consumer subsidy in Bangladesh: the rice in these sacks is food aid from the USA being carried into a government warehouse. It will be sold to selected urban consumers at a price lower than prevailing market prices. Free Traders sought to reduce this subsidy.

or through higher grain output prices. They are intended to encourage the adoption of new technologies such as high-yield seeds (see Chapter 3 in *Bernstein* et al., *1992*), or to maintain farmer incomes (a common aim for subsidy in industrialized countries).

Consumer subsidies reduce or stabilize the price paid for food, in order to reduce poverty or maintain real incomes amongst selected groups. Consumer subsidies hence either force down prices paid to producers, or need to be financed (e.g. from taxation).

According to the Free Trade view, therefore, subsidies waste resources.

Worse, if it is true that competitive markets create an efficient economy, then subsidies, by changing prices, 'distort' this process and create inefficiency. On both grounds, subsidies are therefore argued to be among the causes of ailments of markets.

Overall diagnosis

All of these problems, Free Traders argued in the early 1980s, had interacted to create a multiple crisis in urban food supply, in rural food pro-duction, in overblown inefficient marketing agencies, and in food aid dependence. State marketing boards drained government revenues and producers switched to parallel markets. As turnover dropped in official marketing systems, so unit costs rose. Furthermore, keeping the price of foreign currency low made food imports artificially cheap, and discouraged agricultural exports.

More generally, agricultural marketing monopolies were thought to provide extremely good examples of the conflicts, dysfunctions and inertia of state institutions. Free Traders held to the 'private interest' view of the state. State institutions, they argued, could never be efficient. The institutions had emerged out of years of struggle between classes and interest groups over the redistribution of surplus via the government budget. They had become powerful but inert institutions, functioning in anti-social ways according to their own implicit objectives. Only competition could break this behaviour down.

This diagnosis was made at the beginning of the 1980s: a turning point in the evolution of ideas about development. As Hirschmann (1980) put it, there was an 'ideological shift in the policy cycle'. The 'private interest' view of the state took root among aid agencies and their controlling governments. The ailments of regulated markets, arising from failures of the state, were identified as key features of the generalized crisis.

8.3 The medicine (deregulation) and the prognosis (economic health)

The prescriptions of liberalization and deregulation have their origins in nineteenth century liberalism, a movement favouring free trade, individual liberty and moderate political and social reform. The nineteenth century Free Trade movement played an important role in the growth of the world economy, and the extension of capitalist forms of production. It provided ideas and a rallying cry for the overthrow of social and political barriers to the expansion of world trade (*Allen & Thomas, 1992,* Chapter 8). In the 1980s, the restoration of 'Free Trade' again became a driving force of government action in many parts of the world. As in the last century, the rise of Free Trade ideas has been associated with a marked expansion of capitalist economic organization.

The main elements of the twentieth century Free Trade movement are *liberalization*, *privatization* and *deregulation* (see Box 8.1). As Chapter 7 showed, these processes are central to the design of structural adjustment packages.

In the agricultural sector the deregulation of markets was expected to:

- increase production
- reduce state expenditure
- improve the efficiency of markets
- create more efficient state and private institutions though increased competition.

Box 8.1 The vocabulary of Free Trade

Liberalization involves the removal of government controls on markets and exchange. In grain markets, liberalization involves the repeal of restrictions on prices and on quantities moved or stored. Operationally, it means the lifting of movement and licensing restrictions, the abolition of uniform country-wide prices, the removal of subsidies, and the raising of producer prices.

The transfer of government-owned economic activity to private ownership is known as *privatization*. Ownership change is intended to improve the efficiency of economic activity. Privatization can take place without full liberalization. For example, government-owned bus companies, steel mills or state trading organizations can be sold to private owners without restrictions on markets being removed.

Deregulation is generally used to mean a combination of liberalization, privatization and reform of state institutions. It thus covers removal of restrictions on markets, the sale of state-owned economic activities, and reforms intended to improve the efficiency of state organizations.

Rising grain prices were expected to lead to increases both in production and in the amount of agricultural production marketed. Privatisation of state marketing institutions was expected to reduce the costs of government, and to improve the efficiency of food trading. This combination was expected to lead to lower food prices, thus improving household food security, by making food cheaper, and tackling inflation at the same time (Lele & Christiansen 1990; Rodrik, 1990).

8.4 Market institutions and market regulation: some elements of an alternative view

Before we examine what happened in some cases in Asia and Africa, we should note some possible problems with the Free Traders' diagnosis. In practice, few markets operate simply, uniformly or with equal benefits to all participants.

For example, behind the prescription of deregulation lies one abstract model of competitive market functioning. This model suggests that, if there are large numbers of buyers and sellers in a market, and there is an effective flow of information, then the prices set should adjust rapidly to changes in supply and demand. The actions of any individual seller or buyer should not affect prices. Mackintosh notes, however:

> "Better models recognise that markets concentrate information, and hence power, in the hands of few; that some participants are 'market makers' while others enter in a position of weakness; that markets absorb huge quantities of resources in their functioning [in costly dissemination of information, in storage, transport and processing and in speculation]; that the profits of a few, and growth for some, thrive in conditions of uncertainty, inequality and vulnerability of those who sell their labour power and of most consumers; and that atomized decision-making can have long-term destructive consequences [for example, in the 1943 Bengal famine when trade between rural producers and urban consumers left rural consumers (agricultural labourers) in a state of famine] … "
>
> (Mackintosh, in *Bernstein* et al., 1990, p.50, inserts added)

Markets, in other words, are institutions (as Chapter 7 noted) with varied organizational problems. They do not all operate in the same way but some complexities and difficulties are found in many developing country markets.

Figure 8.4 A petty rice trader is passed by many potential consumers on an urban street in Bangladesh. In the background, sacks of rice are being taken off lorries and into the warehouses of large traders (broker–wholesalers).

Figure 8.5 is an expansion of Figure 8.2 and shows some of the features overlooked by the Free Trade diagnosis. Major differences among traders, producers and consumers, and their relation to markets, are identified. The diagram also identifies common organizational problems and government aims.

Commonly in developing countries, commodity markets have a proliferation of petty traders co-existing with large private firms (see Box 8.2). The large traders usually have better access to information about prices, and much larger buying power. Many of the petty traders are dependent upon them for credit, and for access to information, transport and processing facilities. As a result, petty traders have to buy and sell at given market prices while the large firms can sometimes influence prices. We shall see that this type of market domination influences the outcome of deregulation in Somalia and Bangladesh.

Producers and consumers participating in foodgrain markets may also do so on different terms, with larger producers getting the best prices and the smaller producers and consumers getting the worst. For example, poor peasants may have to sell at harvest time when prices are low, and buy again, once their limited food stocks are exhausted, when prices are high. Rich peasants can choose when, and to whom, to make their sales, so they may be able to get higher prices for their produce.

Agricultural labourers and poor peasant households are usually forced to purchase food at high prices, because they have cash enough only for a few days' supply of food. Larger consumers can buy in larger quantities and choose when and where to buy, and thus buy at cheaper prices.

Price stability, encouraged by some forms of state intervention prior to deregulation, helped to minimize this inequality of participation in markets. Deregulation frequently led to greater price variability, worsening this form of inequality, tending to make the rich richer and the poor more vulnerable. This increased social differentiation occurred in all of our three case studies, but was particularly marked in Somalia.

The reasons for market regulation

All markets are regulated in some way so as to avoid fraud, cheating and chaotic behaviour. At a minimum, rules are established to standardize weights and measures, forms of transaction and systems of payment. Some staple food markets are made up of institutions which regulate themselves collectively, but this is rare.

It is usual for the state to provide a regulatory framework for markets. In India, for example, such a framework was outlined as early as 1898, and the body of regulatory law has been revised and amended ever since. Its stated objectives have been to:

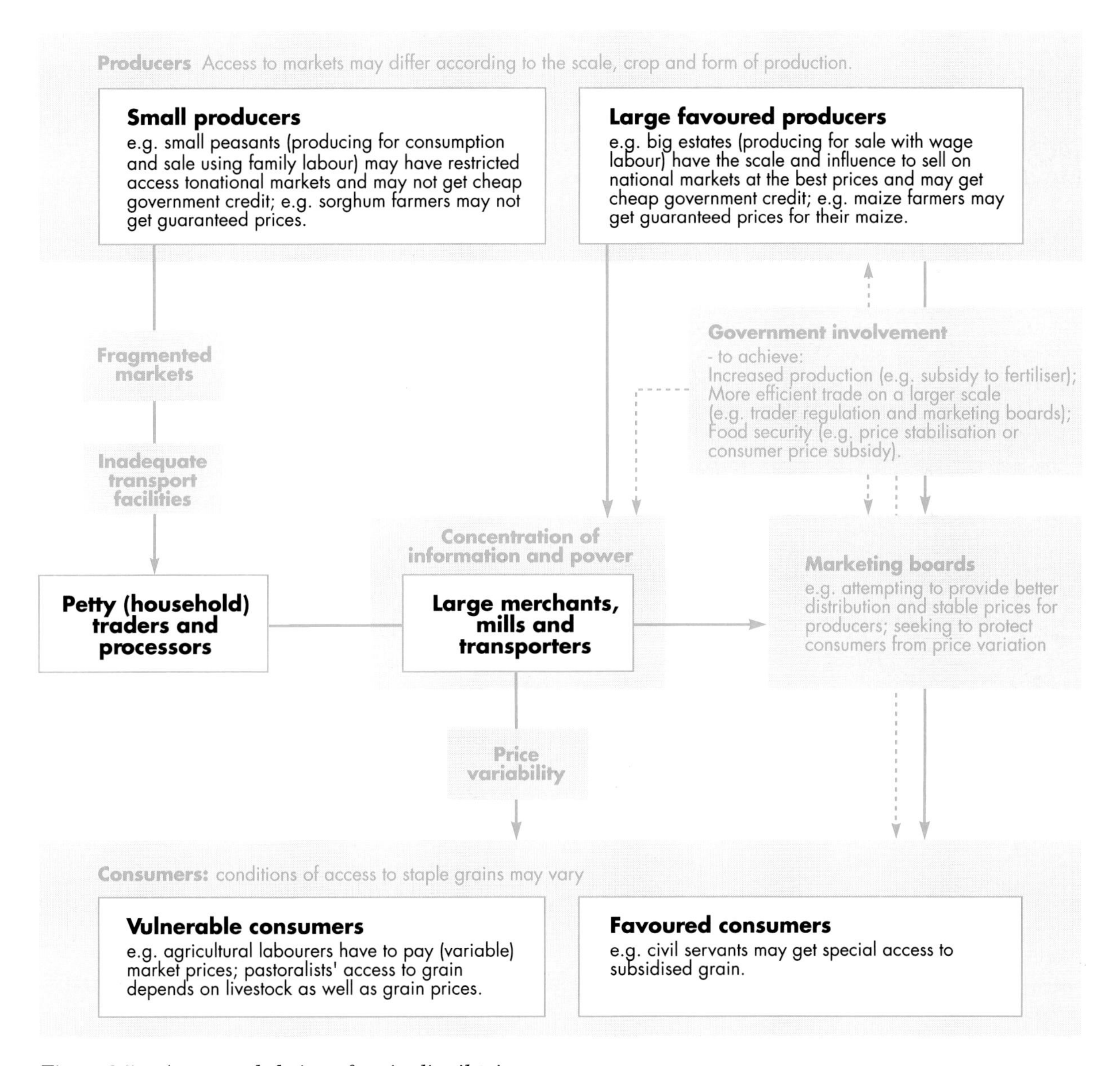

Figure 8.5 An expanded view of grain distribution.

- standardize and decentralize market administration
- supervise and centralize (within its market places) transactions and systems of payment
- control entry into trade and ensure vigilance over traders by means of licences
- provide information to traders and obtain fees from them.

In theory, a market regulated in such a way will benefit from reduced costs of acquiring information (about prices and quantities, and about other people and events in the system) and from reduced costs of transaction (that is, costs involved in creating, adhering to, and enforcing a contract). Trading in such a market will therefore be less risky.

State regulators have assumed that both the

Box 8.2 Market-making Institutions

Markets do not emerge out of thin air: they are constructed through interaction between market-makers, some of whom have considerable power, while others have very little.

Brokers negotiate a transaction between a buyer and a seller for a commission, but may also act as wholesalers.

Petty traders are small, frequently one-person, firms. Often the traders are women. They trade intermittently, and are usually tied (that is, subordinated to larger retailers), indebted, pauperized and unable to accumulate. Such traders are to be found everywhere, but they congregate in weekly or periodic markets.

Wholesalers aggregate crops into consignments for transport. They also commonly operate storage and transport facilities and, less commonly, undertake some processing.

Multinational corporations (MNCs) seek to maximize the profits they send to their shareholders overseas. In agriculture, MNCs are often associated with contract farming, where technical services, extension, credit and inputs are supplied to farmers in return for contracts in which quantities are tightly specified.

Marketing co-operatives are institutions owned by their users, and operated for their mutual benefit. Their efficiency is often hampered by organizational problems, lack of management skills and finance, and subordination of their operations to government priorities.

State marketing boards tend to have high costs, poor accounting, and inadequate finance. Their operations are often hampered when they conflict with other government policies. They tend to have low turnover, and to be over-centralized and over-staffed.

agricultural producers and the consumers would benefit from market regulation. Reduced risks and costs of buying and selling should lead to lower trading margins; that is, higher prices for producers and lower prices for consumers (Harriss, 1984).

There were similar good reasons for the establishment of marketing boards in sub-Saharan Africa. Private trade was much less developed than in South Asia. Throughout much of sub-Saharan Africa, the system of private trade was inadequate to handle planned levels of growth of marketed output from agriculture. Agrarian society had not invested adequately in trade. States had hitherto failed to provide sufficient transport, storage and processing facilities. In some places, ethnic minorities had captured trade, and were thought to act to the detriment of farmers and consumers. Private markets were thought to accentuate regional and social inequalities. Markets were fragmented, and prices fluctuated greatly.

Marketing boards were established, therefore, partly to integrate and develop these fragmented markets. In addition, governments of newly independent countries wanted to achieve rapid rates of economic development. To do this, they needed to be able to raise money and to influence rates of investment and consumption. Marketing boards could help achieve these aims, by influencing food prices and the relation between agricultural and industrial prices.

State marketing boards have since evolved to meet changing objectives. Institutions which some have derided as economic dinosaurs have sometimes in practice demonstrated considerable flexibility. They have managed to

- exploit economies of scale in marketing and transport (where unit costs fall as turnover rises)
- create competition for private monopolies
- succeed at times in stabilizing prices for producers and consumers
- control strategic food security reserves
- provide revenue for other crops and services.

Figure 8.6 (Top) 'some have considerable economic power': a large grain broker–wholesaler with labourers in his warehouse in Bangladesh. (Bottom) '... some have little economic power ...': a petty trader sells rice to a poor woman consumer in Bangladesh.

We can therefore balance this view, which sees *some* regulation of food markets as essential and potentially beneficial, against the Free Traders' view, which focuses on the costs and failures of regulation. An examination of the effects of deregulation in practice will help us to evaluate these two competing approaches to market analysis.

8.5 Assessing the effect of the medicine

The Free Traders noted, rightly, that some state interventions in staple food markets had generated budgetary crises for governments, and some had depressed producer prices below production cost levels. Deregulation was an attempt to address those crises. In their general approach to the revival of agricultural production, however, advocates of deregulation overlooked both the diversity of state intervention and the diversity of market institutions in the different countries in which it was implemented. Nevertheless, there have been some successes; in particular, some government budgets have been set on a safer course. There have also been some admitted disasters.

In general, structural adjustment was associated with deepening crisis in the 1980s. Overall economic activity among the sub-Saharan African recipients of structural adjustment loans, measured by the gross domestic product (GDP), dropped by 20% in the first half of the 1980s. The price of agricultural products relative to other products (that is, the 'terms of trade') went *down* by 34%, in the same countries over the same period, while imports per head shrank by 65% and exports dropped by 30% (Kydd & Spooner, 1990).

This is not at all as predicted. The policy reforms that were implemented in sub-Saharan Africa, as a condition of structural adjustment loans, included food market deregulation. Agricultural prices should therefore have *risen* in relation to other prices. The argument goes that since the diagnosis identified government intervention as the cause of low producer prices, rolling back the state should have allowed them to rise. According to the theory, gross domestic product (GDP) should also have risen after a while, as producers responded to higher prices by increasing their output.

We are therefore forced to draw one of two conclusions.

1 The deregulation has contributed to a deepening crisis, instead of helping to resolve it.

2 The potentially beneficial effects of deregulation have been overridden by external pressures such as world recession, or by institutional inefficiencies in the private food markets themselves.

The diagnosis by Free Traders ignored these factors.

But the experience of countries is very varied. To separate the successes of deregulation from its failures, we need to know more about particular experiences. We need to get behind the average measures of economic change, and to look at the specific differences in the organization of production and markets, and the varied purposes of government intervention in markets. It is these specific differences which explain success or failure.

Scouring the literature for evidence has unearthed a telling lack of information. It would seem that deregulation has been both prescribed and implemented in deep and irresponsible ignorance. What happens when theory, and policy-making based upon it, so far outstrip empirical enquiry? We explore the cases of Somalia, Bangladesh and Malawi, chosen only because they are relatively rich in empirical information, in an effort to find some answers.

In the following three case studies, we ask five simple but telling questions about food market deregulation:

- **Q** What happened to government institutions concerned with market regulation?
- **Q** What happened to prices, and which prices?
- **Q** How did private markets respond?
- **Q** What happened to production (and whose production)?
- **Q** Who were the beneficiaries and who the losers of deregulation?

In other words, we are evaluating the success of deregulation according to the Free Traders' objectives (increasing production, reducing government expenditure, improving institutional efficiency). Our questions also focus on changing market structures, and inequalities among participants in markets. For example, if prices have become more variable, we may find that some merchants and producers are able to benefit from this change, while some consumers and producers may be made more vulnerable.

8.6 Somalia: increased vulnerability, inequality and civil war

Deregulation in Somalia has been associated with increased inequality, increased vulnerability due to price instability, and apparently big gains for some merchants. In one sense, deregulation can be portrayed as a great success (Wehelie, 1989) because production of foodgrains, particularly maize, has increased substantially. However, only a select minority of producers have been able to benefit from this growth. The majority of producers have suffered from lower prices. There has thus been significant social differentiation following deregulation. State bureaucrats, who are also both merchants and maize growers, appear to be the major beneficiaries, and pastoral nomads and deficit farmers (farmers who also buy food), the losers. There have also been other important changes since deregulation.

At the time of writing, the government of President Siad Barre has collapsed and the country is riven by civil war. These events cannot simply be blamed on the deregulation of food grain markets, yet neither can policy reforms which increase social inequalities be separated from social and political tension.

Somalia has an extensive pastoral economy. Sixty percent of the population are nomadic herders. Livestock rearing contributes 35% of GDP and 80% of foreign exchange. A mere 20% of the population are settled farmers and their crops contribute only 8% of GDP.

Most agriculture is located in the southern part of the country, along two rivers, where about half the tilled land is under large-scale irrigation. Here maize, and to a lesser extent rice, are the food staples, grown mostly for local consumption and trade. Bananas are grown for export. In the high-rainfall north-western region of Somalia, similar settled agriculture is possible.

Farming in the rest of the country, however, is rain-fed and drought-prone, so sorghum cultivation predominates. Trading in sorghum is low in volume, localized and dispersed in pattern. By contrast, marketed maize flows to grain deficit regions, and to towns and cities, via traders in the capital city, Mogadishu.

Regulation

During the 1970s, Somalia experimented with a 'scientific socialism' which led to increasing tension with the US Government. The country slid from self-sufficiency in grain to a state of acute food aid dependency. 70 000 tonnes of grain were annually shipped to Somalia by the beginning of the eighties. The system of state trading has been identified as one of the factors responsible for the decline in production.

As part of the state's efforts at rational central planning, and in order to eradicate the 'deplorable exploitation system of man by man' (President Siad Barre), two state marketing boards were created in 1971 to replace private trade:

1 *The Agricultural Development Corporation (ADC)*, intended to control grain distribution within Somalia. This covered maize, because its production was spatially concentrated and easier to control, and, to a lesser extent, sorghum.

2 *The Ente Nationale de Commercio* (ENC: the National Agency for Trade) which distributed imported wheat and rice, increasingly obtained as food aid.

Crisis

Crisis became acute in the late 1970s. Official procurement prices were set so low that producers would not sell grain to the ADC. And the role of the ENC expanded dramatically because Somalia's overvalued exchange rate made grain imports relatively cheaper than internally

Figure 8.7
(Above) Waiting for a grain shipment to be unloaded.
(Right) Once ashore, state-controlled grain distribution frequently became a handout to the bureaucracy.

produced grain. There were also other factors at work. The droughts of 1973–5, together with the resumption in 1977 of armed conflict with Ethiopia over the Ogaden, led to a rush of refugees requiring international assistance in the form of food aid. Let us look in more detail at the role of the ADC in the 1970s, because it sets the scene for deregulation.

In practice, and despite its comprehensive legal monopoly power over domestic grain trading, the ADC could never control more than a small portion of the marketed grain: 31% of maize on average during the 1970s and 24% of sorghum, to a maximum 67 000 tonnes a year, a quantity comparable with the normal *monthly* distribution from the state trading system in some Indian states.

A dual market structure therefore prevailed. One portion of the marketed grain streamed into official channels, while the bulk flowed into residual parallel markets. As the ADC's prices fell increasingly below parallel market prices, its purchases dwindled and the police and the military were brought in to enforce sales to the ADC. Uncontrolled grain trading went literally underground into pit stores. Household food security is thought to have been severely threatened by this forced procurement.

Meanwhile the ADC's sales price was highly subsidized for a narrowly defined group of beneficiaries: a select set of urban, state-employed officials, plus the army and police and hospitals. Because subsidized sales were available only to this group of state employees, state food distribution constituted a handout to the bureaucracy. The government debt incurred in order to finance this subsidy fuelled inflation. The external balance of payments suffered, not only because of the overvaluation of the shilling (the local currency) but also because cattle exports to Saudi Arabia, Somalia's main source of foreign exchange, were banned.

Deregulation

Such were the precarious economic circumstances that in 1979 pitched Somalia into the arms of the IMF, who pronounced the socialist experiment a failure, ripe for dismantling. The IMF made its loan conditional on the immediate implementation of specific stabilization measures: the devaluation of the shilling, the liberalization of external trade and a reduction in public expenditure. By 1982, the World Bank had put together a structural adjustment package involving

- the liberalization of pricing for food staples
- the removal of movement restrictions on inter-regional trade, and of quantity restrictions on private storage
- the removal of controls on private trade
- a reduction in expenditure by (and a change of role for) the ADC.

Deregulation transferred almost all grain trade from the ADC into private hands. Up to 1992, the ADC was continuing to set procurement prices and was trading about 17 000 tonnes of grain, largely maize, to a target group limited to a narrower band of its earlier beneficiaries. A dual pricing structure persisted, but the 'residual' parallel market was now effectively 'the market'.

What happened to private markets?

The structure and functioning of private grain markets varies regionally and seasonally. Firms have complex combinations of trading activity. Entrants to private trade since deregulation are thought to be of two types: young traders with little capital, and experienced traders who fled the grain trade during the 1970s. Trade is unspecialized, with the largest grain traders also being livestock exporters.

There are thought to be few 'entry barriers' to trade. That is, it is easy and cheap to get started. Local markets are not thought to be monopolized by a few traders nor are they dominated by certain ethnic groups, tribes or one gender. As private trading has expanded there has been a noticeable increase in the supply of grain to rural periodic markets.

Grain markets, however, are generally underdeveloped and display a number of problems for society as a whole. One is poor and uneven access to market information, so that those who control information have a great advantage. A second problem is the high and varied costs of storage and transport, so that it is hard to separate legitimate marketing cost differences from deliberate price manipulation. Traders find it hard to get credit to finance their businesses. As a result, only a small number of merchants have the capacity for long term grain storage and for long-distance inter-regional trading, and this may give them some monopoly power (as is known to be the case, for example, in maize storage in the capital, Mogadishu).

Prices

Prices for grain are thought to have increased, with a decline in the influence of low official prices. This was in line with the predictions. The structure of grain prices has, however, changed in favour of maize. Seasonal price variation is now very large, by comparison for example with West Africa. In 1987–89, maize prices varied from 33% of average to 303%. In the same period, sorghum prices varied from 44% to 184% of average. These wild fluctuations are least in Mogadishu, where information, credit and storage are concentrated and the vestiges of state intervention are strongest.

Analysis of the spatial variation of prices shows that, despite large numbers of traders, markets across the country are not well integrated. Price changes in one market may not be reflected in another. This suggests that grain markets are not operating effectively.

One of the functions of a market is to transmit information about extra demand in one part of an economy to another region where there may be available supply. This can only occur if information about prices is transmitted, and goods can follow, from one market to another. Such integration, however, requires transport. The government of Somalia does not have the capacity to build roads, nor does the private sector have the resources to provide the lorries required. Uncertainty about the future role of the ADC may also be contributing to the problems. Private investment in trade may be limited by the fear that a state trading monopoly may be reimposed.

Production

Deregulation has been associated with significant increases in grain production in the early 1980s. Maize production in particular expanded at an average annual growth rate of 18 percent between 1982 and 1986, and by 1989, Somalia was producing a grain surplus.

Beneficiaries

In addition to the traders who benefited from the increase in trade, the 11% of agriculturalists who dominated the cultivation of maize for the market also increased their returns. They were no longer obliged to sell much, or any, of their produce to the ADC at low prices. Hence after deregulation they could store their grain and sell when prices were high, and thus benefit from the big price variation.

Figure 8.8 Despite Somalia's attainment of self-sufficiency in grain production by the late 1980s, deregulation had ambiguous results. Increasing inequality, for example, was among the causes of the civil war.

Much of the increase in grain production was the result of an expansion of the area under cultivation. Newly irrigated land was given on long-term lease to 'influential people' — the very same civil servants and military officers who had access to state food subsidies. It would seem that the major beneficiaries of deregulation were state bureaucrats, who enhanced their power not only as consumers but as traders and producers.

Losers

Nomadic pastoralists and long-settled sorghum farmers are the losers from deregulation. Their selling prices have declined relative to essential purchases. These categories include the majority of the population. Even the majority of maize producers in the Shabelle valley in Southern Somalia were in net food deficit, and therefore were forced to purchase maize on the open market at higher prices than before (Olsen, 1984; Abikar, 1989; Wehelie, 1989).

To sum up, those groups affected by adverse price changes such as those between meat and grain, have suffered, along with poor producers of even favoured grains such as maize, as a result of deregulation.

8.7 Bangladesh: 'slow, modest and highly uneven' progress

In Bangladesh, deregulation is only partly implemented, and has had mixed results. There has been some success in the avoidance of severe consumer price increases, but this has arisen from re-regulation, rather than from a simple liberalization of markets. Attempts to reduce the cost of food subsidies to better-off urban consumers had initial success, but it has not been sustained. Other elements of deregulation appear to have had a limited impact. There has been some increase in agricultural production, along with increasing inequalities.

The 110 million population of Bangladesh makes it the eighth largest country in the world, and the GDP in 1990 of $170 per head put it with sub-Saharan Africa, among the poorest economies. It is a predominantly agricultural economy, with the bulk of livelihoods associated with the production of rice and, to a lesser extent, wheat. Agriculture generates 40% of GDP, and 75% of agricultural income is derived from crop production.

The overall experience of structural adjustment has been disappointing. The World Bank, one of the authors of policy reform, notes that their advice helped to ' ... steer a weak and fragile political/administrative system through a difficult period. However, the pace of change and progress to date have been slow, modest and highly uneven.' (World Bank, 1990b). Despite its role, the Bank comes to pessimistic conclusions about the overall efficacy of structural adjustment reforms: 'At the end of the 1980s, the situation has not changed materially, and Bangladesh continues to face many of the macroeconomic and sectoral issues it confronted in the early 1980s.' (World Bank, 1990b). In other words, structural adjustment reforms, including market deregulation, have had little impact.

Regulation

The structure of government intervention in grain markets in Bangladesh contrasts with the position in sub-Saharan Africa. The greater development of markets in South Asia was noted earlier. This corresponds to a much greater proportion of grain being sold, rather than consumed directly by the producing household. There is, in other words, a greater commercialization of agricultural output, and grain trading is more established. There are many specialized traders and the infrastructure of transport and communications required for distribution and exchange is more extensive and sophisticated.

A second contrast with sub-Saharan Africa concerns the role of government in food distribution. State trading in Bangladesh was initially intended to ensure supplies to city populations, rather than the development of production.

During the Second World War the British Government of India also ruled the territories that are now Pakistan and Bangladesh. In response to the Bengal famine of 1943 it established

Figure 8.9 Market infrastructure and transport facilities in Bangladesh. (Above left) Trucks, shops and warehouses. (Above right) Communications and security. (A grain broker waits for custom with important market facilities to hand — a telephone, a safe, and a ledger). (Left) Boats and market places.

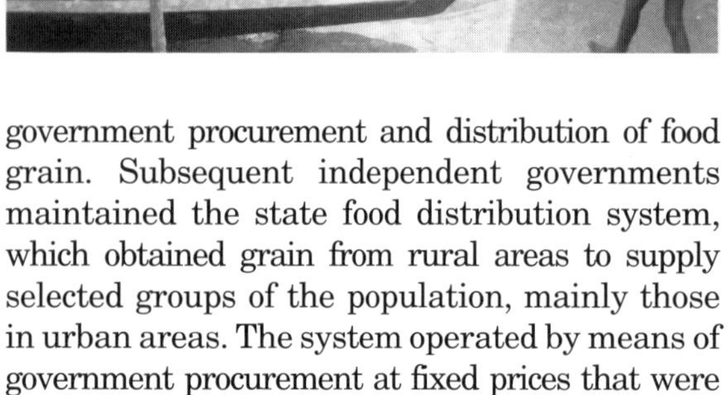

government procurement and distribution of food grain. Subsequent independent governments maintained the state food distribution system, which obtained grain from rural areas to supply selected groups of the population, mainly those in urban areas. The system operated by means of government procurement at fixed prices that were lower than market prices. It was supported by:

- restrictions on the movement of grain from one area to another
- restrictions on the quantity of grain
- restrictions on the length of time traders were allowed to store it
- licensing of traders.

Crisis

During the 1960s and 1970s, local procurement had gradually been overtaken by food aid as the principal source of grain for the public food system. The political sensitivity of food prices had also encouraged an increasing level of subsidy in the ration price. Key groups were supplied with grain at 10–30% below the market price. A notable group was the military (who took part in a series of *coups d'état* between 1975 and 1982 and influenced all changes of government between 1971 and 1991) but also included were the police and government employees.

These changes contributed to the growth of an 'indefensible' government role in foodgrain markets. Nearly 10% of government expenditure was being devoted to the provision of cheap food to these relatively better off groups in the urban population. Almost all of this cheap food was obtained as food aid.

The level of consumer subsidy was also reducing the general price of foodgrain in Bangladesh, including the price paid to peasant producers. About 12–15% of total food consumed in the economy was being sold at a reduced price, and the demand from the consumers for that grain was thus being diverted away from the market.

The state food system was also implicated in one catastrophic breakdown. In the period of adversity prior to a serious famine in 1974, restrictions on grain movement are thought to have exacerbated the crisis, and the managers of the state food system took a decision which almost certainly increased the severity of the crisis. Faced with a reduction in food imports, they decided to cut back the already limited distribution of subsidized food to the rural poor in order to maintain supplies to the cities.

In itself, this decision to support the better-off groups at the expense of the poor is questionable. But in addition, it served notice to private traders that foodgrain distribution in Bangladesh was no longer within government control. The slow rise in grain prices then accelerated. Prices reached unprecedented heights as those with grain held supplies back, in expectation of ever higher price gains.

Deregulation and re-regulation

Market reform has a longer history in Bangladesh than in many developing countries. It started in 1976, with advice from the US Agency for International Development (USAID), partly in response to the famine of 1974. The 1976 agenda laid out the main lines for reforms, which were gradually carried out throughout the remainder of the 1970s and the 1980s. During the 1980s, further rounds of reform were introduced, in order to reduce government storage of grain, and subsidies, and to privatize fertilizer distribution and irrigation pump sales and hire. Most of these reforms were either loan conditions imposed by the World Bank and the IMF, or were conditions imposed upon long term food aid contracts by USAID.

Reform of state intervention included four significant changes:

1 reduction of the state food distribution system, including the ending of grain movement restrictions and compulsory procurement

2 termination of subsidized food distribution to urban populations

3 introduction of Open Market Sales from government foodgrain stocks as a means of stabilizing market prices

4 the proposal to raise foodgrain prices in Bangladesh towards world market price levels.

Whereas items 1, 2 and 4 can be categorized as deregulation, item 3 is more correctly termed re-regulation, because it provides a new role for government in foodgrain markets, as an agent using sales to influence market price levels. In other words, deregulation is incomplete, but the *form* of regulation has changed somewhat.

The overall objectives of policy reform in agriculture were:

> " ... to increase foodgrain production and productivity, to the end of attaining self-sufficiency and a more equitable income distribution. To achieve this overriding objective, the Government's strategy focused on reforms pertaining to input and output pricing policies, improvement of the

> efficiency and equity of the public food grain distribution system, and agricultural credit administration. The Government also recognised the need for greater involvement of the private sector in the delivery of modern agricultural inputs and irrigation equipment."
>
> (World Bank, 1990b, pp.iii–iv)

During the 1980s, the main principles of these reforms were implemented:

- the proportion of food distributed by the private sector increased
- the rate of subsidy on grain distributed by the public sector was reduced
- the proportion of publicly distributed grain going to better-off groups was reduced.

The military were an exception to reduction of public sector subsidy. Their food supplies became the centre of a row between the governments of the USA and Bangladesh, which was resolved on paper whilst retaining the subsidy to the military.

The contribution of food subsidy to government expenditure had been reduced from a peak of 10% in 1982 to 1% in 1987 (World Bank, 1988) but rose again in 1989 (World Bank, 1990b).

Price

In recognition of the increased costs of production imposed by the removal of fertilizer and pump subsidies, the government attempted to support producer prices. It did this by announcing a procurement price at which it was (in principle) willing to buy grain after the harvest (when prices fall to their lowest levels). This is another element of re-regulation. Average prices have risen to about world market levels.

Although fluctuation of grain prices appears to have increased in the period after deregulation (Ahmed & Bernard, 1989), severe price rises seem to have been avoided. Food crises in 1979 and 1984 did not lead to famine, partly because price fluctuations were contained. This, however, can be attributed more to re-regulation rather than de-regulation. One of the proposals included in the 1976 deregulation agenda was for a system of 'open market sales', allowing the government to sell food directly, from the backs of lorries, to traders and to consumers, in order to reduce famine-threatening price rises. Hence the reform gave government greater capacity to influence the market, but changed the form of intervention rather than simply rolling back the state.

This new capacity to influence prices was probably an important factor in the containment of food crises in 1979 and 1984. Clay (1985) argues that improved transport, greater government financial reserves that facilitated rapid ordering of food imports, and generally increased government capacities were equally important elements. A massive expansion of rural credit also gave rural households greater capacity to withstand crisis.

Production

Production of grain has increased in Bangladesh, but the rate of increase of production declined in the latter half of the 1980s (World Bank, 1990b). From a position of dependence on imports and food aid for 12–15% of total consumption, the economy was nearing self-sufficiency in grain in 1990 for the first time since Independence, in 1971. This increase in output is only partly attributable to deregulation of grain markets. Producer prices are still very low at harvest time, particularly in the more remote and backward areas of the country and for poor peasants. Many poor peasants get only a fraction of the market price because they are tied into credit and tenancy arrangements that reduce the price they receive to 50–80% of the market price.

Government attempts to support producer prices have had very limited influence. Instead, the key factor in the increase of output has been the gradual increase in multiple cropping as a result of the spread of irrigation and the adoption of 'Green Revolution' cereal varieties. It is possible that the deregulation of irrigation pump supply gave a boost to the expansion of irrigation at the end of the 1980s, but that is probably all that can be claimed as a direct effect of deregulation on output.

Benefits and losses?

During the several rounds of market change in Bangladesh, some groups of traders were caught out by market changes and forced into bankruptcy. There were new entrants in some roles, using capital from the repatriated earnings of labour migrants to the oil-rich economies of the Middle East. New forms of procurement introduced with deregulation provided huge windfall profits for selected grain millers who were able to enter into contract with the government, and these profits were shared with selected food ministry officials.

Rich peasant producers, unencumbered by tenancy and onerous credit contracts, have been able to increase their returns as a result of increased price levels. Poor peasants and landless labourers, who make up the huge majority of the rural population, have suffered as a result of increased prices, and increased price variation, because they buy more than they sell.

8.8 Malawi: errors of ignorance

Writing about Malawi, Uma Lele, a World Bank economist, admits that serious mistakes were made in deregulation. This, she says, resulted in the impoverishment of poor households and speculative gains by a few traders. There were

> "hasty decisions to dismantle policies and cut back on investments in grain marketing interventions, fertilizer subsidies, and the National Rural Development Program. The implications of these decisions were not adequately considered before implementation."
>
> (Lele, 1990, p.1217)

Figure 8.10 While some private traders benefited from deregulation, there is a consensus that reform has done more harm than good in Malawi.

These mistakes arose from a failure to understand the structure of agricultural production and distribution, and the positive role of government intervention.

According to Lele, Malawi has a 'divided agriculture' with 'a rapidly growing estate sector and a subsector of smallholders mired in poverty'. The smallholder sector, which concentrates on subsistence staples for local distribution, accounts for 80% of agricultural production. The estate sector, in which the holdings of the President's family alone are thought to account for about 20% of production, concentrates on producing export-quality tobacco, sugar and tea. Estates taking up only about 2% of cultivable land account for 20% of production and for 96% of Malawi's foreign exchange earnings.

Lele also notes that virtually all of the early post-independence estates were operated by Malawi's President Banda and a few select members of the élite. These estates arose partly as an attempt to create a middle class. Estates have the right to grow certain export crops, to employ labour, to sell produce at world prices, and to rent out land to tenants. Smallholders did not have the right to grow the most profitable crops.

In recent years, the boundaries between the estate and smallholder sectors have begun to break down and there has been differentiation within the smallholder sector. A small, 'more market-oriented' minority of the smallholders is capable of taking risks, and 'a majority of poor householders living below subsistence ... experience stagnation or near paralysis' (Lele, 1990). Nevertheless, prior to deregulation, the marketing arrangements for the estate and smallholder sectors were distinct. Smallholder agriculture was forced to sell to a state marketing board at low administered prices (between one third and one half of market prices) but benefited more from subsidized inputs. Estate agriculture has always operated under private markets.

Regulation

Prior to deregulation, the state had monopoly control over the marketing of staple foods (and export crops) through its marketing board ADMARC. ADMARC was made responsible for price stabilization, food security and storage, the subsidy of consumer prices and agricultural inputs, and for the creation of financial surpluses. Accordingly it operated stable, guaranteed, pan-territorial prices. Having banned Asian traders, and built up a physical infrastructure for marketing, the Malawian Government did not, nevertheless, encourage the development of an African private marketing system. This is in contrast to Somalian marketing.

Despite ADMARC's multiple objectives, with marketing and development goals, it was remarkably successful during the 1970s. During this period, there was an expansion of agricultural production. ADMARC stabilized prices, taxed the smallholder sector and transferred resources from this sector to the estates. Its financial record was one of the most successful in Africa.

In the early 1980s, however, ADMARC's multiple burden combined with external events to curtail its successful record. Transport costs escalated, primarily because of South African-backed guerrilla activity in neighbouring Mozambique and a consequent need to construct storage facilities in northern Malawi. The costs of increased security reserves of food became overwhelming. World market prices for Malawi's exports were falling, and ADMARC's capacity to repay the necessary loans was compromised.

The organization was also overstaffed. ADMARC's profits fell by 50% in 1983–4 and by 1986–7 had deteriorated into substantial losses. ADMARC's weakened condition, however, was a minor component in the portfolio of reasons forcing Malawi towards the IMF and World Bank in the early 1980s. The major reason was a growing balance of payments deficit.

Deregulation

In the event, familiar adjustment conditions were imposed by the World Bank over the period 1981–3, during which US$224 million was lent.

- ADMARC was to be remoulded, to become a buyer of last resort. The price support role that this implied was, however, to be temporary, and was to be phased out in due course.

- Subsidies to smallholders were to be eliminated.
- ADMARC's financial performance was expected to be turned round by such changes and by employment reforms.
- Smallholder crop purchases were to be privatized, not only to raise prices, but also as a means to improve efficiency in ADMARC, which would have to become a competitor.
- 'Farmers' Clubs' or private traders were to be encouraged in the first instance (that is, in the early 1980s) to perform marketing operations in small, seasonal bush markets, at prices that would compensate them for taking on the role of bulking as agents for ADMARC.
- ADMARC would remain as the buyer of last resort at minimum administered prices.
- Large-scale corporate businesses would be free to export.

What happened to private markets?

Many traders rapidly filled the vacuum left by the retraction of ADMARC's operations in bush markets. Entry into trade seems to have been unproblematic. Trade can easily be combined seasonally with agricultural production. But it is much harder to fund the resources to combine trade with transport or with the ownership of processing facilities. Problems in the evolution of the private marketing system included:

- costly, partial and slow access to information
- underdeveloped storage structures (leading to decentralized, individualistic and locally monopolized markets, sometimes physically located in the traders' homes)
- poor access to credit (whether from banks or informal money lenders) which traders need in order to increase their turnover
- chains of delayed payments undermining traders' finances
- the lack of state capacity to implement the regulatory laws.

Prices

As in Somalia, price fluctuations were much larger than could be justified by the costs and risks of storage. Licensed trade became characterized by unstandardized and manipulated weights and measures, by widespread trade with smallholders on terms outside those legally specified, and by reluctance to file accounts. There was also much unlicensed and unmonitored trade.

Production

Smallholder production has stagnated and there have been increasing bankruptcies among operators of tobacco estates (Lele, 1990). As prices have risen, the 'supply response' from the small-scale agricultural sector has been pitifully low. That is, price rises have not encouraged much more marketed output. Box 8.3 explains why this may happen. 'Non-price' constraints such as the following may often have a greater influence on output than prices:

- legal restrictions on cropping patterns
- the rapid miniaturization of holdings, as land is divided and sub-divided
- the historical shift away from commercial crops towards maize growing, to ensure food supplies
- 'shocks', such as drought, and the impact of war.

Beneficiaries and losers

The beneficiaries of deregulation have undoubtedly been private traders, although they are constrained by lack of credit and information. The larger smallholders in central and southern regions have benefited from increased competition and increased prices.

The losers include smallholders in remote regions where operating costs make private trading unprofitable, and where ADMARC's depots have closed, depriving farmers of alternatives.

Box 8.3 Why agricultural production may not respond even if prices rise

One reason why deregulation may not necessarily increase agricultural output is that agricultural production and sales may not respond to price rises. This is called, 'low price elasticity of supply' and has been observed in many Third World economies (Lele, 1990). It arises because peasant households are not capitalist enterprises, and because peasant farmers find it harder than commercial farmers to obtain inputs.

If a significant part of production is for own consumption, then increased prices may provide little incentive to sell more. If, as in parts of sub-Saharan Africa, women contribute most of the labour for food crop production, but men have a disproportionate say over how the income from sales gets spent, then again, increased prices may not encourage more production. On capitalist farms there may be a direct relation between prices and output, but it may be much more indirect in a peasant household.

Furthermore, factors such as the availability of credit, and of consumer goods to buy with the proceeds of sales, may be more important in determining output and sales than price. Even if peasant households wanted to respond to price incentives, their ability to do so may be limited by the inability of peasant farmers to control natural conditions. Rainfall, frost, insects and soil fertility all have greater influence on output than in the industrialized, irrigated agriculture of large capitalist farms.

Conspicuous losers are the 35% of smallholders who currently operate less than 0.7 hectares, and who cannot meet their own subsistence requirements under any technology. These food deficit farmers have in the past undergone a forced commercialization (described by Ghosh & Bharadwaj, in *Bernstein* et al., *1992*, Chapter 7). They were forced by financial exigencies such as the need for taxes, clothing, ceremonies, and so on, to sell maize immediately after the harvest They then purchased maize for consumption late in the marketing season before the harvest. The abolition of country-wide, seasonally stable prices means that they now have to sell at much lower prices than they can buy at.

Their market participation is not only non-optional and non-speculative, and therefore forced, but also operates on adverse terms of distress. These terms are made worse by the concentration of storage infrastructure in urban centres, as we saw in the Somalian case. There is evidence that such deficit producers are forced to undertake preharvest labour, paid in kind, to the detriment of work on their own smallholdings.

Other losers include the large-scale maize processing companies partly or wholly owned by ADMARC. These have been temporarily starved of their previously secure supplies and their operating margins have been squeezed by the rise in producer prices. Their system of information and contacts on private markets were poor, and the costs of dealing with large numbers of traders are far higher than before. They have tried to solve their problems by finding tied sources of supply, and setting up contracts with estates and large smallholders (Christiansen & Stackhouse, 1987; Scarborough, 1989; Mkwezalamba, 1989; Lele 1990).

Errors of ignorance

There is agreement that the early rounds of structural adjustment and liberalization made mistakes. Thus, World Bank economist Uma Lele writes:

> "The first three SALs [Structural Adjustment Loans] ignored many of the non-price constraints to smallholder growth which had become evident from the rural development experience in the decade preceding the start of the adjustments. The initial SALs achieved their macroeconomic objectives of an improved external and internal balance. However, they had a

disappointing effect on growth and an adverse effect on the poor."

(Lele, 1990, p.1207)

Lele goes further, admitting very serious failings of market liberalization. She identifies three lessons of liberalization in Malawi. First, there were too few traders to take up the grain marketing and distribution role of ADMARC. Those that were able to operate were given free rein to benefit from the extreme price fluctuations following withdrawal of government price stabilization, because the Government had neither the capacity to follow what was happening, nor the ability to do anything about it. Second, the influx of refugees from the war in Mozambique increased demand for maize to such an extent that prices 'went through the roof', with 90% of households in the southern region running out of maize and being forced to sell their labour in order to buy food. Third, it was clear that even without external shocks, such as the influx of refugees from Mozambique, there was an important role for government in protecting the poorest households from price variation. These damaging admissions about deregulation in Malawi are worth quoting:

> "First ... imposing liberalization when a government has little alternative may serve to reduce the Government's role, but may not be accompanied by market development, which takes more time to achieve. Having little access to finance, market information, or transportation, Malawian traders were unprepared for liberalization. ... Indeed, since the mid-1970s, when the Government sharply curtailed the presence of Asian traders in rural areas, donors had mainly encouraged expansion of ADMARC's operations to alleviate the weaknesses in the Malawian private sector. They had not supported the development of the indigenous private sector. ... To meet food security needs of many vulnerable households in periods of shortages, governments also need to have information on traders' speculative activities to ensure that complementary public action is taken in a timely manner. The Malawian Government was unable to obtain such information without either licensing and information-gathering systems or administrative ability.
>
> Second, external shocks such as droughts or other unanticipated disasters can make the timing of liberalization a disaster. In Malawi's case liberalization coincided with a large influx of refugees. As a result, ADMARC ran out of maize stocks and market prices of maize increased sharply, reaching three or four times the official price. ADMARC could then purchase less grain in the market for resale. This adversely affected food security and incomes of a large number of food-deficit households. According to subsequent field surveys, 90% of households in the southern region ran out of their own maize stocks and had little cash for buying food... Traditional support systems amongst households broke down, and the only option for many poor smallholders was to accept employment on more prosperous farms to pay for maize purchases. This short-term solution, however, tended to interfere with the preparation of their own plots, thus jeopardizing their food supply for the next year.
>
> Third, even in relatively normal circumstances free of shocks, grain market liberalization without adequate stocks which the Government can use to defend prices can make food-deficit households extremely vulnerable to fluctuations in market prices and supplies. The Malawian experience has taught donors that these households must be protected through some control of food supplies and prices. ... ADMARC needs to hold stocks and be a buyer of last resort to stabilize prices."

(Lele, 1990, pp.1212–1213)

8.9 Conclusion

The conclusion we draw from these case studies is not that deregulation always fails. What fails are the assumptions:

- that market forces always work well
- that state regulation is generally the cause of economic crisis.

Despite the scarcity of reliable information, the cases reveal that intervention always takes place in the context of real markets, which are complex institutions, with different actors operating with different levels of information and market power. Some of the failings of deregulation arose from ignorance of this context.

Specifically, we can draw the following conclusions from the case studies:

1 Deregulation has more often than not increased inequality. In some cases this has been associated with riots (Zambia) or governmental instability (Somalia).

2 The evidence on increased economic growth is limited: production has stagnated in Malawi, but is rising in Somalia and Bangladesh, though not necessarily or solely as a result of deregulation. The bargaining power and food security of the poor have not increased, especially those dependent on markets.

3 In some cases, as in Malawi, deregulation resulted from a misunderstanding of the function of government intervention. Serious errors have been made in deregulation, simply because of this misunderstanding.

4 In general, deregulation overlooked (and has partly been defeated by) the diversity of market and production conditions. There have been small successes, however, where care has been taken to relate innovation to existing market structures and to needs, as in the case of the re-regulation of Bangladesh markets through open market sales.

Looking back over the experience, it is possible to suggest three wider conclusions:

- that structural adjustment was undertaken in ignorance of local conditions
- that the reformers got it wrong
- and that they are now having to eat their words.

Adjustment in ignorance

There are several ways in which the deregulation of grain markets, and adjustment more generally, appear to have been undertaken in some ignorance.

The first concerns the process of deregulation, which has displayed great *diversity*, reflecting the variety of economic conditions and types of government intervention in different countries. No single model of change predicts the effects of deregulation. The generalizations and prognoses of the theory of deregulation presuppose over-simple, ideal conditions that are rarely found in practice.

Deregulation has, secondly, proved a much *longer* process than was predicted. The rapid improvements expected have not materialized. As the World Bank put it in the case of Bangladesh, ten years later, and much adjustment further on, all the same problems were still there. The process was slow, and was prolonged by the political and economic instabilities of the 1980s.

A third area of ignorance concerns the role of gender in trade. Even though women dominate trade in some areas of the developing world, and are central to agricultural processing in even larger areas, gender divisions of labour and the implications of adjustment for them, are neither mentioned nor addressed by the advocates of deregulation.

Figure 8.11 shows an example of the sexual division of labour in Bangladeshi markets. In contrast to much of Africa, most grain traders are male and the work of women is generally hidden. Until recently, most rural grain processing in Bangladesh was carried out by women within the homestead. The spread of mechanical milling and commercialization of grain exchange has shifted this work from homestead to market place, and from women to men. Small-scale winnowing (separation of rice husks from the grain after milling) is still often carried out by poor women, working either as wage labour or as part of a small household enterprise.

Figure 8.11 Small-scale winnowing in Bangladesh.

Getting it wrong

The central slogan of the twentieth century excursion into free trade reform was 'get the prices right'. In retrospect, we can see that the reformers got so much else wrong, that even where they got the prices right, it did little to achieve what they wanted.

The most glaring misperception was of markets. There was a series of invalid assumptions about markets, most important of which were the following.

1 The skills, investment and infrastructure existed to replace state trading institutions. In many economies they were not there and could not be conjured out of thin air.

2 Markets were somehow independent of their political context. In practice political and policy stability is an important precondition for a rapid evolution of markets.

Would-be beneficiaries were unable to take advantage of deregulation because of:

- lack of information, credit, and transport
- the spatial fragmentation of economies
- the incomplete nature of markets and market participation.

The variable impact of deregulation was not foreseen because the Free Traders focused primarily on the failings of state institutions, and considered only very simplified abstractions about the operation of markets, and the different livelihoods connected to the market. Subsistence producers were not influenced by deregulation at all, except adversely as net consumers of purchased food; small commercialized producers benefited; agricultural wage workers lost. In retrospect, the importance of non-price constraints on agricultural output has had to be recognised.

Eating their words

By the early 1990s, the Free Traders' critique of the state did not appear at all convincing. The proponents of deregulation were themselves less fervent about 'rolling back' the state. Indeed, since deregulation it had become apparent that in certain cases more, not less, state regulation had been required. For example, in Zimbabwe the marketing board had to invest in physical market infrastructure in remote regions of smallholding agriculture.

In many economies where state intervention was curtailed by reform, a need for new forms of intervention rapidly emerged. In other words, deregulation frequently resulted in re-regulation. For example, state trading is necessary if commercial agricultural production is to develop in areas where no private market is possible. In many parts of sub-Saharan Africa, private markets are very slow to develop because there is little capital, no infrastructure, no roads, and transport costs are high.

Finally, Free Traders were having to accept that there are important functions for the state in securing the protection of the public interest during deregulation. Namely:

- It is essential for the reduction of poverty in many developing countries to establish food price subsidies targeted to the most vulnerable groups. In many industrialized countries, too, consumer subsidies constitute an important element of poverty alleviation programmes.
- Price stabilization is an important function for government. Control of price increases may be a necessary part of attempts to prevent disastrous food crises. An attempt to enforce producer price 'floors', with state trading agencies acting as buyer of last resort, may be necessary in order to reduce the risks of new technology, and to protect poor peasants from total impoverishment.
- The regulation of trading, through the provision of market sites, maintenance of trader safety, enforcement of quality standards, provision of standardized forms of contract, and dissemination of information about prices and supplies, may improve efficiency and reduce the costs of trading.

This chapter has shown that the Free Trade view associated with Structural Adjustment is too simple a guide to be the solution to the complex problems of agricultural trade and distribution occurring in the developing world. The deregulation of grain markets, which it informed, has proved to be a more complex process than was anticipated, because grain distribution is more varied and complex than the Free Traders expected, and deregulation has had mixed outcomes. There are some apparent achievements associated with deregulation and re-regulation, but they have to be balanced against the exacerbation of inequality and vulnerability which have also resulted. Overall, we must have reservations about a process which gives a prominent position to reformers driven by so simple a palliative for complex, ill-understood problems.

Summary of Chapter 8

1 This chapter has examined one of the central elements in the structural adjustment programmes of the 1980s: the deregulation of food markets. Attempts at deregulation have been based on the diagnosis that there has been too much state intervention and that the treatment of the problems of food production and supply should be left to the market.

The central arguments of the chapter, however, refute such claims. Food markets are shown to be so diverse and complex that no single prescription can be made. Similarly, the success of food markets appears to depend less upon reducing state intervention than upon finding new and more appropriate ways for the state to intervene.

2 The Free Trade view of state regulated food markets is that they are a drain upon government revenue, they encourage parallel markets, they increase food insecurity, discourage producers, and distort prices. All these ailments were put down to failures of the state. In practice, however, the outcomes of deregulation have been very different. Despite a lack of detailed information, three case studies illustrated the diversity of markets.

3 *Somalia* — deregulation had very mixed outcomes; there was an increase in maize production to self-sufficiency. While this benefited certain producers and traders (principally government bureaucrats), it adversely affected pastoralists and sorghum farmers who together make up the majority of the population. While prices generally rose, as predicted, they also fluctuated wildly.

Bangladesh — Here there is a greater commercialization of agriculture. Reforms began after the 1974 famine and continued through the 1980s. Many of the improvements in production and prices have not come about, however, simply by deregulation. Re-regulation appears to have been the crucial element in reforms. The vast majority of producers and consumers have seen their situation worsen as a result of grain price rises.

Malawi — is an example of what can go wrong with deregulation when specific markets are misunderstood. There were too few private traders to take over from the state. Increased demand for grain sent prices sky high. The removal of government stocks gave no protection to the poorest households. In retrospect, reformers acknowledged that such shock treatment was misguided.

4 The chapter concludes that food markets are too diverse to permit any blanket prescriptions about deregulation. In addition, it seems that deregulation will not succeed without the continued involvement of the state in new forms of market intervention.

In short, food markets are difficult to predict in advance. The experiments in deregulation of the 1980s have shown that there is no one solution to food insecurity. Further, they have shown that deregulation is not the panacea espoused by Free Traders. Rather than being left to their own devices, it seems that food markets need to be understood case by case before food insecurity can be alleviated.

POLICY AS PROCESS

CHANGING PATTERNS OF PUBLIC ACTION IN SOCIALIST DEVELOPMENT: THE CHINESE DECOLLECTIVIZATION

GORDON WHITE

9.1 Introduction

The last part of this book brings together themes discussed in earlier chapters and makes clear what we meant by 'policy as a process'. This will be done by examining two case studies in detail. This chapter deals with the first case: the dramatic events of rural decollectivization in China in the 1980s.

This chapter is about a transition between two different patterns of public action. It traces the change from a highly 'statist' approach to development (commonly referred to as 'communist' or 'state socialist') to a variant of 'market socialism', the state-socialist model of action. In effect, in this model state action is a historical substitute for the role played by private entrepreneurs in the capitalist path to development. Public action replaces private action in two basic ways: first, through the direct involvement of the state in the economy through nationalization of productive assets, and second, through large-scale state sponsorship of collective enterprises, most notably in agriculture.

This highly statist approach to development has been subject to the kind of criticisms and sparked off the kind of debates which have surrounded the general issue of public action, in other types of societies in both the south and the north (as discussed in Chapter 3).

This chapter will analyse:

1 how the impact (economic, social and political) of this particular state-socialist problem of public action has created pressures for change;

2 how new problems of public action have arisen in response to these pressures through programmes of institutional reform;

3 and how these new problems have created their own pressures in turn which have led to the search for fresh policies and institutions.

This is indeed what we mean by 'policy as a process'.

The process of public involvement in development in socialist, as in non-socialist, contexts should be seen in these dynamic terms — as a process of continuing interaction between policies and institutional forms on the one side and their economic, social and political effects on the other side. Options for public action cannot be seen merely as alternative sets of policy choices; they should also be situated within the broader context of social, economic and political change within which they operate both as cause and effect.

9.2 Rural decollectivization in China

These questions will be pursued through a case study of one Third World country, China, which, following the experience of the Soviet Union, adopted the state-socialist pattern of public action after the success of the Communist-led revolution in 1949. Here, we will concentrate on the rural sector where, in China as in other developing countries, the bulk of the population lives and where material and cultural deprivation is the greatest. In the context of the developing world, China is a very suitable case for examining the record of state-socialist strategies of public action since the basic elements of the traditional Soviet system were adopted wholesale after the Revolution in the early and mid 1950s. Moreover, in spite of certain differences which emerged during the Maoist period between the late 1950s and the mid 1970s, the central features of Soviet-style socialism were still in place in both the rural and urban sectors until the beginning of the 1980s.

During the 1980s, moreover, previous forms of public action were subjected to radical criticisms which led to a new strategy of public involvement which sought to reduce the extent and change the character of public action through a sweeping programme of economic reform under the banner of 'market socialism'. In the rural areas this meant that the agricultural producer collectives introduced along Soviet lines in the early years were dismantled and replaced by a system of production based on the household farm. The 'collectivization' of the 1950s was thus reversed by 'decollectivization' in the 1980s. This Chinese experience enables us to trace the dynamic process whereby yesterday's developmental solutions become today's problems and how patterns of public action change in consequence.

We will concentrate on these changes in the nature of public involvement in Chinese *rural* development and ask three main questions.

Q First, how did China's system of collective agriculture operate in the pre-reform era, how effective was it in developmental terms and why did Chinese reformers wish to move decisively towards decollectivization?

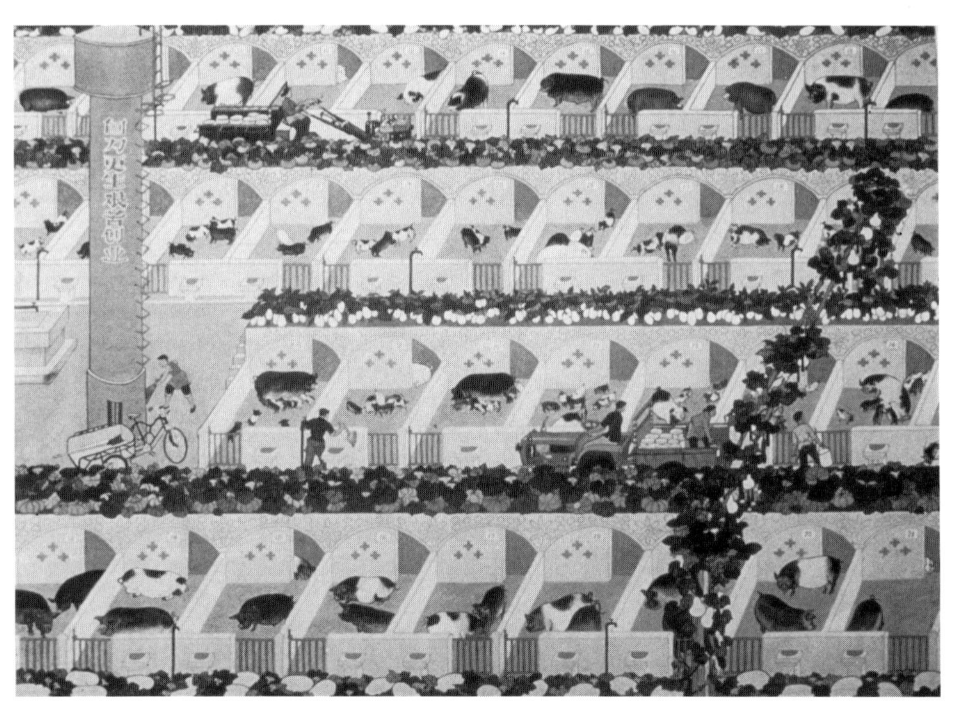

Figure 9.1 Under the commune system, agricultural production was organized collectively by the commune or by smaller scale brigades and teams. (a) A painting of a commune pig farm.

Figure 9.1 (b) Peasants in a production team threshing rice.

Q Second, what form did decollectivization take in the Chinese countryside and how did the new system of agricultural production created by the reforms operate in practice?

Q Third, how successful was agricultural decollectivization in terms of the central problems of developing a poor Third World country, i.e. in stimulating rural economic growth, alleviating poverty and improving the general welfare of the rural population?

9.3 State-socialist models of development: from Stalinism to market socialism

The communist or state-socialist approach to development derived from the experience of crash industrialization and agricultural collectivization in the Soviet Union in the 1930s under the leadership of Stalin. This Stalinist version of socialism became a model for economic development in later 'socialist' societies whether it was imposed by the Soviet Union (as in much of East-

Box 9.1 The Stalinist state-socialist model of development

In this model the state initially establishes its capacity to control and direct economic development by transforming the system of ownership of productive assets through nationalization and collectivization. In the former case, formal ownership of productive assets is vested in the nation as a whole with the state as its representative; in the latter case it accrues in theory to the members of the collective unit. (In agriculture, however, the Soviet Union is distinctive in that land was nationalized, a precedent not followed by other socialist countries in Eastern Europe and the Third World. Indeed, private agriculture predominated in Poland and Yugoslavia.)

In the traditional Stalinist model of state socialism, these two forms of ownership dominated the economy and the ensemble was co-ordinated and directed by a mandatory central plan. The plan was worked out by the leadership of the ruling Communist party and economic planners at the centre, and then issued as binding instructions to the basic productive units, be they state industrial enterprises or collective farms. In such a system private enterprise plays a marginal role in theory (though it may play a more significant role in practice, particularly in agriculture where farmers retain 'private plots') and the operation of markets is kept within strict limits (in consumer goods and to a lesser degree in labour power) with the ultimate intention of displacing markets altogether in a future fully-planned and fully-socialized 'communist' society.

ern Europe) or adopted voluntarily by countries setting out on a fully-fledged revolutionary path of development. In the Third World, for example, this model was adopted with important local adaptations by Vietnam, North Korea and Cuba and was seen by these and other developing nations as a powerful means dealing with their problems of poverty, technological backwardness and national weakness.

Although the Soviet model of state socialism was widely recognized particularly in the poorer countries of the colonial world as a developmental success story worthy of emulation, in the aftermath of Stalin's death in 1953 (and even earlier in Yugoslavia) it became the target of criticism for its economic (not to mention political) failings. Critics in the Soviet Union, Eastern Europe and China argued that the model was problematic in at least two major respects: in terms of macro-economic strategy, its obsessive concern with heavy industry had been detrimental to the development of both light industry and agriculture; in terms of micro-economic efficiency, the predominance of mandatory centralized planning had created systematic waste and poor utilization of productive assets. These problems, it was thought, would get worse as the economy became more complex and, particularly given the need to compete economically with the capitalist West, there was a greater need to increase productivity through organizational and technical innovation rather than mere quantitative expansion by putting more resources and labour power to work.

These criticisms became more influential in the 1960s, particularly in the more advanced economies of Eastern Europe where poor economic performance was giving cause for concern. During this decade and the one that followed, there emerged a new conception of the socialist economy which advocated basic institutional reforms in the traditional model to create what came to be known as 'market socialism'. This was in effect a socialist version of that wider phenomenon of 'economic liberalization' which is explored in non-socialist contexts in Chapter 7 along with its characteristic aspects of privatization and deregulation. This new model was based on a threefold critique of Stalinist economics.

First, previous directive methods of central planning were held to be both organizationally impracticable and economically irrational. They were to be replaced by 'indicative' or 'guidance' planning whereby the state would not intervene directly to instruct enterprises what to produce. Rather it would establish a framework of macro-economic regulation within which enterprises would operate with a high degree of autonomy in ways comparable with capitalist enterprises in a Western 'mixed economy'.

Second, the hostility of the traditional model towards markets was seen as irrational because markets were necessary to deal with the complex problems of co-ordination and exchange in a modern economy and exerted powerful pressures for efficiency in the use of economic resources. Fully-fledged markets should be developed, said the market socialists, first of all in goods and services and ultimately in factors of production such as labour, land and capital. Socialist planners should aim for a productive and complementary relationship between these born-again markets and the new-style planning; indeed the integration of plan and market was the watchword of market socialism.

Third, both the state and collective forms of ownership had significant economic defects and should be reformed if necessary through formal (freehold) or informal (leasehold) privatization of state and collective enterprises. Moreover, greater scope should be allowed for the private sector and other mixed or non-socialist forms of ownership (which might include enterprises set up by foreign capital and joint foreign–local enterprises) which could offer tangible economic benefits to the economy as a whole in terms of greater efficiency, diversity and dynamism. This reformed system could still be defined as 'socialist', it was held, because public ownership (both state and collective) would continue to predominate in the economy as a whole, because central planning would still be retained, albeit in modified form and because the state-socialist political system would continue basically unchanged, notably the dominance of the single Communist party.

Figure 9.2 The cultural revolution was the height of the Maoist period when China's agriculture was organized collectively in 'rural people's communes'. A Red Guard rally in 1966.

This new model of a 'socialist market economy' was the guiding force behind a series of efforts at economic reform in Eastern Europe, the most enduring being in Hungary (in Czechoslovakia the economic reforms were stifled by the Soviet intervention in 1968). Though thinking along these lines had already made some impact in China by the early 1960s, the Cultural Revolution intervened in 1966 to push China in a radically different direction and the ideas of market socialism did not surface again until after the death of Mao Zedong (Mao Tse-tung) and the purge of the radical Maoist 'Gang of Four' in 1976. The new leadership under veteran revolutionary Deng Xiaoping (Teng Hsiao-ping) sponsored a re-assessment of China's economic strategy and at the watershed Third Plenum of the Eleventh Central Committee of the Chinese Communist Party (CCP) in late 1978 launched a new era of market-oriented economic reform. These reforms promised to introduce sweeping institutional changes which were to affect all sectors of the economy including the system of agricultural collectives and it is to their impact on the Chinese countryside that we now turn.

Figure 9.3 Veteran revolutionary, Deng Xiaoping, sponsored China's market oriented economic reforms after the death of Mao. Here he is seen meeting representatives of the European Commission in 1983.

9.4 What was wrong with the commune system?

What is the commune system?

Agricultural producer collectives had been introduced in China in a series of stages in the mid 1950s. Like their Soviet predecessors, the Chinese leadership saw collectives as an economically more productive and socially more equitable way of organizing agricultural production than any hypothetical alternative, whether this be small-scale peasant or large-scale capitalist production. Collectives were also seen as an important means to integrate agriculture into, and subordinate it to, a national strategy of rapid industrialization. While the original agricultural collectives had been enlarged during the Great Leap Forward in 1958–60, the early 1960s saw a series of institutional readjustments which created a system of rural collectives which was to remain unchanged in its essentials until the onset of the reform era in the late 1970s. These institutions are usually referred to as the 'commune system'.

Box 9.2 The commune system

The 'rural people's commune' was a structure with three nested levels: the *commune,* the *production brigade* and the *production team.* The commune was divided into production brigades (on average about nine to thirteen brigades per commune); the brigades were divided into production teams (averaging about seven to eight per brigade). The *communes* were large units (averaging about 15 000 members each as of 1980) and were multi-functional, responsible for local government, party affairs, social welfare (such as secondary education and hospitals), economic planning and management, culture, public security, rural investment projects (such as land reconstruction and irrigation works), local industry and commerce, technical services and so on. The *production brigade* operated under the supervision of the commune and was responsible for political affairs (as the site of the basic-level branch of the CCP), small-scale rural capital construction and industry, and social welfare (elementary schools and clinics). The *production team* under the supervision of the brigades was a small unit composed of between 20 and 40 neighbouring households and operated as the basic unit of agricultural production.

Formally speaking the team owned the land cultivated by its members, organized agricultural and other production and was the 'basic accounting unit', i.e. it calculated revenues and expenditures and distributed income to team members. The income of each working member of the team was a share of the team's net income available for distribution after paying taxes and debts and reserving certain funds for investment and welfare. Remuneration was in cash or kind and was based partly on a basic ration and partly on 'workpoints' awarded to each labourer on the basis of time or task. Incomes varied each year depending on the productive performance of the team, on the amount creamed off for investment and welfare expenditures and on the level of procurement prices and subsidies set by the state.

The communes and their subordinate units were themselves subject to controls and claims exerted by the state, the lowest level of which above the commune was the county (*xian* or *hsien*) government. State controls operated through a dual system of both government and party organization. The commune acted in effect as a dual institution: on the one hand it was a 'collective' responsible to its component parts and membership; on the other hand it was a representative of the state, in effect the basic level of local government, and was therefore responsible to and controlled by higher levels of the state. The commune and the brigades also had party organs which were subordinate to higher levels of the party apparatus. In consequence the autonomy of the communes, brigades and teams was heavily circumscribed and they were under pressure to comply with the policies of the party/state at any given time.

Figure 9.4 In the commune system, daily work was measured in 'work points'. In this painting, an evening meeting discusses how the day's work points should be distributed.

State policies towards agriculture in turn were not merely designed to encourage rural development for its own sake, but to link agriculture and other sectors of the rural economy into a national economic strategy which stressed industrialization, particularly the construction of heavy industry. To this end the state sought to control agricultural production through a system of *procurement* whereby agricultural collectives were obliged to meet mandatory annual output targets, the produce being sold to the state at fixed prices. Since the state wished to hold down the cost of food and other wage-goods in the cities to foster industrialization, it kept down the prices paid to rural collectives for their produce. Moreover, since the bulk of state investment resources were devoted directly or indirectly to industry, investment funds for agriculture were relatively scarce. In consequence, the state imposed a high-investment regime on rural collectives, requiring them to reserve a high proportion of their own incomes for local investment in agriculture, infrastructure, social welfare and small-scale industry. This meant that there were less funds available in the collectives to improve peasants' material living standards.

Achievements and problems

The commune system received a lot of plaudits from foreign development specialists during the 1970s on the grounds that it offered an equitable and productive path towards rural development. Some of their approval, however, reflects the influence of Chinese propaganda about the merits of the communes (systematic field-work was not possible during the Maoist period either by Chinese or foreign scholars and observers had to rely on selective and biased accounts of rural life). The post-Mao backlash in China, by contrast, produced a flood of criticisms from Chinese commentators who painted a very negative picture of the preceding two decades. Particularly devastating was their criticism of the massive famine in 1960-62 which was a consequence of the excesses of the Party's attempt to force the rural population into vast and unworkable communes during the Great Leap Forward from 1958 to 1960 (for a discussion of this famine, see Drèze & Sen, 1989).

But the commune system did have some achievements to its credit which helped to lay the basis for the success of the reform policies which followed in the 1980s. In terms of agricultural production, it performed reasonably well. Between 1957 and 1978 the gross value of agricultural output increased by 3% a year with the all-important grain output averaging 2%, a creditable growth record compared with other Third World countries. The collectives were also able to establish a system of rural health and educational facilities which contributed significantly to an impressive improvement in the welfare of the rural population, comparing very well indeed with other Third World experience as shown in Chapter 1. Collectives also acted to establish a 'floor' income for their poorest members and to restrain income disparities within villages. In terms of rural accumulation, moreover, by organizing labour-intensive public works to build irrigation systems and improve farmland, they significantly improved the level of rural infrastructure; communes and brigades also contributed to the diversification of the rural economy by establishing rural industries and other enterprises.

Though these were areas of significant advance they concealed serious problems. While agricultural output was increasing slowly, the population was growing rapidly, expanding from 646 million in 1957 to 924 million in 1975. In consequence there was very little improvement in agricultural output *per capita* during this period; in fact grain output *per capita* hardly increased at all, rising from an average of 301 kilograms in 1955–57 to only 305 kilograms in 1975–77. This meant that the diet of Chinese peasants did not improve significantly over this period. Since productivity per worker increased only marginally, moreover, peasants' real incomes rose very slowly. Though income differentials within collectives were kept within bounds by egalitarian distribution practices, this nevertheless dampened people's incentives to work and thus hampered growth in productivity. Moreover, equality at the local level co-existed with substantial inequality *between* localities with large areas of poverty still remaining (a Chinese study in 1978 estimated that about 30% of production teams could be considered 'poor', i.e. with an annual collective income of below 50 *yuan* per person).

In general, therefore, the performance of the commune system over the Maoist period from the mid-1950s to the mid-1970s was mixed. On the one hand, it had been successful in building up rural infrastructure, encouraging rural industry and providing for basic material and welfare needs, with very positive effects on health, longevity and literacy. On the other hand, in terms of the vital objective of improving agricultural productivity, a factor on which other improvements in material living standards and social welfare depended, the communes performed disappointingly. In the words of Drèze & Sen (1989) the commune system represented a successful case of 'support-led security' in which considerable progress had been made in improving the standards of living of the rural population through public action (in this case organized through a combination of state and collective provision). The problem was that further advances in the provision of public goods such as education, health and social security rested centrally on an ability to generate more resources by accelerating the rate of agricultural growth. At the dawn of the post-Mao era in the late 1970s, it is to this question that China's experts and policymakers directed their attention.

Where did the problem lie?

If the economic performance of the communes had been disappointing, where did the problems lie? In answering this question, it is useful to recall the two basic dimensions of socialist public action: the 'macro' level of action by the state and the micro level of collective action. This distinction is particularly important when it comes to evaluating criticisms of collective agriculture because it is hard to disentangle whether problems were caused by the nature of collective institutions themselves or by their subordination to unfavourable state policies and controls.

When Chinese reformers came to evaluate the commune system in the late 1970s, they realized that some of the problems they identified were a consequence of state action: the lack of state investment in rural infrastructure; an overly rigid procurement system which forced peasants to grow grain in unfavourable situations thereby preventing them from diversifing into potentially more profitable cash-crops and sideline products; a price policy which kept procurement prices low to support the industrialization programme and offered little in the way of incentives for extra productivity; and restrictions on private production and marketing which imposed a virtual straitjacket on the local economy.

However, strong criticisms were also levelled at the collective institutions themselves. In the day-to-day operation of Chinese agricultural producer collectives (the production teams), two problems seemed particularly intractable:

- how to measure and supervise labour
- how to relate labour to payment.

First, the distinctive nature of much agricultural production (particularly in the context of intensive Chinese agriculture which has been likened to gardening on a vast scale) means that work is highly complex, discontinuous, fragmented and variable. This means that the optimal scale for defining and organizing work is small, smaller even than the production teams which were themselves small units, averaging 25 households or so. Team managers faced formidable problems of measurement and control. Second, this meant that it was very difficult to devise systems of payment which linked the quantity and quality of labour on the one hand to precise payments on the other. The obvious method was through some form of piece-rates (payment according to specific tasks such as, for example, weeding or transplanting seedlings). However, given the complexity of the agricultural labour process, this proved very difficult in practice and the situation was made worse by the fact that the level of accounting skills in many teams was pretty rudimentary. Team managers thus tended to fall back on the cruder time-rate method of payment (essentially a certain number of workpoints for a day's work) which had the virtue of simplicity but did a bad job in linking payment to labour. Time-rates could not adequately capture variations in the quality and quantity of individual labour and thus tended to be more egalitarian than piece-rates, thereby reducing incentives for individual effort. The result was a vicious circle of low labour productivity, low incomes, low incentives, low effort, low productivity and so on.

Q Can one then attribute slow growth in Chinese agriculture during the Maoist era to micro-problems deriving from the operation of the collectives themselves, or to the macro-context of state organization and policy?

Commentators disagree on this question. Peter Nolan of Cambridge University, for example, while recognizing the negative influence of the macro setting, accords this secondary importance and argues that 'it seems unquestionable that a major part of the problems encountered by the rural economy pre-1978 is attributable to the institutions of the collective farm *per se*' (Nolan, 1988). By contrast, Marc Blecher of Oberlin College, Ohio argues that the problems of labour organization and payment methods in the collectives have been exaggerated and concludes that

> "... there is every reason to believe that collectivism could have achieved much better economic results under a different set of state policies on commerce, planning and investment."
>
> (Blecher in Benewick &Wingrove, 1988, p.95)

It is my view that, to understand why the collectives performed in the way they did, they must be placed in their macro-setting. In historical terms they were the means by which agriculture was harnessed to China's dash towards industrialization. One then must consider whether any major policy alternatives were feasible and desirable in the context of the 1950s, 1960s and early 1970s when China faced a hostile external environment under pressure from both the United States and the Soviet Union and given the obvious desire of any Chinese leadership (communist or not) to achieve national power through rapid industrialization. But the key point is that China's agricultural collectives were never allowed to operate as real collectives in the sense that they had a significant degree of independence and with the reality, not merely the form, of collective ownership by their members. Although managers within the communes had some scope for decision and although ordinary members had some opportunity to participate in the making of these decisions, on most basic issues their hands were tied. Production teams were essentially told what to produce, whom to sell it to and at what price, and important economic alternatives were closed off (whether this be the opportunity to sell surplus produce on local free markets, go into business or look for jobs elsewhere, or exploit local natural advantages to produce more profitable crops). It was an important part of the role of officials at the commune and brigade levels to enforce these controls over the teams. This lack of autonomy at the point of agricultural production at the team level meant that collective ownership was largely a sham since the notion of ownership implies the power to control how a productive asset is used, or what the economic policy of an enterprise should be. This lack of real ownership power in turn was a powerful disincentive to effort, since workers were 'alienated' from production: they did not feel a commitment towards collective assets, nor a responsibility to work hard in using them.

In sum, since these were collectives in name only, it is impossible to evaluate their economic virtues or vices as collectives. Moreover, this same point can be made about most of the experience with agricultural producer collectives in other state socialist countries (with Hungary as a partial exception). Even if we were to ignore this problem, however, it is undeniable that producer collectives in agriculture may well have serious inherent problems in terms of economic efficiency, enough to give pause to socialist policy makers aiming at rapid rural development. However, the organization of *production* by no means exhausts the possibilities for and potential of co-operative/collective organization in agriculture. The distinctive role of collective action in rural development may be less in directly organizing production and more in providing ancillary economic services and social welfare. The basic innovation underlying the post-Mao reforms has been to separate these two sets of functions, assigning agricultural production to peasant households and maintaining the collectives for productive services and welfare. We shall examine this major institutional change in more detail in the next section.

Before doing so it is important to recall what was said at the beginning of this chapter about the social and political context of changes in the forms of public action. The commune system was criticized and eventually superseded not merely because a group of policy-makers and economists in Beijing (Peking) had evaluated its developmental shortcomings and chosen a new course of action. This they did, of course, and their decisions were very important for what was to come next. However, they made their choices in the context of a rising tide of social discontent in the countryside directed both against previous state policies towards agriculture and against the communes themselves. Peasants resented state directives about what they could grow, low official prices, state pressure to put a high proportion of their income into local investment instead of allowing them to improve their current living standards, and restrictions on their ability to seek new jobs or trade part of their produce in local free markets. They also resented the exaction of levies to support what many thought was a bloated and unproductive system of local officials at the team, brigade and commune levels, and chafed against payment systems which they felt to be unfair in that they failed to reward merit and subsidized sloth. On the one hand, this peasant discontent translated into lack lustre economic performance contributing further to the danger of rural stagnation; on the other hand, it built up political steam which put pressure on party leaders to change direction.

9.5 The decollectivization of Chinese agriculture

The Third Plenum of the Central Committee of the Chinese Communist Party held in December 1978 marked an important shift in official policies towards the rural sector. Measures were introduced to improve the macro setting of agriculture by raising official procurement prices for grain and other agricultural products, with extra price incentives for above-quota sales to the State; the previous emphasis on grain production to the detriment of cash crops was relaxed; private production was given greater scope and rural free markets were encouraged to expand. The policy environment of agriculture thus improved dramatically, providing more incentives and greater freedom for peasant producers.

The nature of the reforms

The Third Plenum also began a gradual process of agricultural 'decollectivization' which lasted about four years (1979–83) and transformed the rural scene. The new policies rested on two basic principles.

1 There was to be a division of labour between collective institutions and their constituent parts, the peasant households. In the new system, the latter were to take over day-to-day agricultural production. Whereas in the past it had been the team which had been responsible for directly organizing production and meeting output targets, in future this task was delegated to units

Figure 9.5 The reforms provided greater opportunities for farmers to sell their produce in local free markets. (Top) Selling vegetables in Guangzhou, Canton. (Bottom) Selling traditional medicines in Emie Shan, Sichuan province.

smaller than the team. The chosen unit (which in the event was to be the peasant household) was required to deliver a certain amount of output to the team and this responsibility was regulated by a contract.

2 There was the principle of divorcing ownership of productive assets (predominantly land) from control over its use. Even though the peasant households were now to have a great deal of effective control over the land contracted to them, formal ownership was still to reside with the collective. What we have, therefore, is a *de facto* privatization of agricultural production within a *de jure* framework of collective ownership.

At first the contracting unit was a small group within the team in a system called 'contracting output to the group'. The two parties agreed about the output target and the inputs necessary to achieve it (which were supplied by the team) and a certain amount of workpoints was awarded to the group when the target was fulfilled (with incentives for overfulfilment). By 1980, however, the break with collective production became more radical as official permission was given for contracting to the individual household. This took two forms, a more moderate 'contracting output to the household' and a more radical 'contracting work to the household'. In the former system, a household was allocated a piece of land (or rather a number of parcels of land) and a specified proportion of its output was handed over to the team in return for an agreed number of workpoints. Current inputs such as seeds, pesticides and fertilizer were provided by the team or by both partners. Production planning, services such as irrigation and major means of production such as machines or draught animals were still handled by the team.

The second form, 'contracting work to the household', which spread across the country from 1981 on and was nearly universal by the end of 1983 (when it involved 98% of peasant households) was in effect a return to household agriculture. This new system was referred to as the 'household responsibility system' (or often just 'the responsibility system'). Households are now responsible for both production and distribution (workpoints distributed by the team are now obsolete); means of production such as tools,

Figure 9.6 Watering a private plot of land. Under the 'responsibility system', introduced in the early 1980s, Chinese peasants took over cultivation of small parcels of land which had formerly been worked collectively.

draught-animals and mechanical equipment are distributed to the households; the household is still obligated to deliver a certain level of output to the team and must pay state agricultural tax and a contribution to the upkeep of the collective, but after these payments it is free to deal with its extra output as it wishes (consume it, sell it on the local free market, or sell it to the state at higher-than-quota incentive prices). This is essentially a kind of tenantry system with the household paying a 'rent' to the collective landlord. Though the period of the contract was initially a year, this was gradually extended to give the household *de facto* ownership over its allocation of land and, though a formal market in land has not been revived, subcontracting is now widespread, particularly in areas where there are opportunities for off-farm employment.

Box 9.3 The shift towards a two level management system

Though rural collectives are now removed from day to day production, they retain a residual economic role in planning output (by influencing what crops are to be produced for the contract), providing productive services, and encouraging off-farm enterprises. Chinese commentators describe this in terms of *a two-level management system* with a division of labour between the household on the one side and various kinds of co-operative, collective and state institutions on the other. The latter now operate at various levels. The former production team has become a 'co-operative' or a 'villagers' small-group', the former production brigade has been re-incorporated as a 'village', and the former commune has been replaced by a basic level of government called the 'township' (*xiang*). At the village level, the important remaining collective institution is the 'villagers' committee' which presides over a large natural village or a group of small hamlets and is responsible for maintaining welfare services previously handled by the brigade (education, health and aid to vulnerable households or individuals), organizing technical services and rural capital construction to aid agriculture (notably irrigation works), overseeing contracts and problems arising from them and managing, directly or indirectly, village industry. In the last case, the reforms have led to the contracting out of most of the enterprises formerly run by the brigade on the same basis as the agricultural contracting system. Assets are effectively leased temporarily to individuals or groups in return for a cash payment to the village.

At the level of the former commune, the township government has a powerful role, not only in handling the routine business of government (public security, legal affairs, taxation, etc.) but also in stimulating local industry, whether public (run directly by the township), private or mixed. Together with village-level industry, local industry (under the collective name of 'village/township industry') has developed very rapidly during the era of reform, not the least because it has been given a powerful impetus by villagers' committees and township governments. The township is also crucial as a means to maintain state control over the countryside, both directly and indirectly (through its supervision of village organizations). The party apparatus is still intact in the countryside, moreover, with the village party branch taking over the formerly important political role of the brigade branch.

If one wishes to summarize the changed role of collective and state institutions in the countryside, one should first say that the Chinese rural scene is very complex and the situation varies considerably from place to place. In some areas, the combined role of collective, Party and state institutions remains as dominant as before the reforms; in others, collective institutions at the village level have atrophied and rural Party branches have been weakened as Party cadres have defected, lost their effectiveness or succumbed to corruption. This has left the township government alone with the task of getting its locality to conform with state policies, a difficult task when the chain of authority has snapped at the village level. For more analysis of the changing role of the Communist Party and local government in the countryside, see Gordon White (1987) and Tyrene White (in Davis & Vogel, 1990).

Continuity and change

In general, however, from the perspective of the early 1990s, twelve years after the reforms began, much has changed on the rural institutional scene. Agricultural production is now undertaken by peasant households and the private sector has expanded greatly (into transportation, finance, trading and services as well as industry). The economic power of the collectives has declined considerably as they have moved out of direct management of production. The grip of the state over agriculture has relaxed to some extent to allow farmers greater say in decisions about what to produce, where to sell it and on what terms. Overall, there has been a significant redistribution of economic power away from the collective/state and towards individuals and households. This means that the former power of collective institutions to control their members and extract resources from them (for investment or welfare) has declined. To that extent, collectives at the grass-roots level have lost part of their capacity to provide for, in the words of Drèze and Sen, 'support-led security'.

On the other hand, there are important institutional and political continuities. Collective institutions, notably the villagers' committees, do

Figure 9.7 Although the role of collective production has declined, peasants still co-operate to construct and maintain important rural infrastructure. These farmers are digging an irrigation ditch.

retain significant social, economic and political functions in many rural areas. Moreover, though the grip of the state and Party has relaxed, the apparatus of authority and control is still basically intact. For example, though the CCP has gone through the motions (in 1984) of abolishing the official procurement system, it still operates even though its scope and impact may have lessened; though the 'free market' has expanded considerably in the countryside, it is still far from being the dominant principle of China's rural economy, except perhaps in certain highly dynamic areas in the south-east. To a considerable extent, these continuities reflect limitations on the reform process consciously imposed by the CCP leadership. They are reluctant to relax their grip over the rural economy too far because they are concerned about the supply of food to the cities and the contribution of agriculture to their still ambitious plans for industrialization. Given the political problems in the cities over recent years, moreover (reaching a climax in the June 4 Incident in Beijing in 1989), they regard it as vital to retain their political control over the countryside.

But the Chinese countryside is now very different from how it was at the end of the Maoist era, in economic, social, political and institutional terms. What difference has the process of decollectivization and greater reliance on private enterprise and markets made to China's performance in terms of key developmental objectives? What role will the changes in rural society play in conditioning the nature of, and prospects for, continuing public involvement in shaping the developmental future of the Chinese countryside? It is to these questions that we now turn.

9.6 The impact of decollectivization on Chinese rural society

In Section 9.3 we investigated the performance of the previous system of rural people's communes in achieving certain basic developmental objectives. Here we shall be asking similar questions of the new system of agricultural organization in an attempt to compare it with its predecessor. Views of commentators vary considerably on the effects of the reforms. Some, such as Nolan, applaud the new policies for achieving a breakthrough in productivity and growth which has brought significant welfare benefits for the rural population. Others, such as Drèze and Sen, while admitting that significant economic progress has been achieved, have been concerned about its sustainability over the longer term and about certain negative social consequences, such as growing inequality and a decline in welfare services resulting from the atrophy of previous systems of collective provision. This debate clearly relates closely to the central themes of this book.

Q To what extent, if any, have the economic reforms, by weakening the role of previous collective institutions, reduced the capacity of a society to solve the problems of mass poverty and insecurity?

Q Alternatively, have the new policies produced a new balance between collective and private, state and market, which has enhanced China's developmental performance and contributed to the welfare of its rural population?

Let us address these questions by discussing certain basic developmental objectives in turn.

The impact on economic growth

First, in economic terms there seems to be a fair unanimity among commentators that the reforms have brought a very significant improvement in agricultural output and productivity. The gross value of agricultural production rose by an annual average of 6.2% in the decade from 1979 to 1988, about double the growth rate during the collective period from 1957 to 1978. Land productivity rose substantially: output of grain crops per *mu* (a Chinese measure of land area, equal to about one fifteenth of a hectare) rose from 169 kg per *mu* in 1978 to 239 kg per *mu* in 1988. Moreover, at least in the early period of the reforms, real output per farm worker rose

Figure 9.8 In the 1980s many Chinese farmers branched out into small scale industry. Many of these operated in the household. (Top) Household pasta production, 1986. (Bottom) Making teapot moulds, 1986.

rapidly, to the tune of nearly 60% over the six years from 1978 to 1984. There was also a surge towards the diversification of the rural economy, both within the farming sector (towards cash crops such as cotton and other activities such as aquaculture, tree products and animal husbandry) and towards rural industrialization (industry's share of total rural output rose from

19.5% in 1980 to 38.1% in 1988). The share of rural workers in agriculture and other farming activities dropped from 90 to 78% between 1978 and 1988; this had an important effect on incomes since peasants could earn more by working outside agriculture (for example, in 1987, a person could earn 6.5 *yuan* a day in grain production and 16.5 *yuan* in local industry).

In spite of these happy results, the rural reforms have encountered serious economic problems which became increasingly apparent from 1985 on. While performance in the early 1980s was very impressive (a growth rate of about 7.7% a year from 1979 to 1984), it decreased quite sharply after that, averaging only 4.1% annually in the four years from 1985 to 1988. Output of grain was a particular headache: from a base of 300 kg. per capita in 1978, it had only risen to 359 kg per person by 1988, an increase of only 16.6% in ten years. There was also concern that the stock of rural infrastructure (particularly irrigation works) which had been built up during the Maoist period had been allowed to deteriorate in a new economic climate in which the collective organization which had built and maintained them had been substantially weakened. For example, the area of land under irrigation fell from 45 million hectares in 1979 to 44.5million in 1988. Concern was also voiced about certain adverse environmental consequences of the new stress on private enterprise and markets (decline in soil quality caused by 'mining' the land for short-term profit, invasion and destruction of forests and pollution by unregulated rural industries).

Some Chinese policy-makers and economists felt that the early years of spectacular economic progress had been at least partially a 'one-off', i.e. the result of institutional and policy changes which increased output dramatically but temporarily and which had relied heavily on an accumulated stock of rural capital built up over the previous two decades. For example, a commentator in the party organ *People's Daily* in early 1987 argued that 'we must realize that the ultra-high growth rates of recent years are mainly due to an outburst of long-restrained productive forces triggered by reform. The problem of the underdeveloped state of agriculture has not been solved'. For many Chinese observers, the resulting slowdown could only be tackled by decisive new policy measures: for instance, an increase in state investment in agriculture, higher prices for agricultural products, particularly grain, and more radical reforms within the agricultural sector itself (such as the complete abolition of the state procurement system, or the creation of a market in land). As we shall see later, however, there are certain political obstacles to the adoption of these proposed policies and the economic problems of agriculture continue to defy solution.

Welfare and social provisioning

Let us turn now from economic performance to the impact of the reforms on the material and social welfare of the rural population. *Per capita* consumption by the agricultural population rose by an annual average of 8% in the decade from 1979 to 1988, a dramatic improvement over the commune period. Not only were peasants eating better, but there was also a virtual explosion in house-building and a spread of durable consumer goods, such as bicycles, sewing-machines, radios, watches, TVs and electric fans, which greatly improved the quality of rural life. This overall improvement was accompanied by some increase in inequality of incomes and wealth, both within villages and between regions. Some successful entrepreneurs became rich quickly and often displayed their wealth ostentatiously through conspicuous consumption, causing their less fortunate neighbours to catch the 'red-eye disease' of jealousy. However, intra-village inequalities were limited by the fact that land was distributed under the 'responsibility system' on an egalitarian basis (according to the number of people or workers in a family), thereby providing each family with a basic 'floor' income.

More worrying was an increase in differentials between regions as more fortunate areas (near cities or in the more developed eastern provinces) were better situated to take advantage of new market opportunities. However, while some regions prospered more than others, there was also a very significant improvement in the lot of the poorer areas. Their inhabitants

Figure 9.9 Farmers' rising incomes enabled them to purchase a wide range of consumer goods which they could never have afforded before.

benefited from increasing opportunities to diversify agricultural production or move into non-agricultural work in their own locality or elsewhere. The improvement of previously poor areas was also aided by government policies: for example, the reduction of quotas for mandatory grain sales to the state, tax exemptions or reductions, or subsidies to local governments with fiscal deficits.

Though these benefits in material welfare can be counted as a major success for the reforms, other aspects of rural welfare appear to have fared less well. Enrolment in primary schools dropped as parents saw greater economic advantage in sending their children out to work earlier and this may have affected literacy levels in the countryside; nor is there much evidence to suggest that the quality of rural education improved significantly (rural teachers were notoriously badly paid and were often poorly trained). There were also problems in the provision of, and access to primary health care. With the weakening of collective institutions, most of the previous systems of co-operative health insurance collapsed and most of the previous 'barefoot doctors', one of the proudest achievements of the Maoist era, were privatized. While health care was still available in the villages, it was on a fee-for-service basis and its quality left much to be desired due to a decline in the level of supervision by higher levels of the medical system.

Overall, though the problems of rural education and health-care do give cause for concern, we need to put them in context. China's level of provision in rural education and health-care continues to be excellent when compared with other poor and populous Third World countries. Moreover, there is a great deal of variation between richer and poorer areas: many of

Figure 9.10 Many of the 'barefoot doctors' previously supported by the collective have gone private during the 1980s.

the former have ploughed back profits from increased agricultural productivity and local industry to improve the quality of local welfare services while the latter still lack the resources to do so. In overall terms, however, it does seem to be the case that the capacity of grassroots collectives to deliver welfare benefits to the village population has declined and to this extent peasant households have been thrown back on their own resources. One can argue that levels of collective welfare provision have lagged behind agriculture's growth performance and the increase in the material living standards of rural households. However, to the extent that the resources available to households have increased, this counterbalances the decline in public provision to some degree.

To summarize, one must conclude that the impact of the reforms, in terms of certain basic aspects of development, has been positive. Particularly impressive has been their impact on economic growth and on the real incomes of China's peasant majority. While distributive inequalities appear to have increased, they have been kept in bounds to some extent by retaining a provision for floor incomes within the villages, and by the continuing (though weakened) role of collectives and government. There has also been significant progress in raising the material level of the poorest areas, through a combination of government action and increased economic incentive and opportunity. At the same time, levels of collective welfare provision do not appear to have improved *in tandem* and give cause for concern. This is one of the key problems which China's policy-makers must face as they tackle the challenge of rural development in the nineties; it is to this prospect that we now turn.

9.7 Conclusions

Growth-led improvements

In the terms used by Drèze and Sen, the improvements in popular welfare which have taken place in the era of decollectivization have been largely 'growth-led' rather than 'support-led'. In other words they were the result of accelerated growth in the rural economy and a dramatic increase in the real incomes of peasant households. Though the capacity and effectiveness of public provision has declined, by contrast local collective and state institutions are still generally in place. As a result, some of the negative consequences most usually associated with an economic strategy which relies on economic growth alone have been avoided or contained:

- Egalitarian methods of allocating land and other productive assets to peasant households have served to guarantee a basic income and restrain intra-village differentials.
- Village collectives remain to protect vulnerable households and groups (such as the old and the disabled).
- The continuation of a state procurement system provides a guarantee of protection against insecurity which might arise from fluctuations in markets for agricultural goods.
- Village collectives and local governments have helped to organize a dynamic process of local industrialization which has served to absorb labour no longer needed in agriculture.

The central and local governments have also retained the capacity to redistribute economic resources to poorer areas, either on a continuing basis or in times of crisis. Although rural health care and education are areas of concern, the pattern of provision is very uneven. In some aspects, for example in preventive medicine and in the provision of doctors and hospital beds in rural towns, the situation may have improved to some degree. For more detailed analysis of the impact of the reforms on the economic and social welfare of the rural population, see Bramall *et al.* (1990). It is also worth comparing China's levels of popular welfare with other low-income countries by investigating the information provided by the World Bank (1991a).

Growth-mediated security

To this extent, the era of decollectivization has not been one in which crucial developmental objectives, such as providing basic security and reducing poverty have been sacrificed to a strategy of 'growth first'. The strategy could thus be described as one which continues to provide 'growth-mediated security' in the sense that, while accelerating the rate of economic growth is a central priority, the mechanisms still remain (at both state and collective levels) to guard against vulnerability and provide a basic level of welfare for the bulk of the rural population.

In the context of the late 1980s and the early 1990s, however, China's rural sector faced increasingly serious problems which needed decisive action. The rate of growth in agricultural production and rural incomes had slowed substantially; there was a critical need for more investment in agriculture which the central government was unwilling or unable to supply; agricultural production is held back by fragmentation of land holdings, a shrinking area of usable land, the deterioration of rural infrastructure, the desire of many peasants to move out of agriculture, stagnant prices for agricultural products and rising prices for agricultural inputs such as fertilizer and pesticides, and growing unemployment caused by the austerity policies of the late 1980s which had put many rural industries out of business.

Further reforms needed?

If the earlier success in stimulating rural economic growth is to be repeated and maintained, many Chinese economists feel that the rural reforms must be taken further. They propose that:

- the state procurement system must be reduced or removed to allow more space for a market in agricultural goods
- any remaining procurement prices for agricultural produce should be increased substantially to provide peasants with incentives to increase production (particularly of grain)
- peasants should be guaranteed formal ownership title to their land (this would reduce their incentive to 'mine' it for short-term gains)
- a market in land should be allowed to pave the way for its increasing concentration in the hands of more efficient producers
- much greater freedom should be given to the private sector.

Taking this route is far from easy since the government's ability and willingness to implement further reforms is limited by broader political considerations:

- Raising the prices paid for rural produce might increase the cost of living in urban areas which are politically volatile.
- A rapid expansion of rural credit might fuel the inflation which has dogged the last few years of the reforms.
- The state's capacity to increase spending on rural investment is limited by its chronic fiscal deficit, the improvement of which depends on more thorough-going reforms in the state industrial sector which is proving resistant to change.
- A further spread of markets and private power might undermine the state's ability to exert political control over the countryside.

If the route towards further market-oriented rural reform is to be followed, however, it may lead to some problems. Sen and others point out the potential negative effects of a growth-led rural development strategy which privileges private enterprise and the role of markets and pays inadequate attention to the role of public action to counteract the negative impact of unfettered markets and private enterprise. These include losses in human welfare resulting from economic insecurity caused by market fluctuations, growing inequality, landlessness and rural unemployment, and a weakening of the capacity of local co-operative and governmental institutions to tackle these problems.

Balance between market growth and welfare

If market-oriented economic reforms are to be taken further (and there are some strong economic arguments for so doing), the above problems can be anticipated and countered at least partially through public action. In essence this means an attempt to achieve a balance between on the one side, policies which aim to accelerate economic growth by encouraging private initiative and freeing markets; and on the other side, policies which provide increasing levels of public security and welfare, both by countering the deficiencies of the market and by providing fundamental guarantees of social provision for the rural population as a whole. This balance is important for any poor society in the early stages of development. But it is even more important for an avowedly socialist country committed to principles of co-operation, social equality, full employment and economic security.

Much more than in the 1980s, policymakers' concern about how to stimulate growth should be accompanied by equally careful thought about the kind of public action necessary to achieve these social goals. The aim should be to establish a system of rural institutions, both collective and governmental, which will facilitate economic growth and construct a humane social context for it. This involves experimentation with various kinds of economic co-operatives while retaining the 'responsibility system' by and large for agricultural production; strengthening the role of village committees (particularly by giving them greater autonomy in relation to their superiors in the state apparatus); and reorganizing local government at the *xiang* level to put the provision of public services on a firmer organizational and financial basis. It is this task of achieving both economic growth and institutional reform which formed the major challenge confronting Chinese reformers in the nineties.

Political feasibility

It is, however, fruitless to utter such prescriptions without considering whether or not they are politically feasible. Much has changed since

Figure 9.11 June 1989, student demonstration in Tiananmen Square, Beijing: a major turning point in China's modern political history.

the onset of the reforms. Reformers have become stronger within the party leadership. The power of local party, state and collective officials has been undermined to some degree; and the rural population is now more diverse and differentiated than previously, not the least through the emergence of a new elite of relatively well-off rural entrepreneurs with considerable potential influence on the local political scene.

The future form and force of developmentally oriented public action in the countryside depends on the balance of power between these different forces: on the relative power of radical *vis-à-vis* moderate reformers within the party leadership; on the continuing strength of the communist party-state at the local level; on the business fortunes and political influence of the new rural economic elite; on the countervailing ability of poorer peasants and areas to exert pressures for redistributive provision in their favour; and on the potential alliances between these contending political forces.

Clearly there are conflicting pressures on public action and, in the context of the post-Tiananmen era, there is a political deadlock which is not at all conducive to resolving the complex issues of balancing public and private, state and market, which we have discussed above. Any clear answer to the question must await the breaking of that deadlock, an event the nature and consequences of which are, at the time of writing this chapter, unpredictable.

Summary of Chapter 9

1 Public action for development is a process of continuing interaction between policies and institutional forms on the one side and their economic, social and political effects on the other side. This chapter investigated this process in the context of the Chinese experience of decollectivization in the 1980s.

2 Chinese rural development prior to the economic reforms relied on the 'commune system' which provided a framework for support-led security but also imposed real constraints on economic growth and development.

3 The emphasis on industrialization meant that rural communes were never allowed to operate as real 'collectives' with significant autonomy of their own. This lack of real ownership and autonomy was a powerful constraint on effort which impeded agricultural growth and innovation.

4 The economic reforms created a two level management system between households on the one hand and various co-operative, collective and state institutions on the other. The reliance on market forces was intended to enhance productivity and incentives.

5 The economic reforms altered the character of the rural economy and population which became more diverse and differentiated, and strengthened the position of reformers within the leadership. But public action is in a continuous state of flux as the balance of power changes.

10

SOCIAL NEEDS AND PUBLIC ACCOUNTABILITY: THE CASE OF KERALA

GITA SEN

10.1 Introduction

This chapter is about the social definition of human needs, and the associated structures of public accountability. In previous chapters of this book we have tried to argue that what a society comes to view as the legitimate field for public action and public policy is a product of complex, often contradictory, social and historical processes. What constitutes the public space is, more often than not, subject to conflict and renegotiation. This is especially true in times of social or economic crisis.

In a way this is not surprising because all human societies are formed by sets of social relationships that structure people's needs, rights, obligations and responsibilities. How any person fits into that structure, and how the society in turn is viewed by that person, is shaped by a number of social forces, not all of which are obvious on the surface. (Note that the term 'individual' is not used here, because in current usage it carries too many layers of ideological meaning, most of which are inappropriate in this context.)

The central arguments

There are a few points about the meaning of the public sphere that are worth emphasizing here:

1 What today we may think of as the appropriate and proper role for governments is not necessarily what was considered so a century ago, or even a couple of generations ago. It is not even the case that all countries or regions within countries hold the same views at the same time as to the legitimate space for public action. While there are sometimes trends in the political climate which shape the boundaries of the public sphere and appear to encompass large sections of the world, local conditions clearly also have considerable influence on the way in which those ideas are appropriated in particular contexts.

2 Struggles over the nature of the public sphere and what is included in it are also struggles over accountability. Accountability in this context includes both its positive sense of responsibility for doing something, and its negative sense of who is to blame if things do not work as they are supposed to and who should therefore set them right. It is important to distinguish between these two ways of thinking about accountability, because the one does not automatically imply the other. For example, a Ministry of Health may be charged with the task of child immunization and may set up the infrastructure and budgets for tackling the task. But the Ministry may take no responsibility for the fact that large numbers of children continue to fall ill from preventable diseases, claiming instead that the fault lies with people's superstition or ignorance.

3 Making governments and local functionaries accountable in the negative sense is much more difficult. A society's ability to enforce such accountability often depends on the political mobilization and awareness of local communities, groups and people generally. Thus, the success of an immunization programme may depend, among other things, not only on how much money a government allocates to primary health centres in its budget, but also on:

- whether such centres as are set up actually function to deliver the services they were set up to deliver
- whether the staff do indeed come to work regularly
- whether supplies are made available in adequate quantity and quality and on time.

But whose responsibility is it to ensure that all this happens? Or, more to the point, who is capable of ensuring that it happens?

4 The concept of the public sphere we are using here is one where we can think of people participating along with governments; in defining needs, in making choices appropriate to those needs, and in enforcing accountability. This is a far cry from a notion of the public sphere as including those activities that have to be carried out socially or collectively because they are too costly to be done individually; that is, a notion of governmental action as a substitute for individual action by persons. Defining a good as 'public' in this latter sense means taking it out of the hands of persons. It is a short step from this to a welfarist approach to the provision of public services, which come to be seen as something the government does for people or on their behalf, not something that people can have any direct say over on an ongoing basis.

In the mid-1970s, such a 'welfarist' notion of public action came to dominate much of the thinking about, and many of the development programmes focusing on, basic needs provision in the Third World. These programmes were laudable in their aim of focusing directly on requirements like basic nutrition, health, education, shelter, sanitation, etc. But most of the programmes in these areas were dominated by a way of thinking about public action that effectively precluded participation by people in the process of defining action, rather than being only its recipients. Not surprisingly, the programmes often suffered both in terms of appropriateness and accountability.

Another area which is a classic example of this way of thinking about public action is population policy. At least one influential approach is guided by the belief that poor people in the Third World, if left to themselves, would want to have more children than would be socially desirable. Governments, therefore, it is argued, have little choice but to impose population control measures on people. Typically people and particularly women, who tend to be most affected, are rarely even consulted about appropriate policies or alternative possibilities, let alone actually allowed to make decisions. Experience shows that such approaches seldom work on a long-term basis, and this perhaps was one of the reasons for the growing influence of the view that 'development is the best contraceptive' which was first articulated at the World Population Conference held in Bucharest in 1974. This view represented a step forward in that it identified the satisfaction of people's basic needs for survival with dignity as a prerequisite to their wanting to have fewer children. But it stopped short of recognizing the value of people's involvement as actors in the process.

The case study

We shall illustrate the above arguments by looking at the experience of Kerala, a small state in India. During this century, Kerala's people have had quite remarkable success in improving their health, raising their levels of literacy and education, and bringing down the birth rate. And all this at a level of per capita income that has been lower than that for the country as a whole. Recognition of these achievements by policy-makers and researchers has led to a view of Kerala as representing one type of 'model' for social development and fertility reduction.

Figure 10.1 Despite Kerala's remarkable success in social development, it remains a poor region where life is hard work. (Above) Making thatch from palm leaves. (Left) Child worker sorting spices.

Some of the conclusions of this view tend at times to be a bit simplistic, insofar as they indicate a direct cause–effect relationship between government expenditures in education and health on the one hand, and declining fertility and mortality on the other. This chapter aims to show that a prior issue is what makes governments willing to commit themselves to improving health and education for the majority of the people they are supposed to be governing. This question is not a simple one. It involves precisely those processes identified thus far as critical in shaping and defining the public sphere. In examining some of these processes for the case of Kerala, the chapter

Figure 10.2 Stereotypical images of Kerala. (a) Tea pickers.

shows that the region does indeed constitute a kind of 'model' for development activity, but one whose lessons are more complex than is commonly supposed. It is not simply a story of a government and its relation to its people, but one about the shaping of the relationship of governance itself. This is what we are going to be looking into more closely below.

A final point: this chapter only deals with the origins of Kerala's remarkable social development. The region's development problems, particularly issues of economic growth, productivity and employment, are much more complex, but are not covered here. One area of intersection between the two sets of issues is the question of a possible trade-off between economic and social development as it manifests itself in government spending priorities, especially in a time of fiscal stringency. This question is both complex and unresolved at the present time. Nor do we address current health concerns such as the causes of high morbidity in Kerala.

10.2 Kerala: the land and its people

Situated in the southwest corner of India, the state of Kerala lies between northern latitudes 9 degrees and 12 degrees, and eastern longitudes 71 degrees and 74 degrees. The state has a long coastline of about 600 kilometres, but its width is nowhere greater than about 120 kilometres. This narrow strip of land accounts for roughly 1.18% of the total area of the country. It is bordered on the west by the Arabian Sea and on the east by the Western Ghats, a mountain range which runs down the length of the west coast of peninsular India. With an average annual rainfall of around 300 centimetres, and with as many as forty rivers running from the Ghats to the sea, Kerala has traditionally been blessed with a more than adequate supply of water. However, this picture has changed in recent times with growing deforestation in the high ranges of the state and growing demand for hydroelectric

(b) *Coconut harvesting.*

power. A chain of backwaters connected by canals runs parallel to the sea. The topography of mountains, rivers and backwater lagoons divides the land naturally into three regions; the highland, the midland and the lowland. The land is fertile and there is a considerable range of crops cultivated, including paddy (rice), coconuts, bananas, varieties of tropical fruits, spices, tea and rubber. The average temperature is around 32°C (90°F) with a high humidity level.

The economy of the state is largely agricultural with much of industry being agriculture-based. The share of the tertiary or services sector in the net state domestic product grew rapidly in the 1960s and again in the 1980s, so that it now accounts for the single largest component. Despite the relative importance of agriculture and the favourable agro-climatic conditions, Kerala has been a food deficit region since around the beginning of this century. In part this is because a lot of the cultivable land has been used to grow commercial and export crops which are only partially consumed by the local population. In part the reason lies in the relatively low ratio of cultivable land to population, which currently is under a quarter of an acre per person. Kerala's population is currently around 29 million, with a density of over 655 persons per square kilometre, as compared with a density of around 230 persons per square kilometre for the country as a whole. This gives Kerala a population density similar to that of Bangladesh.

The region's contacts with lands overseas goes back many centuries. One of the world's earliest Christian churches was established here by the apostle St. Thomas in the first century AD. In

Figure 10.3 The harbour at Calicut, northern Kerala, in the 15th century.

later centuries towns along the Kerala coast became regular ports of call on the trade routes between China, Southeast Asia, and the Arab world. Early Portuguese and Dutch settlers of the modern colonial era also found their way to Kerala, and their descendants like those of the Arab traders can be found in parts of northern Kerala. The region served not only as a natural harbour, but its narrow hinterland contributed considerably to the lucrative spice trade.

Kerala's social structure

Extensive foreign contacts notwithstanding, Kerala's social structure was for a number of centuries preceding this one, extremely rigid and hierarchical even for India. Swami Vivekananda, a religious and social reformer from Bengal, who visited Kerala in the late nineteenth century commented on the 'lunacy' of a caste system so rigid that there were groups who were not only untouchable but even unseeable and unapproachable. Clearly, long-standing contact with the external world, and the very early presence of Christian missionaries had done little to alter that structure at least until very nearly the end of the last century. In the modern period, Hindus constitute about 59% of the population, with Christians accounting for 21% and Muslims for 20% (Figure 10.4).

By the early nineteenth century the caste hierarchy was well entrenched, with the Namboodiri Brahmins (who never accounted for more than around 1% of the population) at the top; they were traditionally the large landowners. They were followed by the Tamil Brahmins who came into Kerala in the eighteenth century and were absorbed into senior positions in government service. The Nairs came next in the caste order, and were important tenants and landowners in their own right, as well as the main military caste. They were also often tax collectors, and functionaries in the lower levels of the government bureaucracy. Below them were the castes that were in one way or other considered 'polluting' by the ritual order. Highest among these were the Ezhavas who were smaller tenants, toddy-tappers, coir-workers and providers of various services. At the bottom were the Pulayas and Parayas who were largely the landless agricultural labourers and providers of menial services, and who were so-called agrestic slaves until slavery was abolished in various regions between the middle of the nineteenth century and the early twentieth century.

The Nairs and Ezhavas together accounted for 35–40% of the total population. This is a point of some significance to our story. The Nairs as an upper caste and the Ezhavas as an upwardly mobile caste probably had a symbiotic relationship that had profound implications for the social evolution of the region, and in particular the almost unique social position of Kerala's women. The Nairs were matrilineal and matrilocal; all property descended from mothers to daughters, and women stayed on in their mothers' families even after marriage. Although the official head of a Nair household was usually a man who was charged with managing the property and economic affairs, the matrilineal/matrilocal system gave women considerable autonomy in their personal lives and in decisions around the home (Box 10.1). Certainly daughters were prized among these castes, and there was none of the over-weening preference for sons that characterizes patrilineal and patrilocal societies. What is striking about traditional Kerala society was its extreme rigidity with respect to the caste hierarchy as against the flexibility of its gender relations. It is not often in history that one finds such a striking contrast.

Politically, during much of the period we are considering, from the early nineteenth century up to Indian independence from colonial rule in 1947, Kerala had been divided into three distinct regions. During the British period, there were three different governments in these three regions. The regions of Travancore and Cochin in the south were so-called princely states under their own Maharajas. Such states were not directly ruled by the British, but were overseen by a British 'resident' appointed by the colonial administration. The region of Malabar in the north was directly ruled by the British. It was a part of the larger Madras Presidency. These differences in political and administrative arrangements had a significant impact on the

Figure 10.4 *To the present day, Hindus, Muslims and Christians coexist in Kerala. Temples, mosques and churches, such as these in Trivandrum, are often only a few streets apart.*

Figure 10.5 One of the things which has contributed to the Kerala story is the relatively strong position of women in society: Malayali women and children near Cochin. The women with the earrings are probably Nair caste.

way in which the health and education story evolved in Kerala as we shall see. After Indian independence in 1947, there was considerable internal political reorganization in the country. The current state of Kerala was formed in 1956 by amalgamating the erstwhile princely states of Travancore and Cochin with the linguistically compatible region of Malabar.

Important to our story is the nature of popular political participation in the state. The state claims to have had the world's first elected Communist government in 1957. Before this went decades of intense social and political ferment, including political organization, social reform movements, caste organization, and trade union activity. Some of this was linked to the nationalist movement for independence from colonial rule, but a great deal was not. It is probably no exaggeration to say that no other region in India experienced so thoroughgoing a transformation of ideas and institutions as did Kerala during the period we are considering.

10.3 Social Development: Kerala's story

Probably the simplest way to tell this story is to begin from where we stand now and work backwards. Comparing Kerala's position in health

Box 10.1 The system of *marumakkathavam*

Matrilineality and matrilocality or some mixed variant of it have been estimated to have extended over more than half of Kerala's population until the 1930s. Property was inherited by and through women, and a typical household consisted of a male head (who was its manager), his sisters, and their children and grandchildren. Husbands typically had visiting rights, and may in some instances have chosen to stay in their wives' households. Girls had considerable say over the choice of marriage partners, and the rules for divorce and remarriage for women were quite liberal, particularly when viewed in contrast with the rest of the subcontinent. Partly for this reason, the mean age at first marriage for women was as high as 17 years as early as 1901, when the average for the country was only 13 years.

Matrilineality and matrilocality do not mean that women dominate in household decision-making or power. Often it is still a man who takes the major decisions affecting the household. But they do mean that women come to have greater say over their own lives, and are valued more by the household and by society at large. This is strikingly evident in Kerala in contrast to much of the rest of India (barring some hill tribes). For example, the sex-ratio (i.e. the number of women per 1000 men in any given population) in Kerala has always been favourable to women. In most societies that have solved basic problems of physical deprivation, the life expectancy of women is longer than that of men. In India, as in a number of other developing countries, the reverse is true, partly because of high rates of maternal mortality caused by poor medical care for women in the reproductive ages, and partly because girl babies and girl children generally get poorer healthcare and nutrition because of their parents' preference for sons. Girls and women in Kerala have not, by and large, suffered from these disadvantages.

An exception to the above is the traditional position of Namboodiri women. Since the eldest son in a Namboodiri family inherited all its land, only he was allowed to marry formally into the Namboodiri caste. Younger sons formed liaisons with women from Nair families, and these were generally looked upon with favour. However the sexual and caste 'purity' of Namboodiri women was protected. So large numbers of them remained unmarried and without any sexual relationships. They were sequestered within the home, and were known as '*antharjanam*', or 'inner person'.

and education with that of the rest of the country brings out the sharpness of the differences. Table 10.1 not only does this, but it also compares Kerala with an average for countries classified internationally as belonging to the upper middle-income group. These include some of the most developed Latin American countries such as Brazil, Mexico, Argentina and Venezuela, some of the East Asian 'tigers' like South Korea, Singapore and Hong Kong, and a number of southern and eastern European countries such as Greece, Yugoslavia, Portugal, Poland and Hungary among others.

The comparison between Kerala and the group of upper middle-income countries was chosen to point out that while Kerala's per capita income is currently around one-ninth the average for this group of countries, yet on almost every indicator of performance in basic health and education the state's performance is at least as good if not better.

Q What accounts for this remarkable performance?

Kerala is not unique in combining low per capita income with excellent social development; Sri Lanka is another such case. But these cases are rare enough to have drawn considerable attention from policymakers and researchers in the last decade and a half. In Kerala's case, improvements in both health and education date back to the early part of this century, and in some instances even to the latter part of the nineteenth century. A closer look at the history of the

Table 10.1 Social development indicators, 1989

	Kerala	*India*	*Upper middle-income countries**
GNP per capita (US$)	<200	290*	1890
Crude birth rate	22.4	33.6	27.0
Crude death rate	6.2	11.9	8.0
Total fertility rate	2.1	4.4*	3.5
Infant mortality rate	27	96	50
Maternal mortality rate	1.3	3.5	1.2
Life expectancy at birth			
(male)	67	55	64
(female)	70	54	70
Sex-ratio (women per 1000 men)	1032	931	
Literacy rate#	70.0	36.0	
(male)	75.0	47.0	
(female)	66.0	25.0	
Population per registered doctor	1994*	2529*	1380#

[Data sources: Government of Kerala, 1989; World Bank, 1989; *Census of India*, 1981]

* 1986 data

\# 1981 data.

nineteenth century is necessary if we are to understand why the statistics for today's Kerala are as good as they are.

Three factors in Kerala's success

Three factors appear to have contributed significantly to Kerala's health and education experience.

The first is the relative autonomy of the government in two out of the three main subregions during the colonial period. The fact that the subregions of Travancore and Cochin were ruled by native rulers positively affected the ability of the government to spend on health and education, and to use relatively innovative methods. It also enabled it to build on, rather than destroying, the existing system of local schools, and the best of the indigenously developed medical systems and healthcare practices.

Interestingly, in the matter of laws and regulations for both education and health, Travancore and Cochin followed the lead of the neighbouring, British-ruled Madras Presidency. However the implementation of the laws was much better in these native states, and accounted for their strikingly better performance.

One of the reasons for this was the traditionally unusual position of women in a society where, as mentioned before, an important social caste was matrilineal and matrilocal. The position of women in Kerala society ensured that when health and education expenditures were made, they did not exclude women either as workers or as recipients. This is the second factor behind the region's performance. Women's status with respect to land-holding, inheritance and the family system made it possible for the society to absorb health and education expenditures more effectively.

The third factor was the surge of social and religious reform movements that shook the foundations of the region's social structure during the latter part of the nineteenth and the early part of the twentieth centuries. As it functioned in the earlier part of the nineteenth century, the caste structure must have excluded large sections of the people from participating in defining social needs or affecting the way in which rulers responded to their aspirations. The reform movements brought about a democratization of society through which many of the hitherto unheard groups in the population could articulate their demands and legitimize their needs. In so doing they completely transformed the meaning of the public space and beliefs about the proper role of government. They ensured also that the benefits of social development percolated down the social hierarchy to include much of the population.

Processes of change

These three factors were the main ones behind Kerala's exceptional health and education story. Conditioning them was the growing process of commercialization during the previous century that had linked the region ever more strongly to the world economy. As we saw earlier, the region had for many centuries been involved as a producer and as a natural harbour in the spice trade. With the growing consolidation of British rule in India, British capital became increasingly involved in the expansion of plantations and agro-processing. Most of this was production for export. Thus, tea, rubber, spice and cashew plantations grew and produced largely for the export market. In Travancore, a major industry based on processing coconut fibre into coir rope also expanded in this period, as did the manufacture of tiles.

The expansion of the coir industry was important in Kerala's history in at least two ways. It was for many decades, and continues to be one of the major sources of industrial employment in the region and particularly so for women, who constitute the bulk of this labour force. The industry was also important because its expansion improved the economic fortunes of the Ezhavas who were the traditional workers with coconut produce. Thus, it raised the aspiration levels of this traditionally lower caste, and made them a powerful force for social and religious reform.

It has also been argued that, as job opportunities in the plantations and commercial sectors increased, there was a growing demand for basic education so that people could handle bookkeeping, contracts, mortgages, and so on (Tharakan, 1984). It is difficult to know what weight to attach to this explanation, especially if it is viewed as a factor behind the relative importance given to *primary* (as distinct from higher) education in Kerala when compared to the directly ruled parts of British India. Very probably, however, commercialization did play some such role.

Another important set of changes provided the background for the working of the three factors

Figure 10.6 A worker baling coir for export.

identified above. The nineteenth century saw very significant moves towards land reforms in the regions of Travancore and Cochin. These came on the heels of a process of political centralization (through annexation) which, in the first half of the eighteenth century, combined the many independent principalities of the south into the unified state of Travancore. A half century of extended peace and stability followed. The annexation of petty chiefdoms also consolidated almost two-thirds of the cultivated area under direct state ownership. Early in the nineteenth century the state began to encourage the cultivation of waste lands by passing a Government Regulation in 1818 which made such cultivation tax free for ten years, allowed claims for the cost of improving the land, and assigned waste lands to those who were willing to cultivate them. A royal proclamation in 1865 conferred ownership rights on tenants on state lands. By the end of the century, tenants on private lands were guaranteed security of tenure. By all these acts the state of Travancore helped to strengthen the class of middle and small peasant proprietors.

A number of the beneficiaries, especially of the waste lands legislation, were Ezhavas and Syrian Christians who would not traditionally have had access to land. The reforms therefore worked to improve these people's economic position; raising their desire and ability to struggle for social reforms and for a greater share in the rewards of commercialization and in the expansion of government employment. In particular, the Syrian Christians who had already begun to accumulate capital through trade, took advantage of the waste lands legislation and of the ownership rights conferred on government tenants to make rapid strides in commercial agriculture. They became the most important group engaged in the cultivation of commercial and plantation crops in the region.

The native state of Cochin followed the lead given by Travancore in the field of land reform, and enacted similar legislation by the end of the nineteenth century. The region of Malabar in the north provided a sharp contrast to both Travancore and Cochin in this regard. For various historical reasons no process of political centralization took place in Malabar until the British absorbed the region into one of its Presidencies. British rule did not break the hold of the large landowners and chiefdoms. Instead it consolidated and strengthened their control over land by converting their traditional rights into formal ownership. The relationship of these owners to the temples gave them further control over the considerable amount of land held by the temples. Thus, direct rule by the British strengthened a landed oligarchy in Malabar, as it did in many other parts of India. Consequently, a strong middle group of aspiring peasants did not emerge in the same way in Malabar. This may well have contributed to Malabar's relative backwardness in social development when compared to the other two subregions.

A third set of processes that began to accelerate in the nineteenth century social background were the changes within the Christian community. As mentioned earlier, the Syrian Christian church is among the oldest in the world, and dates back to the first century A.D. Before the arrival of Portuguese and Italian missionaries, the religious life of the community was organized in relatively autonomous self-governing units with the local churches providing the focal points. The local priests were chosen by the community and were maintained by them. The arrival of Portuguese Jesuits and Italian Carmelites increasingly eroded the power of the local communities over the churches, as priests began to be appointed by the Bishops who were themselves European. This centralization of power had occurred by the nineteenth century among both the Syrian Catholics and the Syrian Mar Thomites who were largely Anglican.

The advantage of hierarchical organization in the churches was that it was easier to consolidate resources and channel them to chosen activities. But it also had a negative influence in that the European Catholic prelates, for quite a long time, did not believe in the value of secular education (Tharakan, 1984). The growing prosperity of the Syrian Christian community in the nineteenth century generated a demand for education, as well as pressure for entry into the upper levels of

Figure 10.7 A bishop and priest of the Syrian Church, by a Malabar Coast artist, circa 1828.

the church hierarchy. Both these forces worked to make the Catholic churches extremely active in the educational field in the latter part of the century. This process was helped along by healthy competition from the Protestant missionaries who had always recognized and valued education. Anglican reliance on the reading of the Gospels meant that literacy was a prerequisite for religious conversion. It was the Anglicans who provided the leading edge to the rapid spread of educational institutions among Christians in this period.

The three processes discussed above (commercialization, changes in land ownership and organizational changes within the Christian churches) worked together synergistically to create an economic and social background that was open to pressure for improvements in education and health. In what follows we will first see in greater detail what happened in the field of education and then go on to use that discussion to tell us more about health.

10.4 Education: vernacular and primary

As was true in much of the rest of India, there was a fairly well developed system of popular and higher education in traditional Kerala prior to our period. For centuries, higher education (largely scriptural learning in Sanskrit) had been the almost exclusive preserve of the Brahmins. In the two or three centuries preceding the nineteenth, other upper castes such as the Nairs had begun making inroads into that preserve. In part this may be seen as the unsurprising result of the growing economic and political eminence of that caste. It also points to the continuing importance of religious authority, since higher education was still largely education in Sanskrit. Institutions of higher learning were supported by gifts and land grants from the rulers.

There was also a network of popular schools which catered to the needs of other castes. The existence of this network probably owes a great deal to the early Buddhist and Jain influences in Kerala. Both of these religions, developing as they did in reaction to some of the perceived hierarchies of Hinduism, placed considerable emphasis on popular learning. These schools were largely maintained by the support of the local communities themselves. They taught basic reading, writing and arithmetic, in addition to a little medicine, astrology, the arts and ethics. Some schools were supported by the local temples or the wealthy of an area.

It is hard to come by reliable estimates of the extent of literacy at the start of our period. Although it is clear that literacy and numeracy extended considerably beyond the Brahmin caste, it is probably the case that most of the lower castes, the majority of the untouchables, and the majority of women were excluded from both popular and higher education. However, the existence of a network of popular schools that were locally supported and funded did provide an opening for economically rising castes and communities to bring literacy to their people when the demand arose. Thus, for example, there were educated Ezhavas even before the mass movements for social reform that transformed that caste in the late nineteenth and early twentieth centuries.

Traditional education clearly fulfilled two functions.

1 Its higher levels met the requirements of religious ritual, philosophical discourse, literature and the arts.

2 Its popular level incorporated sections of the people who were at the middle of the social order into the emerging economic and administrative systems. It also opened to them such higher service functions as medicine and astrology in which each village was largely self-sufficient.

Popular education also brought the religious order within the purview of non- Brahmins, albeit only in its epic- and parable-telling aspect, and not in its ritual aspect.

The emergence of 'modern' education

Although the traditional system of education was fairly widespread, it was in the nineteenth century that rapid strides were made in 'modern' education. There was a growing demand from both government and commercial circles for people trained in the newly emerging school system. The political centralization of Travancore in the eighteenth century was followed by its coming under British overrule in the early nineteenth century. The British began to introduce a series of administrative changes which were, no doubt, intended to facilitate their control over the state. In the process, they laid the foundations of the secretariat system of governance and of the setting up of distinct administrative divisions. Large numbers of locally trained clerks and accountants were demanded by the new system, and the schools were expected to produce them.

The judicial system was separated from revenue collection, and judicial regulations were spelled out. This was interesting. Elsewhere in British India where the British ruled directly, the (revenue) Collector of the district was its chief administrative officer as well as the district magistrate. Obviously the British did not want such a consolidation of powers in the native states.

Elsewhere in India, in the regions that were ruled directly, such changes usually meant an increase in higher education in the English language. In Kerala, presumably because administration continued to be conducted in local languages under native rule, there was not so strong an emphasis on the English language. Largely for this reason, the existing system of schooling was not totally destroyed as it was in many other parts of the country.

An active policy of encouraging primary education was enunciated early in the century. A royal proclamation in 1817 promised financial support to schools; among other things the salaries of two teachers in each school would be paid. These would be teachers of Malayalam, Tamil, Arithmetic, and Astronomy. Although the British administrators elsewhere in India pronounced similar support for primary and vernacular

Figure 10.8 (Above) Education in Kerala in the eighteenth century was more widespread than this print of a medical missionary and students would suggest. (Left) A street vendor reading as she waits for customers.

education, they made little headway, since the weight of job creation in the administration and judiciary favoured education in the English language. Furthermore, there was little active support for the traditional schools in the areas of direct rule. As a result, popular education all but disappeared from the purview of large sections of the people in much of the rest of British India.

Clearly, even the traditional schools in Kerala had to undertake considerable changes in curriculum to match the requirements of the new jobs that were being created. The schools started by Christian missionaries stepped into the breach. Rising aspirations within the Christian community, combined with the missionary attempt to bring literacy to backward castes and women who had largely been ignored by the traditional education system, led to a rapid expansion of missionary schools. These schools took advantage of the available financial support from the government, and vied with each other for pupils.

In the first half of the century the government did provide some support to vernacular education, but this was not very strong. In the second half, the Travancore government started a number of vernacular schools which initially faced stiff opposition from the older indigenous schools. Through a flexible policy of co-opting the old teachers by appointing them to teach traditional subjects (astrology, music, religion) in the new government schools, the government was able to soften their opposition (Jeffrey, 1987). Systematic attempts were made thereby to integrate the indigenous schools into the new system.

The government started a separate department of Vernacular Education, provided grants-in-aid (based on the number of pupils on a school's rolls) to vernacular schools, and organized both the preparation of new textbooks and translation of books from Sanskrit and English into the vernacular. Teacher training was also organized under government auspices. All this is not to say that English education did not grow by leaps and bounds in the region. It did. But the shift to English language education was undertaken largely by those who aspired for higher positions in commerce and administration. The trilinguality (Malayalam, Tamil, English) of administration and commerce meant that lower level and even middle level jobs could be filled by those whose education had been in the vernacular.

Although in all probability girls had traditionally gone to school much less than boys, the rapid changes of the nineteenth century did not pass them by. With the active support of the government, girls started going to the new vernacular schools in large numbers. The traditionally high status and autonomy enjoyed by Nair women meant that they shared in the growing education of their community. There were few taboos against girls being seen in public spaces even after puberty. In contrast, such taboos militated against female education (and continue to do so even today) in many parts of northern India. Jobs opened up for educated women in the newly active departments of education and health. The Nairs as a caste had always been emulated by those below them; this held good in this case also. As a result, the female literacy rate in 1941 was 36% in Travancore, and 31% in Cochin, as compared to only 7% at the all-India level.

Kerala's early inroads in education, particularly female education, gave it a head start which the rest of India has not caught up with till today. According to the 1981 census, the state's overall literacy rate was 70% compared to 36% for the country as a whole. Female literacy was 66% while male literacy was a little higher at 73%. For the country as a whole, only 25% of women were literate. Clearly, Kerala's performance owes much to the fact that the female half of its population has not been left out in the cold. Indeed, if Kerala's women had been only as literate as their sisters in the rest of India in 1981, the overall literacy rate in Kerala would have been only 49%, i.e., it would have been 21 percentage points lower than what it actually was.

Movements for social reform

Movements for social reform played no small role in bringing education within the purview of the poorer members of castes like the Ezhavas and even the Nairs. These movements started initially in the late nineteenth century with demands for access to government employment. Even the Nairs found at this time that, despite their growing educational achievements, the higher levels of government jobs continued to go to Tamil Brahmins. Similar grievances on the part of the Ezhavas and the Syrian Christians led the three communities to join together to petition the Maharaja of Travancore in 1891,

Figure 10.9 Female literacy in Kerala is significantly higher than anywhere else in India.

through the so-called 'Malayali Memorial' which was signed by over 10 000 people. The petition requested entry into government jobs. As a result, some jobs were opened up to Nairs and Syrian Christians, but the Ezhavas continued to be excluded. In 1896 a separate 'Ezhava Memorial', signed by 13 000 people was presented but it was also rejected.

These setbacks only served to unite the Ezhava community; under the leadership of Sri Narayana Guru, who was a spiritual mystic as well as a social and religious reformer, the Sri Narayana Dharma Paripalana Yogam (SNDP) was formed in 1903. Its express purpose was the social uplifting of the Ezhavas and other lower castes, reform of the religious order, and political change. One of the principal tools used by the SNDP was education. Another was the setting up of temples and related institutions which were similar in modes of worship to caste Hindu temples, but which were open to the lower castes, and in which the priests themselves came from such castes. This democratization of the Hindu religion and the spread of education among the Ezhavas raised their self-image and provided a much stronger basis for their continuing struggle for economic and political inclusion.

In 1906, three years after the formation of the SNDP and fifteen years after the submission of the 'Malayali Memorial', government schools in Travancore began to be opened to Ezhavas and other lower castes. The Ezhavas also began agitating directly against the taboos of the caste system. Prolonged agitation for the opening of temple roads culminated in the Vaikom Satyagraha in 1924 in which Gandhi and the Indian National Congress played an active role. Although this particular struggle ended in a compromise, the movement against the caste system only gained momentum. The 1931 census in Cochin noted the refusal of the lower castes to accede to untouchability, and called the social changes that had taken place 'revolutionary'.

The social and religious struggles of the first quarter of the twentieth century were followed by political struggles in the 1930s. Ezhavas, Christians and Muslims joined together in the 'abstention movement' and refused to vote for the state legislature in Travancore in which they did not have representation. They also demanded government job allocations proportional to their representation in the population. The extent of social ferment in the region in the first half of this century clearly played an important role in altering the attitudes of both lower and upper castes towards the entire social and political order. Although the castes below the Ezhavas were not so successful in improving their lot, they too emulated the Ezhava organization, and achieved a measure of success in both education and health. According to the 1981 Census, the literacy rate of the formerly untouchable, so-called 'scheduled caste' population was 56% in Kerala, as compared to 21% for the country as a whole.

Improvements in literacy in Travancore and Cochin began, as we have seen, in the early part of the nineteenth century. The major movements for social reform occurred, on the other hand, only towards the end of that century and the beginning of the next. But the early expansion of education benefited most those castes like the Nairs who already had a dominant position in the social and political order, and the Syrian Christians who had a command over economic resources and access to missionary institutions. It was the movements for social reform that helped those lower on the social and economic ladder to give voice to their aspirations, and to bring education within their reach. In a sense, this illustrates the point that government or missionary activities in the field of education will largely work within the bounds of the existing social order, unless pressure from below forces an expansion beyond those bounds. Hence although Protestant missionaries had been trying to expand education among untouchable castes much earlier, it was really the internal struggles against the caste order that finally made that expansion possible on a significant scale.

10.5 Health: public and private

We have already seen in Table 10.1 that the crude death rate in Kerala is currently 6.2 per 1000, which is just over half the rate for the country as a whole. The infant mortality rate at 27 per 1000 is around a quarter that for the country, and life expectancy is 12 years longer than the country's average for men, and 16 years longer for women. Data on morbidity (i.e. rates of illness rather than death) present a more mixed picture, and Kerala does not appear so strikingly superior to the rest of the country. This is partly explicable by the fact that health service institutions are used more by people in Kerala. The recorded data may thus capture more actual cases of illness in the state than in other regions where people do not have access to such institutions, and where much illness is therefore never counted. However the high morbidity data may also be pointing to a more serious problem. That is, a good healthcare system can only prevent deaths from disease up to a point, and in a situation of very low per capita incomes people may still be quite malnourished and disease-prone.

Bearing that caveat in mind, we can look at the development of public health and healthcare institutions in the region,which is still rather remarkable. Similar forces to those that led to the spread of education; that is, the position of women, the fact of native rule, and social reform movements appear to have been at work in health but with some differences of degree and emphasis. An additional and important factor was the spread of education itself, which played no small role in the growing demand for healthcare, and growing local participation in determining the quality of public and private health facilities.

The role of native rule

It is not clear how much one ought to emphasize the fact of native rule *per se* in affecting health in Travancore and Cochin. Certainly, as in the case of education, the same public health codes were adopted in these two subregions as in Malabar, the part of the state which was under direct British rule. But they were implemented much better in Travancore and Cochin. Indirect evidence of this is the estimate that the infant mortality rate may have been twice as high in Malabar in 1956 when the modern state of

Kerala was formed. Considerable emphasis had to be given during the post-1956 period to improving both public health amenities and availability of medical institutions in Malabar in order to bring its facilities up to the level prevalent in the other two subregions. But, was this difference between the subregions due to native rule in itself, or was it due to the greater spread of literacy and education and the impact of the latter on health in Travancore and Cochin? In the latter case, the impact of the institution of native rule would have been indirect.

There was, however, one important direct benefit from native rule. There was greater governmental support for systems of traditional medicine such as the Ayurvedic system. Thus, the government of Travancore began to give grants-in-aid to Ayurvedic dispensaries at the same time as it did for allopathic institutions, and set up an Ayurveda College in 1895. As a result, a system of medicine that is in many ways complementary to the allopathic system did not die out as it did in much of British India. In fact, till today, Kerala's people, rich and poor, tend to use a combination of Ayurvedic and allopathic treatment to good effect.

Whatever the direct or indirect benefit of the institutional framework of native rule, there is no question but that particular native rulers in Travancore and Cochin provided strong backing to health efforts, both public and private. Significant missionary activity in this field began with the setting up of a dispensary at Neyoor in Travancore by the London Missionary Society in 1854. This was later expanded and upgraded to become a major hospital. Other dispensaries and hospitals were opened by different Christian groups in the following decades. Health care provision, probably even more than education, was a big attraction especially among the poorer people, and in the climate of competition among different churches, medical institutions providing curative care based on the allopathic system of medicine grew rapidly. The Travancore government began to give grants-in-aid to medical institutions in 1895, and as in the case of schools, this provided a fillip to the expansion of these institutions. Such institutions not only provided curative care, but in many instances worked closely with the public health institutions of government e.g., in the area of vaccination. They also performed an important function in training women as midwives and nurses to work in rural areas. In effect, the early beginning that Kerala made in training women as nurses led to the region's becoming a major supplier for hospitals all over India. At a time when taboos of both caste and gender acted against female nurses elsewhere, Kerala's young women migrated as nurses to quite distant corners of the country.

Figure 10.10 The signs of Kerala's history of health care are abundant in all parts of the state: doctors' signs in Kottayam, 1991.

Public health

The first major activity of the modern period in the field of public health began in the middle of the nineteenth century with a thrust in the area of vaccination and public sanitation. A new government department of vaccination was set up in the 1860s in Travancore with strong backing from the royal family. It must be remembered that under the rigid caste order of the time, vaccination involving physical contact was a tricky business and faced considerable opposition from the upper castes. The importance of the Maharaja's support must be judged in this light, backed as it was by a royal proclamation in 1878 making vaccination compulsory for all public servants, hospital patients, convicts and students. In 1888-89 a vaccine depot was set up to produce calf lymph as a substitute for the practice hitherto of hand-to-hand vaccination, and this helped to soften the caste opposition to vaccination. House to house vaccinations were also started in this period so as to increase the coverage of the programme.

Public sanitation measures started in the 1880s in the towns of Travancore — Trivandrum, Quilon, Alleppey and Kottayam.

- Public toilets were built.
- Tanks and feeder canals were cleaned.
- Fish markets were relocated.
- Burial and cremation grounds were brought under closer supervision.
- Town Improvement Committees were set up whose job it was to oversee all sanitation activities in the towns.

Rural sanitation activity began in 1888-89, and was given a boost by the cholera epidemic of the 1890s. In 1895 a separate Sanitation Department was set up which gave considerable emphasis to sanitation during fairs and religious festivals, traditionally the site of the spread of epidemic diseases. Both sewage facilities and cleaning and chlorination of wells and other water sources received attention.

After 1927 a new thrust was given to Travancore's public health by the entry of a Rockefeller Foundation mission at the invitation of the state government. Vector-borne diseases such as filariasis, malaria and hookworm, and their entomology began receiving attention. Greater emphasis was also given to the collection of vital statistics and to training, and to measures to improve maternal and child health.

The state of Cochin followed closely on the heels of Travancore in introducing both vaccination and public sanitation. However, the system in Cochin was more decentralized, with greater reliance on rural councils (*panchayats*) for both physical and financial support for measures such as water protection. This openness to popular participation was one of the strengths of the public health system in Cochin.

Education campaigns were obviously of great importance in spreading an awareness of public health issues. In this, the expanding educational system of the region and growing literacy played a significant role. In 1928, when the Rockefeller Foundation team assessed Travancore's potential for spreading public health measures, its report noted that there were as many as 110 publications which could be used to spread the message of public health. Educational institutions were also an obvious and important site for spreading awareness, and for actual public health measures such as vaccination. Today, among ordinary people in Kerala, the extent of knowledge of diseases and their basic causes is quite remarkable.

Women and public health

Public awareness was also aided by the participation of women in the emerging health system. In most societies women are responsible for meeting the routine healthcare needs of their families. As such, they are usually the repositories of traditional knowledge about health maintenance, disease prevention and cure. With the professionalization of healthcare, medical knowledge has often gone out of women's control, but they still remain the care providers of first resort in most instances, especially for the routine health needs of family members. Thus the importance of women's participation for the advancement of primary healthcare and of public health cannot be stressed enough.

In the nineteenth century in Travancore and Cochin, women had to learn the value of vaccinations for the health of both children and adults, and also to learn to use the dispensaries and hospitals for medical care that were being set up. Learning the value of the new allopathic medical practices had to take place both through publications and through the school system. Literacy was clearly an important prerequisite for this, but according to the 1901 census, the combined (male and female) literacy rate was under 15%. Clearly then, there must have been a synergy generated by the schools and by public debate around the new health measures espoused by the government which acted beyond the simple numbers of those literate. Perhaps also, as often happens, children taught parents and other adults what they were learning in school about health and hygiene.

And what children were learning in school was a great deal. In the area of health particular attention was paid to the education of girl students, whatever we may think of such gender biases in education today. For instance, girls could replace the study of compulsory subjects such as geography and history by the study of domestic science and hygiene in Travancore's schools. In Cochin after 1900, housekeeping and nursing were required subjects in the final examinations of all girls' schools. The study of nursing included both theory and practice at local hospitals. In Travancore, teacher training for both English language and Malayalam high schools came to include hygiene as a required course. Learning about hygiene through the schools must have been important in fostering ideas about public sanitation at a time when the local governments were making major attempts in that area.

Nursing on the other hand would have been of considerable importance in the newly emerging system of medical care. The first trained midwives in Travancore date back to the 1870s. By 1901 there were 32 licensed midwives, 6 nurses and 8 vaccinators, all women, attached to government departments in Travancore. Women health workers helped reduce rates of infant and maternal mortality, in addition to providing preventive and curative care. Nor were they only present at the lower levels of the health hierarchy. In the 1930s, well before the rest of the country, there were women surgeons in Cochin. In 1924 a woman doctor became the chief medical officer of the Travancore government, a post she held for 18 years. Syrian Christian and Nair women appear to have been particularly well represented in the ranks of Kerala's women health workers.

In addition to their importance in advancing public health and medical care, the fact that Kerala's girls began entering school in growing numbers had a direct impact on their own health. Unlike girls elsewhere in the country, Kerala's girls benefited like their brothers from the programmes of vaccination and school sanitation that became mandatory in the schools. In all these ways, one

Figure 10.11 The importance of women's education has had positive repercussions in healthcare provision: a girls' classroom in Kerala.

might argue that women not only made Kerala literate (Jeffrey, 1987) but also healthy.

Public awareness

In addition to the role played by the institution of native rule and native rulers, and by women, public awareness around health issues became a major factor determining the availability and quality of healthcare. Literacy and government campaigns raised public consciousness, which was then further promoted by means of social reform movements and the growth of the trade union movement and of political parties. We have already discussed the importance of some of these mechanisms. Kerala also has a relatively long history of militant peasant and worker movements which fed into its strong communist parties.

In the early part of this century, much of the organization of the peasantry in the state took place in Malabar, which was under direct British rule, and thus did not experience any of the changes in landholding patterns witnessed in the other two subregions. As we have seen, Malabar lagged behind the others in the matter of education and health. It appears plausible, however, that although the militancy of its rural organizations may not have had an impact on Malabar's own social development, the general political ferment in a relatively small geographical area would have affected the other two regions, where improvements in health and education were already in progress.

Public awareness appears to have affected the developing healthcare system in three ways. The first was by improving compliance with the new requirements in such things as sanitation and vaccination. Visitors to Kerala are always struck by the remarkable level of public sanitation and awareness thereof in a region with such a low per capita income. The second was by making people willing to question and, if necessary, challenge the priorities and methods of the emerging system of healthcare. A good early example of this was the response to the Rockefeller Foundation's activities in the state of Travancore.

As mentioned, the Foundation had come into Travancore at the request of the government of Travancore in 1927, and it had made significant contributions to reshaping the priorities of the

Figure 10.12 Evidence of a tradition of political militancy in Kerala. (a) The Communist Party logo on a Trivandrum street 1957.

government's healthcare programme in needed directions. But some of its programmes were also criticized. For instance its hookworm eradication campaign was questioned for not being combined with an adequate programme of rural sanitation in the area in which it was conducted. Critics argued that permanent eradication of the disease could only be guaranteed by adequate sanitation. The Foundation was also attacked for the model healthcare unit it insisted on setting up at Neyyatinkara. The Foundation's representatives clearly wanted this model to become a showpiece for rural health, but the demands the unit made on both finances and on trained personnel were considered exorbitant by some members of the Travancore Legislative Assembly. They argued that Travancore had neither the funds nor the personpower to extend the 'model' to the rest of the state. Hence funding it on such a scale was not only fruitless but also was drawing resources away from more urgent needs elsewhere. The Foundation had also been insensitive to the emerging politics of the social reform organizations which were trying to break the monopoly of the dominant castes over government employment. When the term of the Foundation's contract ended in 1937, it was not renewed, and the Foundation left Travancore a bare ten years after it had begun to work there. Whatever be the merits of the case on each side, the history of the Foundation's presence in Travancore points to the fact that the growing literate public, the reform movements, and the legislature all took a strong interest in priorities and methods in the field of social development.

(b) *A bookshop in Trivandrum, 1991.*

A third way in which public awareness affected the healthcare system is through accountability enforced by people at the grassroots level. A typical problem faced by publicly funded primary healthcare in countries with large rural populations is one of accountability. Often only the building designated as the primary health centre stands; neither doctors nor nurses visit, medicines and equipment are rarely available, and mandated services are not provided. Although the reasons for such a state of affairs are many, one of the surest ways to cure it is through mobilization by the people at the base. In Kerala, widespread literacy and social and political organizing have made people willing to demand accountability from both the education and health systems. Observers have noted that services are, in general, much better in Kerala than in the rest of the country, and this state of affairs is due to the willingness of people to join together to demand accountability from the systems and their employees. Public awareness has contributed in no small measure to broadening and democratizing the definition of social needs in the region, and to enforcing accountability.

10.6 Conclusion

In the development literature today, the 'Kerala model' of social development has achieved some recognition. The policy prescription flowing from it appears to suggest that emphasis on education and health in government policy can pay handsome dividends despite low levels of per capita income and in the absence of rapid rates of economic growth. A closer look at Kerala's history provides a more complex picture. Certainly the role of government was important in this history. But government did not always function as an initiator of social change. Government action was as often as not a response to the pressures of social movements in a context that was in many ways favourable to change.

At the start of the nineteenth century, the remarkable aspects of Kerala were

- the growing importance of commercial production for export
- the unusual position of the women in one of its dominant castes
- the rigidity and extreme oppressiveness of its caste structure.

The legitimate field for public policy and the definition of public needs were clearly determined traditionally by the latter two attributes, though not perhaps without resistance from below. At the end of the century, Kerala still held these attributes, but the rumblings of change had already begun to be heard.

Fifty years later Kerala society had been completely transformed. A process of democratization had altered the meaning of the public sphere, with the result that large numbers of people could participate both in the definition of social needs, and in assuring accountability for their fulfilment. The mechanisms by which accountability got built into the system included

- political agitation and organization on a large scale to increase participation and access
- local activity aimed at lower-level health and education institutions, to ensure that once they were set up, they did function as they were expected to.

Accountability, in both the positive and negative senses identified at the start of this chapter, became part of the structure. If there is a moral to the so-called 'Kerala model', it is that government action, in order to be successful, requires popular participation. The definition of the public sphere evidently must incorporate both government and people.

Summary of Chapter 10

1 The 'Kerala model' of social development is *not* the outcome of a simple cause-and-effect relationship between government expenditures in education and health on the one hand and fertility and mortality declines on the other. Instead, this chapter argued that, if there is a moral to the so-called 'Kerala model', it is that government action, in order to be successful, requires popular participation.

2 Kerala's success, therefore, was the outcome of protracted historical processes in which large sections of its population came to participate both in the definition of social needs, and in assuming accountability for meeting those needs.

3 These historical processes involved the transformation under colonial rule of Kerala's traditional society, characterized by extremely rigid caste hierarchy but flexible gender relations. Native rule and the relative autonomy of the local states, the unusual position of women in society, and the weakening of the caste hierarchy under the impulse of social and religious reform movements, laid the basis for a democratization of society which enabled large sections of the population to articulate their demands and legitimize their needs.

4 This combination of a responsive state, public awareness and social reform movements rooted in popular participation led to a remarkable spread of education and public health. Pressures on the state made governments willing to commit themselves to spend money on education and health; public awareness and participation made them effective. The unusually strong position of women in society ensured that when health and education expenditures were made, they did not exclude women either as workers or as recipients.

11

CONCLUSION: DEVELOPMENT POLICY AS PROCESS

MARC WUYTS

The aim of this book was to examine development policy as a social and economic process. To this end, we have developed analytical concepts and explored concrete examples. Throughout the book, we have stressed that policies operate, and therefore have to be understood, within specific historical contexts. In the Introduction we contrasted this approach with one which treats policy as prescription. In this Conclusion we want to deal with the most common criticism of the 'policy as process' view: that is, that no general lessons can be drawn from such analysis. On the contrary, we shall argue that we can only draw general lessons for participants in the policy process if we face the fact that policies do not operate in a social vacuum, but, instead, are the outcomes of complex social and economic interactions.

This conclusion therefore draws together some general threads of argument developed in this book and, in this way, seeks to clarify some connections between its different chapters. Consequently, it does not summarize the arguments in the various chapters of this book. The concern here is rather to establish some key points which connect chapters, and establish the outlines of a broader argument about policy as a process. The conclusion is therefore focused on five major lessons which can be learned from the various contributions in this book.

11.1 The public sphere as socially constructed

The first point concerns the definition of the public sphere itself. This book challenges the prevailing view that the definition of the public sphere, its content, its organization and its financing, is a *technical* matter which can best be left in the hands of economists, public administrators, or experts in general. Chapters 1 and 3 showed that this view is prevalent in orthodox economic theory, which argues that the definition of public goods, and the public sphere which surrounds them, is a technical issue based in categories of market failure. Similarly, Chapter 7 argued that the IMF and the World Bank tend to present their structural adjustment policies and their implications for the public sphere as a matter of technical economic necessity and efficiency. In parallel fashion, Chapter 10 noted that the 'Kerala model' of social development and fertility reduction is frequently used by development and population experts to prescribe a greater emphasis within government expenditures, on education (particularly of women) and on health.

This idea that the public sphere is a matter best left to experts also crops up in ways which are less obvious, or even rather unexpected. Chapter 5, for example, shows that the notion of a new distinctive NGO model of development based on 'empowerment' often boils down to a matter of adopting a correct (technical/organizational) approach based on alternative sets of prescriptive rules developed by experts, for experts. Within this conception, therefore, the place of NGOs within the public sector is defined in terms of their specific expertise in working at the grassroot level.

All these examples have in common that the scope and dimension of the public sphere is seen as a matter of expertise, normally put forward in the shape of a prescriptive model which can be fruitfully applied in different historical and geographical settings. Policy analysis, therefore, concerns what ought to be done (prescription) and how to do it (implementation). Policy mistakes result from the pursuit of inappropriate models or from the failure to implement correct policies.

Instead, in this book, we argued that the public sphere is socially (and variously) constructed under the impulse of public policy and public action. Seemingly similar policies, therefore, do not necessarily lead to similar outcomes. For example, spending more on health and education in a country with low per-capita income by no means implies that social development is necessarily enhanced or that fertility declines. As Gita Sen puts it in Chapter 10 '... if there is a moral to the so-called "Kerala model" it is that government action, in order to be successful, requires popular participation'. But popular participation, she argues, is not something which can be easily switched on or off. On the contrary, it is protracted processes of social development which progressively build up, or destroy, a capacity for public action.

Conversely, similar social processes do not by themselves generate similar public responses. For example, Chapter 1 showed that rapid urbanization in nineteenth century Europe brought in its wake the development of urban sanitation as a public good. But it went on to argue that this does not mean that similar processes of urbanization elsewhere provoke a similar response to the plight of the urban poor, as is illustrated vividly by the removal of the black population in Johannesburg under apartheid. Understanding policy as a process implies that we examine how public policy and public action respond to tensions and crises within society and, in turn, bring about transformations in society. These transformations feed back upon people and their livelihoods, hence upon poverty and deprivation. In times of economic or social crisis, the scope and organization of the public sphere is subject to tension, conflict and renegotiation.

11.2 Morality and economics

Societies change under the impulse of public action. This book has shown that such changes concern not only the organization of the economy, but also the way people feel and think about economy and society. Those who strive to effect changes, or to defend the status quo, invariably seek to occupy the moral high ground to give legitimacy to their purpose. This is the second point.

Tom Hewitt's chapter, for example, shows that, in Brazil, public action over the plight of street children varies from outright neglect, containment or even physical violence, to charity or solidarity and that these responses are associated with definite moral positions. These positions picture the children as victims of poverty, petty thieves living at the margins of crime, or as young people trying to assert some independence in extremely adverse circumstances. Another example of the conflict over the moral high ground discussed in Chapter 6, is the way in which 'motherhood' as a concept served both as a controlling and as a politicizing mechanism in Pinochet's Chile. In this case, the regime's drive to control society and to suppress opposition was tinged with a strong moral flavour in which appeals to the values of 'motherhood' played an important part. However, Chilean women organizing in defence of their own, turned the concept on its head and used it as a weapon instead, so that motherhood became an organizing tool and a justification for action.

These types of appeal to morality, values, and feelings in defence of specific public actions matter a great deal in the debates which surround the definition of the public sphere. Furthermore, these appeals to morality relate closely to the development of social theory. A clear example of how morality and theory interact, discussed by Maureen Mackintosh in Chapter 3, concerns the radical shift in the 1980s in orthodox economic theory of the public sphere. The earlier 'public interest view' provided the theoretical foundations of a welfarist notion of public action. Inherent in this view was the premise that the state was capable of defining public need and organizing the delivery of public goods on behalf of citizens. This particular view tended to set aside conflict and renegotiation in society over its public space, and to ignore the way specific interests might dominate the organization, financing, and control of public goods. However, on the positive side, implicit in this view is the defence of a space for public action, mainly state action, as distinct from private action. This was not just a matter of theory, but it incorporated a particular moral stance: the state conceptualized as responsible for provision which markets could not handle. For example, the state was seen as legitimately and justly responsible for securing a high level of employment, and providing for basic health care and education. Consequently, people came to expect and depend upon this provision.

The 1980s swept aside the theoretical foundations of this view on state action, but also, in the process, challenged it on moral grounds. The axiom that the state is capable of defining public needs and of providing desirable public goods was challenged by the 'private interest view' of the state. Bureaucratic monopoly, the dangers of the Leviathan state, and rent seeking became the new concepts which questioned the established notions of the public sphere. This new view postulated that people *should* not depend on the state to define and provide for their needs. Instead, they should rely less on state support and be given the opportunity to take initiatives and make choices, based on individual preferences. Thus, developments in economic theory, as argued in the Introduction to this book, were reinforced in the language of morality: terms like choice, freedom, self-respect, individual rights assumed dominance in this newly assertive moral language of freedom and individual responsibility. Once more, theory and morality intertwined to legitimize a particular concept of public sector reform and the public space.

Both theories, the public and the private interest views of the state, are essentially prescriptive in design, and ahistorical in nature. Each theory defends its respective policies as universal norms: abstract in formulation, generally valid and, as such, unquestionably desirable, independently of the concrete context in which they are applied. This is not without its dangers, and Joshua Doriye, in his chapter on Tanzania, illustrated this point vividly.

In the 1980s, under structural adjustment, Tanzania witnessed a series of market based reforms of the public sector which were inspired by the new orthodoxy in economic theory. These reform sought to reorganize the public sector on the basis of increased commercialization and privatization of public services, in pursuit of greater efficiency and market-based accountability. Such a process encounters real limits when the pursuit of private gain enters areas such as policing or taxation, where problems cannot be resolved through market solutions. The chapter points to a real danger: public sector reform premised on private interest behaviour may lastingly undermine the probity of public administration, and complete the destruction of a broader, socially constructed concept of the public purpose, public morality, and public accountability.

The changing nature of public debates, the consequent shifts in emphasis in social theory, and the related appeals to morality are all inherent elements of the process of policy making. Public action, morality, and its accountability emerge from the social processes which structure the public sphere. Consequently, they cannot be understood independently of their historical context.

11.3 State action and public action

This book argues that public action cannot be reduced to state action alone. But the third lesson of this book is that public action is not a simple *addition* of state action and of other forms of non-governmental public action. The two forms of public action do not simply run parallel, so that one may be substituted for the other. They interact in ways which have implications for understanding accountability and autonomy of public action.

Public expenditure, to be effective, requires popular participation. This point, quoted above, was vividly argued in Chapter 10 on social provisioning in Kerala state, India. The author showed that, historically, large sections of the population came to participate in the definition of social needs and in assuring accountability in their fulfilment. The state does not define public need for people or on their behalf. Rather, public need is defined through active participation, or exclusion, of various groups in society. This was also a central argument in Chapter 1. Deprivation is surrounded by conflict and tensions in society over how it is defined, and what is to be done about it. This involves, therefore, struggle between the rich and the poor, the powerful and the powerless. The basic lesson from the first chapter was that little or nothing will be done about the poor and the deprived if they are excluded from the processes which structure public need.

Hence the Kerala example implies that accountability of state action cannot be assumed but *can* be fought for. Relations between people and the state depend on broader public action and other public institutions which effectively put pressure on the state. Constructing and defending these public institutions is a complex historical process.

Accountability, therefore, cannot be reduced to market discipline. Nor, however, is the World Bank's appeal for the creation of 'intermediary institutions' such as NGOs to act as a countervailing force against the abuse of state power sufficient to resolve the problem. As Chapter 4 argued, grassroots institutions which reflect the view of local communities subject to the breakdown of law and order may well resort to taking the law in their own hands rather than act as the World Bank expects. Public pressure will only be effective where it succeeds in modifying the operation of the state itself.

In fact, the question of violence, the abuse of public office, and narrow-minded local response to deprivation is a disturbing theme which cropped up at various instances in this book, particularly in the chapters on Brazil and Tanzania. In Tanzania, the breakdown of the cohesiveness and morality of state action resulted in the police using their public function for private gain. This, in turn, often provoked community responses through the creation of local security groups which took the law in their own hands. In Brazil, violence against street children is a response by certain sections of the local communities and members of the police to the perceived threat posed by these children. Both cases illustrate how violence crops up when state action is seen to be hollow, ineffective or corrupt and when accountability breaks down.

Local public action, therefore, like state action, does not operate in a vacuum. It can be constrained, twisted or enhanced by state action and often also by the actions of better funded, resource-rich NGOs from the industrialized countries. One example, discussed in Chapter 5, concerns the ASSEFA (Association of Sarva Seva Farms) movement in India which aimed to develop self-reliant communities based on lands donated for use of the landless through the *Bhoodan* (land gift) system. This movement was rooted in local action, but its successes were dependent on broader processes outside the sphere of direct local action by these communities. These included positive support by the state and central governments and large foreign NGOs, as well as the willingness of landowners to forego their property rights on such lands.

Hazel Johnson discusses another example which concerns women's organizations in Chile. During Allende's time, women's organizations, particularly those who sympathized with Popular Unity, encountered real difficulties in developing their own capacities for independent action. The state and party were supportive to these women's organizations but in various ways constrained their independent action. In contrast, the anti-authoritarian struggle against the Pinochet regime had the effect that women's action gained in strength and autonomy. This example suggests that local action and state action are clearly interdependent but that the character of the interplay is by no means straightforward. Public action outside the realm of the state, therefore, cannot be seen as merely running parallel to state action in an additive fashion.

In some instances, however, NGOs have adopted strategies of operating in precisely in such a parallel fashion. In Mozambique, as described in Chapter 5, several cases existed where NGOs effectively competed with statutory state services in the delivery of welfare and relief services and even set up parallel administrations. Here the interdependencies were denied or concealed, but the effect was to undermine the basis for improved state services in the future. In contrast, in other cases in Mozambique, NGOs reinforced the effectiveness of state services through active support, while, at the same time, pursuing their own agenda for action.

In a different context, Chapter 9, on economic reforms in China, also raised this issue of autonomy of local public action. The commune system, with its remarkable achievements as well as drawbacks, played a key role in China's development strategy during Mao's era. However, the growth in agricultural production remained stunted and there was a definite problem with incentives at the commune level. Gordon White argues that this was because communes were collectives in name only and *de facto* had little effective autonomy in action; as a result they were reduced to being mere appendages of state planning. The reforms sought to effect change by pursuing a particular solution to the problem. This solution was not to give greater autonomy to local public action in the form of the collectives, but implied a move towards devolving greater responsibility in production towards households while maintaining tight state control over the countryside in the context of the two level management system. It appears, therefore, that the reforms were in fact motivated by keeping central political control over rural transformations in China while devolving some economic processes to the market based upon the household responsibility system. Consequently, local public action lost ground in the process.

11.4 Structural adjustment as social transformation

The fourth point I want to make is that structural adjustment is about social transformation which can only be understood in our framework of 'policy as a process'. State action works through and is structured by *institutions* which include markets (although we do not usually think of markets as social institutions). These institutions are influenced by public action and, in turn, provide the means through which this action is sustained or modified. Hence, public action builds upon, but also seeks to overturn, existing social institutions which are themselves the living products of earlier social settlements.

Policies, therefore, change the character of the economy and society inasmuch as conscious public action continuously restructures existing social relations and institutions. This is the view put forward by Mahmood Messkoub, in Chapter 7, in his analysis of structural adjustment. Policies, he argues, do not operate in a vacuum nor do similar policies necessarily lead to similar results. Rather, specific forms of state action operate in a definite institutional context, which acts as a filter through which such policies effect change. For example, as shown in Chapter 7, policies of structural adjustment often alter the balance between paid and unpaid labour to the detriment of the latter, normally performed by women. Some state provision, for example, may substitute for women's unpaid work: the sick may be nursed by paid nurses in a hospital rather than nursed at home (generally, by women). Consequently, cuts in public expenditures or the introduction of user charges by reducing the availability of state services, add to women's unpaid work, the cost of which is effectively treated as zero under structural adjustment. These supposedly 'hidden' costs of adjustment are not just a question of the social impact of adjustment within an otherwise invariant institutional context. Instead, the argument is that households and gender relations form part of the institutional fabric of society and are consequently shaped and transformed by policies.

Markets too are social institutions. Orthodox economics projects an image of market forces which, left on their own, will blossom naturally into competitive structures to, by and large, enhance the welfare of society. The market, therefore, is seen as unproblematic, in sharp contrast with cumbersome state interventions and actions. But, as argued by Ben Crow and Barbara Harriss in Chapter 8, real markets are more complicated than the simple models suggest. The case of World Bank-induced deregulation of food markets in Malawi vividly illustrated the disastrous consequences of the failure to take effective market structures into account. The process of deregulation fuelled speculative profits of traders, failed to induce growth in production and had adverse effects on the poor. In this case, therefore, deregulation effectively fuelled sectional interests, particularly those of speculative traders. The lesson is not that deregulation is necessarily bad or undesirable. The point is that markets are socially constructed and cannot be assumed to operate according to some abstract logic independent of the social context in which production, distribution and exchange takes place.

In sum, adjustment policies are not neutral in content. Instead, they effect changes in the social fabric of society as manifested in its institutional make-up. Social transformations involve institutional change and alter the possibilities for, and constraints on, further public action. In the conclusion of Chapter 9, Gordon White makes the same general point in the context of the economic reforms in China. Under the impulse of economic reforms, the rural population became more diverse and differentiated, a new élite of relatively well-off rural entrepreneurs emerged with considerable potential influence on the local political scene, previous 'barefoot doctors' were privatized, labour markets created different opportunities in the local community or elsewhere, and the power of the local party, state and collective officials has been undermined to some extent. These changes alter the institutional mechanisms which condition business fortunes as well as redistributive policies, and, in turn, exert pressure for further transformations.

11.5 The relevance of policy analysis?

Finally, I shall end with some interlinked questions. If the whole process of policy depends on the place and time, what then is the point of policy analysis? Is it merely to understand policies or should we be concerned with how to change them? That is, how can we meaningfully contribute to policy? In this respect, I think, the message of this book is twofold.

First, quick solutions do not exist for complex social and economic problems. Policy prescriptions often convey the impression that such solutions are available, precisely because the prescriptions are often abstractions of the process of policy itself. In contrast, we argue that policy analysis can and should be done, but not by removing it from the very essence of what gives it life — processes of social transformation propelled by public action.

Take, once more, the case of the Kerala 'model'. Chapter 10 argued that, insofar as a lesson can be drawn from the Kerala experience, it is that governmental action, in order to be successful, requires popular participation. But, as the chapter shows, this did not come easily. Rather, it involved a long process of democratization which transformed society and the public sphere within it. This was not just a matter of spending money on education or on public health. On the contrary, the Kerala example shows how wider public action rendered state expenditures more effective. The Kerala 'model' can be more appropriately characterized as a *virtuous circle* of social processes which define and structure the public purpose, its morality, the character of its public goods, and the nature of accountability.

Similarly, Chapter 4, on Tanzania's public sector reforms in the 1980s, illustrates the same argument, but, in this case, it operates in reverse. Here, a *vicious circle* appears to be at work, which involves a process of erosion and deconstruction of the public sphere. Increasingly, public office became the vehicle for private gain. Once more, this is not just an economic question concerning the size and the rationale of the public sphere.

Rather, it involves a process which affects, not only how people work and live within and outside the public sector, but also how people perceive the public purpose, its accountability and its morality. Not surprisingly, the conclusion of this chapter is gloomy and seeks no refuge in a simple, unilineal policy prescription.

Secondly, to put the emphasis on the importance of social processes does not mean that nothing can be done to change policy. Rather, the very concept of policy as a process implies the *conscious* activity of people and organizations in effecting change. Viewed in this way, this concept of policy can be empowering, not discouraging. Policy analysis matters because it helps us to act effectively. The purpose of this book is to invite participants in the public sphere to think realistically about their objectives, place and role in the policy process.

Here are some examples of what this might mean. When Alan Thomas, in Chapter 5, challenges the idea that grassroot empowerment through the action of NGOs could be the basis of a new development model, the argument is not that NGOs cannot play a positive role in empowering people. In fact, the chapter gives some examples where this is the case and contrasts these with others where NGO activity was, at best, questionable. Whether NGOs play a positive role or not depends largely on the place they occupy within broader processes of social change, and the political tensions, conflicts and alliances associated with them. NGO action, therefore, does not merely act upon a process; it is part of the process itself. This point is also illustrated in Chapter 6 which shows that women's organizations in Latin America underwent considerable changes, as they both responded to and influenced the political and social context in which they operated. Similarly, Chapter 8 does not argue that deregulation of markets is necessarily either good or bad, in terms of stimulating output growth or reducing vulnerabilities. Thus, the example of deregulating food markets in Malawi, shows that the impact of deregulation depends upon the way production is organized, and how specific markets work in practice. To see this point, however, requires a concept of markets as part of the social fabric of society, which in turn is historically influenced by past and present policies.

Policy analysis, then, should help us to understand the role and influence of different policy 'actors' within specific historical and institutional contexts. It is not as paradoxical as it seems to suggest that this should help to induce both modesty (in recognising what is possible) and greater effectiveness (in explaining the bounds of what is possible).

References

Abikar, A. (1989) *Intertemporal and spatial market price analysis in Somalia,* Food Studies Group, Oxford.

Agarwala, A. & Singh, S. (1958) *The Economics of Underdevelopment,* Oxford University Press, Oxford.

Agnelli, S. (1986) *Street Children — a growing urban tragedy,* Report for the Independent Commission on International Humanitarian Issues, Weidenfield and Nicholson, London.

Ahmad, E., Drèze, J., Hills, J. & Sen, A. (1991) *Social Security in Developing Countries,* Clarendon Press, Oxford.

Ahmed, R. & Bernard, A. (1989) *Rice price fluctuation and an approach to price stabilization in Bangladesh,* IFPRI Research Report 72,Washington DC.

Allen, T. & Thomas, A. (eds) (1992) *Poverty and Development in the 1990s*, The Open University/ Oxford University Press, Oxford.

Amnesty International (1990) *Brazil: torture and extrajudicial execution in urban Brazil*, Amnesty International Briefing, London.

Annis, S. (1987) 'Can small-scale development be a large-scale policy? The case of Latin America', *World Development*, 15 (Supplement) pp. 129–134.

Baguma (1990), quoted in the *'Daily News'*, 25 April 1990, Tanzania.

Bates, R. (ed.) (1988) *Toward a Political Economy of Developmen,* University of California Press, Berkeley.

Benewick, R. & Wingrove, P. (eds) (1988) *Reforming the Revolution: China in transition*, Macmillan, London.

Bennett, R. (ed.) (1990) *Decentralization, Local Governments and Markets: towards a post welfare agenda*, Clarendon Press, Oxford.

Bequele, A. & Boyden, J. (eds) (1988) *Combating Child Labour*, International Labour Office, Geneva.

Berlin, I. (1958) 'Two concepts of liberty', Inaugural Lecture, Clarendon Press, Oxford.

Bernstein, H., Crow, B. & Johnson, H (eds) (1992) *Rural Livelihoods: crises and responses*, The Open University/Oxford University Press.

Bernstein, H., Crow, B., Mackintosh, M. & Martin, C. (eds) (1990) *The Food Question,* Earthscan, London.

Bhatt, E. (1989) 'Toward empowerment', *World Development*, 17(7), pp.1059–1065.

Bienen, H. *et al.* (1990) 'Decentralization in Nepal', *World Development*, 18(1), pp.61–75.

Bienefeld, M. (1983) 'Efficiency, expertise, NICs and the accelerated development', *IDS Bulletin*, 14(1).

Borega, E. (1990) 'Polisi Mugumu Wanaomba rushwa', A letter to the editor, *Uhuru*, 4 April 1990, Tanzania.

Bramall, C., Nolan, P., & Sender J. (1990) 'Food and freedom: A. K. Sen on China', Faculty of Economics and Politics, Cambridge University, Cambridge, UK.

Bratton, M. (1989) 'The politics of government–NGO relations in Africa', *World Development*, 17(4), pp.569–597.

Brett, E.A. (1990) 'Competence and accountability in the voluntary sector: organization theory, adjustment policy and institutional reform', Institute of Development Studies, Brighton.

Catalyst (1990) 'From love to need. Latin American women call for a new set of values', *Catalyst,* No 4 (July–Sept.), pp.14–15.

Cernea, M. (ed.) (1985) *Putting People First: Sociological Variables in Rural Development*, New York, World Bank/ Oxford University Press, Oxford.

Chambers, R. (1983) *Rural Development: putting the last first*, Longman, Harlow.

Chambers, R. (1988) 'Bureaucratic reversals and local diversity', *IDS Bulletin* 19(4), pp.50–56.

Charlton, S. E. M., Everett, J. & Staudt, K. (eds) (1989) *Women, the State and Development*, State University of New York Press, Albany.

Christiansen, R. E. & Stackhouse, L. A. (1987) *The privatization of agricultural trade in Malawi,* UNDP.

Clarke, J. (1991) *Democratizing Development: the role of voluntary organizations*, Earthscan, London.

Clay, E. J. (1985) 'The 1974 and 1984 floods in Bangladesh: from famine to food crisis management', *Food Policy*, 10 August.

Colburn, F. D. (ed) (1989) *Everyday Forms of Peasant Resistance*, M. E. Sharpe, New York.

Colclough, C. (1983) 'Are African governments as unproductive as the Accelerated Development Report implies?' *IDS Bulletin*, 14(1).

Cook, P. & Kirkpatrick, C. (1988) *Privatisation in Less Developed Countries,* Harvester Wheatsheaf.

Corcoran-Nantes, Y. (1990) 'Women and Popular Urban Social Movements in São Paulo, Brazil' *Bulletin of Latin American Research*, 9(2), pp.249–264.

Cornia, G. A., Jolly, R. & Stewart, F. (1987) *Adjustment with a Human Face: protecting the vulnerable and promoting growth, a study by UNICEF*, Clarendon Press, Oxford.

Cornia, G.A., Jolly, R. & Stewart, F. (eds) (1988), *Adjustment with a Human Face: country case studies*, Volume II, Clarendon Press, Oxford.

Davis, D. & Vogel, E.F. (eds) (1990) *Chinese Society on the Eve of Tiananmen*, Harvard University Press, London.

De Swaan, A. (1988) *In Care of the State: health care, education and welfare in Europe and the USA in the modern era,* Polity Press, Cambridge.

De Waal, A. (1990) 'A re-assessment of entitlement theory in the light of recent famines in Africa', *Development and Change,* 21(3), pp. 469-490.

Deere, C. D. & León, M. (eds) *Rural Women and State Policy: feminist perspectives on Latin American agricultural development*, Westview Press, Boulder and London.

Dimenstein, G. (1991) *Brazil: War on Children*, Latin American Bureau, London. Originally published in 1990 as: *A Guerra dos Meninos: Assassinatos de Menores no Brasil*, Editora Brasiliense, São Paulo.

Dornbusch, R. & Helmers, F.L. (eds.) (1988) *The Open Economy: tools for policy makers in developing countries,* World Bank/Oxford University Press, Oxford.

Drèze, J. & Sen, A. (1989) *Hunger and Public Action*, Clarendon Press, Oxford.

Ekins, P. (1986) *The Living Economy: a new economics in the making*, Routledge & Kegan Paul, London and New York.

Elson, D. & Pearson, R. (eds) (1989) *Women's Employment and Multinationals in Europe*, Macmillan, London.

Elson, D. (ed.) (1991) *Male Bias in the Development Process*, Manchester University Press, Manchester.

Ennew, J. & Milne, B. (1989) *The Next Generation: lives of Third World children*, Zed Books, London.

Evans, A. & Young, K. (1988) *Gender Issues in Household Labour Allocation: The Case of Northern Province, Zambia*, ODA/ESCOR Research Report.

FitzGerald, E.V.K. & Wuyts, M. (1988) *Markets within Planning: socialist economic management in the Third World,* Frank Cass, London.

Foster-Carter, A. (1985) *The Sociology of Development*, Causeway Books, Ormskirk.

Fowler, A. (1988) *Non-governmental Organizations in Africa: Achieving Comparative Advantage in Relief and Micro-development*, IDS Discussion Paper 249.

Freire, P. & Shor, I. (1987) *A Pedagogy for Liberation*, Macmillan, London.

Freire, P. (1972) *The Pedagogy of the Oppressed*, Penguin, Harmondsworth.

Fyfe, A. (1989) *Child Labour*, Polity Press, Cambridge.

Galeano, E. (1988) *O Século do Vento III: Memória do Fogo*, Editora Nova Fronteira, Rio de Janeiro.

Galheigo, S.M. (1991) 'The formation of social policies in Brazil: the case of FUNABEM', unpublished research paper, University of Sussex.

Garrett-Schesch, P. (1975) 'The mobilization of women during the Popular Unity government', *Latin American Perspectives*, 2 (1), pp.101–103.

Ginneken, W. van (1988) *Trends in Employment and Labour Incomes*, International Labour Office (ILO), Geneva.

Government of Kerala (1989) *Health Profile, Kerala.*

Hanlon, J. (1991) *Mozambique: who calls the shots*, James Currey, London.

Harriss, B. (1984) *State and market,* Concept, New Delhi.

Helm, D. (ed.) (1989) *The Economic Borders of the State*, Oxford University Press, Oxford.

Hewitt, T., Johnson, H. & Wield, D. (1992) Industr*ialization and Development,* The Open University/ Oxford University Press, Oxford.

Hirschmann, A.O. (1980) *Shiftings, involvements: private interests and public action,* Princeton University Press, Princeton, New Jersey.

Hoggett, P. & Hambledon, R. (eds.) (1987) *Decentralisation and Democracy : Localising Public Services*, SAUS Occasional Paper No. 26, School for Advanced Urban Studies, Bristol.

Hogwood, B. & Gunn, L. (1984) *Policy Analysis for the Real World,* Oxford University Press, Oxford.

Holloway, R. (1989) *Doing Development*, Earthscan, London.

Hulme, D. (1991) 'Social Developmnet Research and the Third World', Discussion Paper No. 26, Institute for Development Policy and Management, University of Manchester.

Iliffe, J. (1987) *The African Poor*, Cambridge University Press, Cambridge.

Jaguaribe, H. (1990) *Brasil 2000: Para um Novo Pacto Social*, Paz e Terra, São Paulo.

Jaquette, J. S. (ed) (1989) *The Women's Movement in Latin America: feminism and the transition to democracy*, Unwin Hyman, Boston.

Jayawardena, K. (1986) *Feminism and Nationalism in the Third World*, Zed Books, London and New Jersey.

Jeffrey, R. (1987) 'Governments and culture: how women made Kerala literate', *Pacific Affairs*, 60 (3).

Jelin, E. (ed) (1990) *Women and Social Change in Latin America*, Zed Books, London.

Khan, M.S. & Knight, M.D. (1985) *Fund-supported Adjustment Programmes and Economic Growth*, IMF Occasional Paper No. 41, November, International Monetary Fund, Washington DC.

Killick, T. (1989) *A Reaction Too Far: economic theory and the role of the state in developing countries*, Overseas Development Institute.

Kirkwood, J. (1983) 'Women and politics in Chile' *International Social Science Journal*, XXXV (4), pp.625–637.

Korner, P., Mass, G., Seibold, T. & Tetzlabb, R. (1986) *The IMF and the Debt Crisis: a guide to the Third World's dilemmas*, Zed Press, London.

Korten, D. (1987) 'Third-generation NGO strategies: a key to people-centred development', *World Development*, 15 (Supplement) pp. 145–160.

Korten, D. (1989) 'The U.S. Voluntary Sector and Global Realities: Issues for the 1990s', Mimeo, Institute for Development Research, Boston, Mass.

Kydd, B.H. & Spooner, N. (1990) 'Agricultural market liberalization and structural adjustment in sub-Saharan Africa', *Agrarian Studies*, Oxford .

Lal, D. (1983) *The Poverty of 'Development Economics'*, Institute of Economic Affairs, Hobart Paperback 16, London.

Larrain, C. (1982) 'Catrasto de organizaciones femininas de gobierno', Documento de trabajo, Instituto Chileno de Estudios Humanisticos, Santiago de Chile.

Lee-Wright, P. (1990) *Child Slaves*, Earthscan, London.

Leeds, R. (1989) 'Malaysia: genesis of a privatization transaction', World Development, 17(5), pp.741–756.

Lele, U. & Christiansen, R.E. (1990) *Markets, marketing boards and cooperatives: issues in adjustment Policy*, MADIA Paper, World Bank,Washington.

Lele, U. J. (1990) 'Structural adjustment, agricultural development and the poor: some lessons from Malawian experience', *World Development,* 18(9), pp.1207–1219.

Letelier, I. M. (1989) 'Women's action in Chile' *Women's Studies International Forum,* 12(1), pp.125–127.

LEWRG (London-Edinburgh Weekend Return Group) (1979) *In and Against the State*, Black Rose Press, Edinburgh.

Lindblom, C.E. (1977) *Politics and Markets*, Basic Books, New York.

Loxley, J. (1986) *Debt and Disorder: external financing for development,* Westview Press, USA.

Lutsig, N. (1990) 'Economic crisis, adjustment and living standards in Mexico, 1982–85', *World Development*, 18 (10), pp.1325–1342.

Machado Neto, Z. (1982) 'Children and adolescents in Brazil: work, poverty, starvation', *Development and Change*, 13(4).

Madihi, M.C. (1990) *Survey of Total Accrues of Civil Servants,* June 1990, Tanzania.

Mamdani, M. (1976) *Politics and Class Formation in Uganda*, Heinemann, London.

Maziku, W.K. (1990) 'Nurses' views are unacceptable', *Sunday News*, 15 April 1990, Tanzania.

McGaffey, J. (1983) 'How to survive and become rich amidst devastation, the second eonomy in Zaire', *African Affairs*, 1.

Meier, G. (1984) *Leading issues in Economic Development*, Fourth Edition, Oxford University Press, Oxford.

Mkwezalamba, M.M. (1989) *The liberalization of smallholder agricultural produce marketing in Malawi and implications on household food security,* Economics Department, Chancellor College, Zomba, Malawi.

Molyneux, M. (1985) 'Mobilization without emancipation? Women's interests, state and revolution in Nicaragua' *Feminist Studies*, 11(2).

Moser, C. (1989a) 'Gender Planning in the Third World: meeting practical and strategic gender needs' *World Development*, 17(11), pp.1799–1825.

Moser, C. (1989b) 'The impact of recession and structural adjustment policies at the micro-level: low income women and their households in Guayquil, Ecuador', *Invisible Adjustment,* 2, UNICEF.

NCVO (National Council for Voluntary Organisations) (1990) *New Times, New Challenges: voluntary organisations facing 1990*, NVCO.

Niskanen, W. (1971) *Bureaucracy and Representative Government*, Aldine, Chicago.

Nolan, P. (1988) *The Political Economy of Collective Farms*, Polity Press, Oxford.

Nunberg & Nellis (1990) 'Civil service reform and the World Bank', PRE Working Paper Series, WPS, 422, May, World Bank, Washington DC.

Nyerere, J.K. (1969) Speech by the President to the T.A.N.U. Conference, 28 May 1969, in United Republic of Tanzania.

Nyerere, J.K. (1970) 'Arusha Declaration Parliament', Address by the President to the National Assembly, Government Printer, Dar es Salaam, 6 July 1970.

Nyerere, J.K. (1985) 'Speech by the President to Parliament', Dar es Salaam, 29 July 1985.

ODI (1986) *Adjusting to recession: will the poor recover?,* ODI Briefing Paper, November, Overseas Development Institute, London.

Olsen, W. (1984) *Food grain marketing in Somalia: a case of partial liberalization*, Hertford College, Oxford.

Onimode, B. (ed.) (1989) *The World Bank and African Debt: the economic impact* (2 volumes), Zed Press, London.

Open University (1985) *Caring for Health,* U205 Book VII, The Open University Press.

Panikar, P.G.K. (1975) 'Fall in mortality rates in Kerala: an explanatory hypothesis', *Economic and Political Weekly*, 22 November.

Poulton, R. & Harris, M. (1988) *Putting People First: the NGO approach to development*, Macmillan, London.

Rodrik, D.S. (1990) 'How should structural adjustment programmes be designed?' *World Development,* 14(3), pp.429–439.

Roemer, M. (1986) 'Simple analytics of segmented markets: what case for liberalization?' *World Development* 18(7), pp.933–47.

Sachs, J.D. (ed) (1989) *Developing Country Debt and Economic Performance,* The University of Chicago Press.

Salole, G. (1991) 'Not seeing the wood for the trees: searching for indigenous non government organizations in a the forest of voluntary self help associations', *Journal of Social Development in Africa*, 6(1), pp.5–17.

Samoff, J. & Wuyts M.E. (1989) 'Swedish public assistance in Tanzania', Educational Division Documents No. 43, SIDA, Stockholm.

Sandbrook, R. (1986) 'The state and economic stagnation in tropical Africa', *World Development*, 14 (3).

Scarborough, V. (1989) *Food marketing post liberalization in Malawi and Tanzania,* Wye College, Ashford, Kent.

Schumacher, E. F. (1973) *Small is Beautiful: economics as if people mattered,* Abacus, London.

Sen, G. & Grown, C. (1987) *Development Crises and Alternative Visions: Third World women's perspectives*, Earthscan, London.

Shivji, I. (1973) *The Silent Class Struggle*, Tanzania Publishing House, Dar es Salaam.

Slater, D. (ed) (1985) *New Social Movements and the State in Latin America*, CEDLA, Amsterdam.

Stinton, W. (1982) *Community Financing of Primary Health Care,* American Public Health Association,Washington DC.

Streeten, P. (1981) *First Things First: meeting basic needs in developing countries*, Oxford University Press, Oxford.

Tanzi, V. (1991) *Public Finance in Developing Countries,* Edward Elgar.

Tendler, J. (1989) 'Whatever happened to poverty alleviation?' *World Development*, 17(7), pp.1033–1044.

Tharakan, P.K.M. (1984) 'Socioeconomic factors in educational development — the case of nineteenth century Travancore', *Economic and Political Weekly*, Vol XIX (nos 45 &46 , 10 & 17 Nov.).

Thomson, A. & Smith, L. D. (1990) *Privatization of inputs, services and marketing of agricultural outputs,* Paper prepared for UN FAO, Rome.

Timberlake, L. (1985) *Africa in Crisis*, Earthscan/ International Institute for Environment and Development, London and Washington DC.

Titmuss, R.M. (1970) *The Gift Relationship*, George Allen and Unwin, London.

Toye, J. (1987) *Dilemmas of Development*, Blackwell, Oxford.

UNDP (1990) Human *Development Report 1990*, United Nations Development Programme, Oxford University Press, Oxford.

UNICEF (1989) The State of the World Children 1989, Oxford University Press, Oxford.

United Republic of Tanzania (1969), *Tanzania Second Five-Year Plan for Economic and Social Development, 1st July 1969–30 June 1974, Vol. 1: General Analysis.*

United Republic of Tanzania (1975) *The Annual Economic Survey, 1974–75.*

Van Ryneveld, P. (1989) *An Analysis of How Combi Taxis compete with Bus Services in South Africa,* MA Research paper, Institute of Social Studies, The Hague.

Van der Heijden, H. (1987) 'The reconciliation of NGO autonomy, program integrity and operational effectiveness with accountability to donors', *World Development*, 15 (Supplement),pp.103–112.

Vargas, V. (1990) 'The Women's Movement in Peru: rebellion into action', Working Paper Sub-series on Women, History and Development: Themes and Issues, No 12 (April), Institute of Social Studies, The Hague, Netherlands.

Vargas, V. (1991) 'The feminist movement in Latin America: between hope and disenchantment (notes for debate)', Paper presented at Workshop on 'Rethinking Emancipation. Concepts of Liberation', 30 Jan. – 1 Feb. 1991, The Hague, Netherlands.

Wallace, T. & March, C. (eds) *Changing Perceptions: writings on gender and development*, Oxfam, Oxford.

Wehelie, Y. J. (1989) *Grain market liberalization and price policy reform: some empirical experience in Somalia,* Paper for the seminar on food pricing and marketing reform, UNRISD, Geneva.

Wellings, P. (1983) 'Making a fast buck: capital leakage and the public accounts of Lesotho', *African Affairs.*

White, B. (1982) 'Child workers and capitalist development: an introductory note and bibliography', *Development and Change*, 13(4).

White, G. (1987) 'The impact of economic reforms in the Chinese countryside: towards the politics of social capitalism?', *Modern China*, 13 (4), pp.411–440.

White, G. (ed.) (1988) *Developmental States in East Asia*, Macmillan , London.

Whittaker, A. (1987) 'Brazil's slave-like practices', in *Anti-slavery Reporter*, The Anti-slavery Society, London.

World Bank (1981) *Accelerated Development in Sub-Saharan Africa: an agenda for action*, World Bank, Washington DC.

World Bank (1983a), *World Development Report*, World Bank, Washington DC.

World Bank (1983b) *China, Socialist Economic Development,* Vol I, World Bank, Washington DC.

World Bank (1986), *Poverty and Hunger*, World Bank, Washington DC.

World Bank (1987) *Tanzania; an agenda for industrial recovery,* World Bank, Washington DC.

World Bank (1988a), *World Development Report, Oxford* University Press, Oxford.

World Bank (1988b) *Bangladesh: adjustment in the eighties and short term prospects*, Report 7105-BD, Vol.II Statistical Appendix, World Bank, Washington DC.

World Bank (1989) *Sub-Saharan Africa: from crisis to sustainable growth*, World Bank, Washington DC.

World Bank (1990a) *World Development Report*, Oxford University Press, Oxford.

World Bank (1990b) *Bangladesh: review of the experience with policy reforms in the 1980s,* Report No 8874, World Bank Operations Evaluation Department, Washington DC.

World Bank (1991a) *World Development Report,* Oxford University Press, Oxford.

World Bank (1991b) *The Africa Capacity Building Initiative*, World Bank, Washington DC.

List of acronyms, abbreviations and organizations

ADC	Agricultural Development Corporation (Somalia)
ADMARC	a state marketing board (Malawi)
ASSEVA	Association of Sarva Seva Farms (India)
BRAC	Bangladesh Rural Advancement Committee
CCP	Chinese Communist Party
ENC	national agency for trade (Somalia)
FUNABEM	National Foundation for Child Welfare (Brazil)
GDP	gross domestic product
GNP	gross national product
GRID	Grass Roots Integrated Development (Thailand)
IBASE	*Instituto Brasileiro de Anâlise Social e Econômica* (Brazilian Institute for Social and Economic Analysis)
IMF	International Monetary Fund
JAPs	*Juntas de Abasticimiento y Control de Precios* ('grassroots' associations for the control of supply and prices, Peru)
LDCs	less developed countries
MNC	multinational corporation
MOMUPO	*Movimento do Mujeres Populares de la Zona Norte* (movement of poor women from the northern zone, Chile)
NCVO	National Council of Voluntary Organizations (UK)
NGOs	non-governmental organizations
OECD	Organization for Economic Co-operation and Development
OEPs	*organizaciones economicas populares* (popular economic organizations, Chile)
PAR	participatory action research
SEWA	Self Employed Women's Association (India)
SINAMOS	National Support System for Social Mobilization (a state initiative in Peru)
TECs	Training and Enterprise Councils (UK)
TNC	transnational corporation
UN	United Nations
UNDP	United Nations Development Programme
UNICEF	United Nations (International) Children's (Emergency) Fund
USAID	US Agency for International Development
VDAs	village development animateurs (Thailand)
VOs	voluntary organizations
WHO	World Health Organization
WWF	WorldWide Fund for Nature

Acknowledgements

Grateful acknowledgement is made to the following sources for permission to reproduce material in this book:

Text

Chapter 1: De Waal, A., 'A reassessment of entitlement theory in the light of famines in Africa', *Development and Change*, 21(3), Institute of Social Studies, by permission of Sage Publications Ltd; *Chapter 2:* Rocha, J. (1991) 'Introduction' in Dimenstein, G. (ed), *Brazil: war on children,* Editoria Brasiliense Portugues, São Paulo, Brazil.

Tables

Table 1.1: Drèze, J. & Sen, A.K. (1989) *Hunger and Public Action*, reproduced by permission of Oxford University Press.

Diagrams

Figure 4.9: data source Odunga *et al.* 'Employment systems and need for policy in the civil service' (project on redeployment of human resources in Tanzania.

Photographs and cartoons

Section title pages: p.11: © Julio Etchart; *p. 5:* © Sarah Errington/Panos Pictures; *p.115:* © Ivan Strasbourg/Hutchison Library; *p.173:* © Camera Press, photo by Claude Jacoby; *p.229:* © Julio Etchart/Reportage;

Figure 1.4: Oxfam; *Figures 1.5 & 1.6:* Mark Edwards/Still Pictures; *Figure 1.7:* Sally & Richard Greenhill/Society for Anglo-Chinese Understanding; *Figure 1.8 (top):* The Mansell Collection; *Figure 1.8 (bottom):* Walter Holt/Oxfam; *Figure 1.9:* Maggie Murray/Format Photographers; *Figure 1.10:* Format Photographers;

Figure 2.1: Alan Hutchison Library; *Figures 2.2, 2.6 & 2.7:* © Julio Etchart; *Figure 2.3 (top):* Oxfam; *Figure 2.3 (bottom);* Susan Cunningham; *Figure 2.4 (top)*; B. Gysembergh/Camera Press; *Figure 2.4 (bottom):* Wellcome Institute Library, London; *Figure 2.5:* Museu de Arte de São Paulo, photo by Luiz Hossaka; *Figure 2.8:* John Ryk/Hutchison Library; *Figure 2.9:* Carlos Reyes-Manzo, ©Andes Press Agency; *Figure 2.10:* Jenny Matthews/ Womankind (Worldwide);

Figure 3.1: ©Nick Thorkelson; *Figures 3.2 & 3.5:* R.K. Laxman, in *The Times of India*; *Figures 3.3, 3.10 (bottom), 3.12 & 3.13(top):* Mark Edwards/ Still Pictures; *Figure 3.4:*© Tom Hanley; *Figure 3.6:* © *Financial Times,* photo by Trevor Humphries; *Figure 3.10 (top):* M & V Birley/ Tropix; *Figure 3.11:* ©David Austin; *Figure 3.13 (bottom);* © Olivia Graham/Oxfam;

Figures 4.1, 4.3, 4.5 & 4.7: © Maggie Murray/ Format Photographers; *Figures 4.2 & 4.11:* Hutchison Library; *Figure 4.4:* R.K. Laxman, in *The Times of India*; *Figure 4.6:* © Liam White Photography; *Figure 4.10:* drawings by Robert Hunt in *Beware of Debtspeak*, Institute of Development Studies, University of Sussex, © 1988 Mike Faber and Robert Hunt;

Figure 5.1: © Tom Hanley; *Figure 5.4:* © Maggie Murray/Format; *Figure 5.5:* © Jenny Matthews/ Format; *Figure 5.6 (above):* © Geoff Sayer/Oxfam; *Figures 5.6 (left) & 5.9:* © Oxfam; *Figure 5.7:* © Tom Learmonth/Panos; *Figure 5.8: Jeremy Hartley;*

Figure 6.1: Julio Etchart; *Figure 6.2:* Andes Press Agency; *Figure 6.3 (top):* Andes Press Agency; *Figure 6.3 (bottom):* Carlos Reyes-Manzo/Andes Press Agency; *Figure 6.4 (top)* Camera Press, photo by Lynn Pelham; *Figure 6.4 (bottom)*: Camera Press, photo by Christian Belpaire; *Figure 6.5:* Julio Etchart/Reportage; *Figures 6.6 & 6.7:* Julio Etchart; *Figure 6.8:* Mark Edwards/Still Pictures; *Figure 6.9:* Carlos Reyes-Manzo/Andes Press Agency.

Figure 7.1(left):© Liam White; *Figures 7.1 (right) & :* © Carlos Guarita/Reportage; *Figure 7.2:* © Ron Giling; *Figure 7.3:* drawings by Robert Hunt in *Beware of Debtspeak*, Institute of Development Studies, University of Sussex, © 1988 Mike Faber and Robert Hunt; *Figure 7.4:* © *Financial Times,* photo by Terry Kirk; *Figure 7.5:* ©Aldin le Garsmeur/Panos Pictures; *Figures 7.8 (left) & 7.9* :© Alison Evans; *Figure 7.8 (right):* Ines Smyth;

Figure 7.10: © Julio Etchart;

Figures 8.7 & 8.8: © Liam White Photography; *Figure 8.10:* © Margaret Waller/Link;

Figures 9.1, 9.5 (top), 9.6, 9.7, 9.8 & 9.10: © Sally & Richard Greenhill; *Figures 9.2, 9.3 & 9.4:* Sally & Richard Greenhill/Society for Anglo–Chinese Understanding; *Figures 9.5 (bottom) & 9.9:* © Melanie Friend/Format Photographers; *Figure 9.11:* Sam Greenhill;

Figure 10.1: © Mark Edwards/Still Pictures; *Figure 10.2(a):* © Douglas Dickens; *Figures 10.2(b), 10.6 & 10.11:* ©Tom Hanley; *Figure 10.3:* The Mansell Collection; *Figures 10.4, 10.8 (left), 10.10 & 10.12(b):* © Mary Thorpe; *Figure 10.5:* © Douglas Dickens; *Figure 10.7:* Oriental and Indian Office Collections; *Figure 10.8(a):* Oriental and Indian Office Collections, from Mateer, S. (1871) *Land of Charity,* John Snow; *Figure 10.9.:* © John & Penny Hubley; *Figure 10.12 (a):* Associated Press.

Index

D

E

F

M

N

O

P

Y

Z